AF678746

Crystallization Process and Technology

Edited by

Rose Marie O. Mendoza

ARCLER PRESS

www.arclerpress.com

Crystallization Process and Technology

Rose Marie O. Mendoza

Arcler Press
2010 Winston Park Drive,
2nd Floor
Oakville, ON L6H 5R7
Canada
www.arclerpress.com
Tel: 001-289-291-7705
001-905-616-2116
Fax: 001-289-291-7601
Email: orders@arclereducation.com

© 2018 Arcler Press

ISBN: 978-1-77361-237-9 (Hardcover)

This book contains information obtained from highly regarded resources. Reprinted material sources are indicated and copyright remains with the original owners. Copyright for images and other graphics remains with the original owners as indicated. A Wide variety of references are listed. Reasonable efforts have been made to publish reliable data. Authors or Editors or Publishers are not responsible for the accuracy of the information in the published chapters or consequences of their use. The publisher assumes no responsibility for any damage or grievance to the persons or property arising out of the use of any materials, instructions, methods or thoughts in the book. The authors or editors and the publisher have attempted to trace the copyright holders of all material reproduced in this publication and apologize to copyright holders if permission has not been obtained. If any copyright holder has not been acknowledged, please write to us so we may rectify.

Notice: Registered trademark of products or corporate names are used only for explanation and identification without intent of infringement.

Arcler Press publishes wide variety of books and eBooks. For more information about Arcler Press and its products, visit our website at www.arclerpress.com

About the Editor

Dr. Rose obtained her PhD in Chemical Engineering from the University of the Philippines-Diliman in 2013. She is also a Professor in the Graduate School Department under the Master of Engineering Program at Adamson University since 2006, and is a Visiting Research Fellow at the Department of Environmental Engineering and Science, and the Department of Environmental Resource Management in Chia Nan University of Pharmacy and Science, Tainan Taiwan since 2010. She is now taking up her Post Doctorate Degree in Green Power: Hydrogen Generation and Fuel Cell Development at the University of Philippines-Diliman under the Department of Mining, Metallurgy and Materials Engineering and the University of California, Merced, USA under the Department of Mechanical Engineering.

Contents

Preface

Crystallization is a key technique for separation and purification of substances as well as for product design in chemical, pharmaceutical and biotechnological process industries. It is vital in many aspects of food preparation and storage, as well as in producing and refining the ingredients for foods and everyday use products. Crystallization is also employed to purify sugars and fats, in addition to change the texture of things such as ice cream, fondant, fudge, and chocolate. Controlling crystallization is important when freezing foods. Crystallization occurs in nature like water freezing and snowflakes formation, stalactites/stalagmites, rings formation, natural (mineral) crystal formation, removing sucrose from beet solutions, removing KCl from an aqueous solution.

This book gives a clear, concise, balanced and up to date presentation of crystallization and solid-liquid separation of the crystalline product. Addressing topics that reflect key applications and recent trends of crystallization; it contains separation and formulation of drug or nutrition components in pharmaceutical and food industry, crystallization issues in medical applications, crystallization process design in production of bulk chemicals. A number of topics are concerned with particle and shape engineering by additives or controlled nucleation, displaying the ongoing interest in defining a specific product quality by crystallization. Ultimate problems of crystallization such as quantifying and understanding phase equilibria and admixture effects, elucidating growth mechanisms, and altering the growth behavior are addressed. The aim of the book is to outline the vital aspects of crystallization theory and practice, together with some closely associated topics.

Editor

CHAPTER 1

INTRODUCTION TO CRYSTALLIZATION

Materials are found in one of the three states of matter; gaseous, liquid or solid state. These three states are characterized by thseir atomic or molecular mobility which determine compressibility and other physical and chemical properties that can be used to predict its behavior under different processes. In gaseous state, the material can take the shape of any container and also becomes compressible for high degree (for example compressed air in car wheels) due to random and dynamic molecular motion. In the liquid state, random molecular motion is more restricted than gaseous state. As a result, the volume occupied by a liquid is limited and the liquid only takes the shape of the container with a flat surface. Liquids with high viscosity such as wax and paraffins have specific properties as compares to the known properties of other liquids like water and oil. These high viscous liquids have solid state appearance but their structure tend to easily get deformed under pressure. In the solid state, the materials have rigid structure and are capable to resist compression. Thus, solids are described to be impenetrable. At high applied pressures, solids became fractured as a sign of deformation. However, fractures fragments still retain portions of the ordered crystalline lattice. Molecular motion in solid state is limited to its oscillation about a fixed position or point of rotation (Mullin, 2001). ß Remove citations like this since you are writing a textbook, but make sure all sources are included in the references. Materials in solid states (solids), can be found in crystalline or amorphous forms. The crystalline state differs from the amorphous state in many aspects such as the regularly arranged constituent molecules, atoms or ions have fixed and rigid pattern known as a lattice. Solids with crystalline lattice exhibits a regular 3D pattern, while amorphous solids do not. Amorphous solids have what is seemed to be a random aggregate of molecules. This surface characteristics of solids can be seen through an X-ray analysis. The X-ray analyses revealed that even in many of the substances previously found to be considered amorphous, some degree of regular molecular arrangement exists. Still, the term 'crystalline' is most commonly used to refer to a high degree of internal regularity, leading to the development of definite external crystal structure (Mullin, 2001).

As the molecular motion in a gas or a liquid is characterized by being free and random, the physical properties of these fluids show high degree of similarity no matter what direction they are measured. That means they are isotropic. True amorphous solids like glass and metals are considered isotropic, as a result of the random arrangement of their constituent molecules. On the other hand, most crystals are anisotropic; their properties such as mechanical, electrical, magnetic and optical properties differ according to the direction in which they are measured. Thus, wood and composites are anisotropic materials. However, crystals belonging to the cubic system are not anisotropic. Their exceptionally symmetrical internal arrangement renders them optically isotropic. Anisotropy is most readily detected by performing refractive index measurements as will be showed in the following chapters (Mullin, 2001).

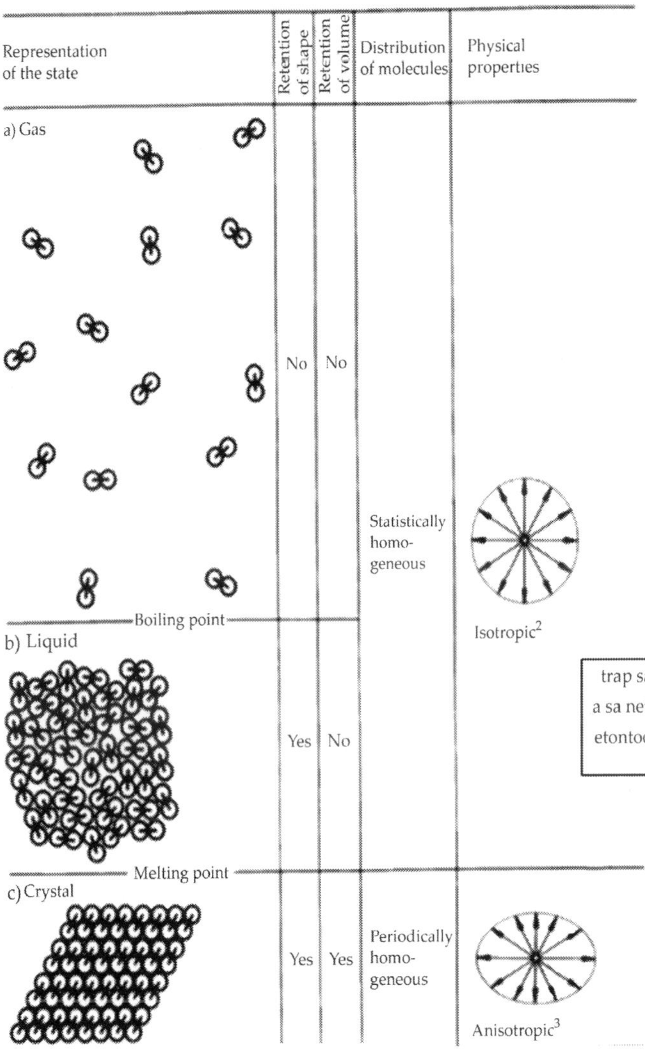

trap sa devomer eb tsum sih T
a sa nettirwer dna erugif eht fo
etontoof

Figure 1.1: Schematic representations of the states of matter. (a) gas, (b) liquid, (c) crystal. (Borchardt-Ott , 2011). Remove this citation. Email Borchardt and acquire permission to copy this material to be used as an illustration to a book entitled “ ”. Then include in the footnote above “With Permission from Author’s Name/s, Book/Journal Article, Publication details”

LIQUID CRYSTALS

In general, liquids are isotropic. However, some 200 cases of these substances are known to exhibit anisotropy in the liquid state at temperatures just above their melting point. These liquids are called 'liquid crystals'. Although this term is widely held, it is irrelevant; since the word 'crystal' points toward the presence of a particular space lattice. Though lattice formation is not possible in the liquid state, certain types of molecules under different circumstances can have variable form of molecular orientation. Accordingly, the term 'anisotropic liquid' is preferred over "liquid crystal". The name 'mesomorphic state' is used to refer to anisotropic liquids, since they are considered in between the true liquid and crystalline solid states (Mullin, 2001).

Liquid-crystalline polymers (LCPs) are characterized by having highly ordered, stiff, crystalline regions in the liquid phase. Therefore, a LCP is very different from the usual polymer material whose macromolecules are disordered during melting. The LCPs are capable of forming a liquid-crystalline intermediate state, called mesomorphous, between the liquid and the solid state. The stiff crystalline regions of a chain molecule are known as mesogens. If a LCP consists only of mesogens in the main chain, it cannot be processed like a thermoplastic since it melts at 400–600°C (750–1110°F) only (i.e., above its decomposition temperature). Thermotropic LCP chains, which melt at 250–400°C (480–750°F), can be obtained through selective creation of points of disturbance between the mesogenic regions of the macromolecule (DROBNY, 2012).

There are many examples of anisotropic liquids, the better-known of these examples include p-azoxyphenetole, p-azoxyanisole, cholesteryl benzoate, and ammonium oleate and sodium stearate. Despite the fact that these substances exhibit a sharp melting point, they melt to form a turbid liquid. With additional heating, the liquid abruptly turns out to be clear at some fixed temperature. With cooling, the reverse processes come about at the same temperatures as formerly. Anisotropy is revealed in the turbid liquid stage. The changes occurring in the physical state corresponding with change in temperature for the case of p-azoxyphenetole are:

$$Solid(anisotropic) \overset{137^{\circ}C}{\leftrightarrow} turbid\ liquid\,(mesomorphic) \overset{167^{\circ}C}{\leftrightarrow} clear\ liquid\,(isotropic)$$

The simplest representation of the phenomenon is provided by Bose's swarm theory, according to which orientation of molecules into a number of groups in parallel formation as illustrated in Figure 1.2. Regarding many respects, this demonstrates a rather similarity to the behavior of several logs floating down a river. Frequently, the substances that can be in the mesomorphic state are organic compounds, usually aromatic, with elongated molecules.

Regarding the mesomorphic state, it can be conveniently divided into two main classes. Firstly, the *smectic* (soap-like) state that is characterized by possessing an oily nature, and the flow of such liquids takes place by a gliding movement of thin layers over one another. Secondly, the *nematic* (thread-like) state that is characterized by normal flow of liquids like normal viscous liquids, however mobile threads can repeatedly be observed within the liquid layer. Furthermore, a third class, in which exhibition of a strong optical activity is noticed, is known as the *cholesteric* state; some workers consider this state as a special case of the nematic. The term of cholesteric state comes from the fact that most of the known examples are made up of cholesteryl compounds (Mullin, 2001).

(a) (b)

Figure 1.2: Isotropic and anisotropic liquids. (a) Isotropic: molecules in random arrangement; (b) anisotropic: molecules arranged into swarms (Mullin, 2001).

SOLID CRYSTALS

The true solid crystal consists of a rigid lattice of molecules, atoms or ions, which are characterized by having locations that are specific to the substance. Because of this regularity in the internal structure of the solid body, the crystal possesses a distinctive shape; smooth surfaces or faces develop as the crystals grow, and the planes of these faces are parallel to atomic planes in the lattice. However, in very rare situations, two crystals of a certain substance may appear matching; in fact, any two given crystals often look quite different in both size and external shape. In a way, this is not very surprising, as many crystals have grown under different conditions, principally the natural minerals. Still, few natural crystals have grown "free"; mostly under some limitation leading to stunted growth in one direction and much increased growth in another (Mullin, 2001).

CRYSTAL SYMMETRY

Crystals can be classified in terms of their crystal symmetry. Many (but not all) of the geometric shapes that look in the crystalline state are readily known to achieve some level of symmetry. The three simple elements of symmetry that can be recognized are:

- A *center* of symmetry: symmetry about a point
- An *axis* of symmetry: symmetry about a line
- A *plane* of symmetry: symmetry about a plane.

For example, the cube has 23 elements of symmetry; 1 center of symmetry, 13 axes of symmetry and 9 planes of symmetry (Figure 1.3). For instance, an octahedron has 23 elements of symmetry and is accordingly crystallo-graphically linked to the cube. A additional element of symmetry known as "compound, or alternating" symmetry about a "rotation-reflection axis" or "axis of rotary inversion" can be displayed by some crystals (Jones, 2002).

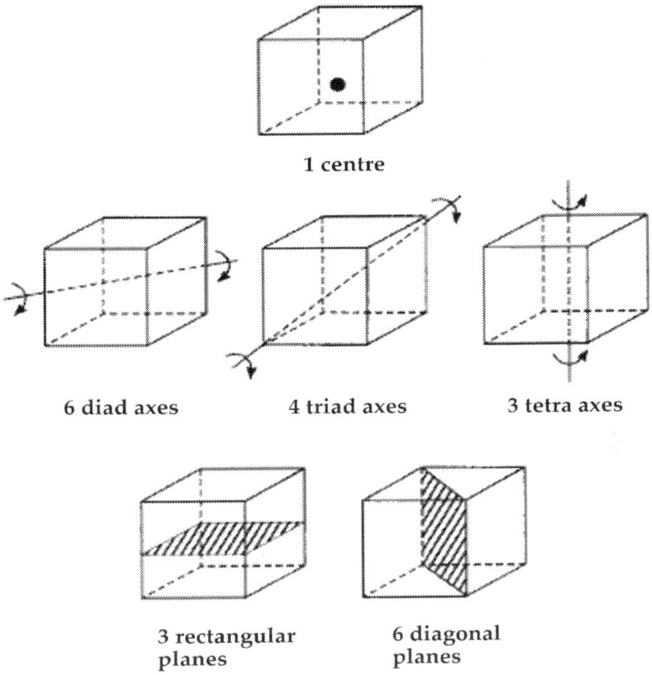

Figure 1.3: Symmetry in the cube (Jones, 2002).

Only 32 combinations are possible to obtain the above-mentioned elements of symmetry (including the asymmetric state), and these combinations are the 32 classes or point groups. For simplicity, these 32 classes are grouped into 7 systems categorized by the angles between their x, y and z axes, which are identified by the following names: regular (5 possible classes), tetragonal (7 possible classes), orthorhombic (3 possible classes), monoclinic (3 possible classes), triclinic (2 possible classes), trigonal (5 possible classes) and hexagonal (7 possible classes). The first six of these systems can be defined referring to the three (3) axes, x, *y* and *z*. The direction of *z* axis is vertical, and the x axis has a direction from front to back while the *y* axis is from right to left, as showed in Figure *1.4a*.

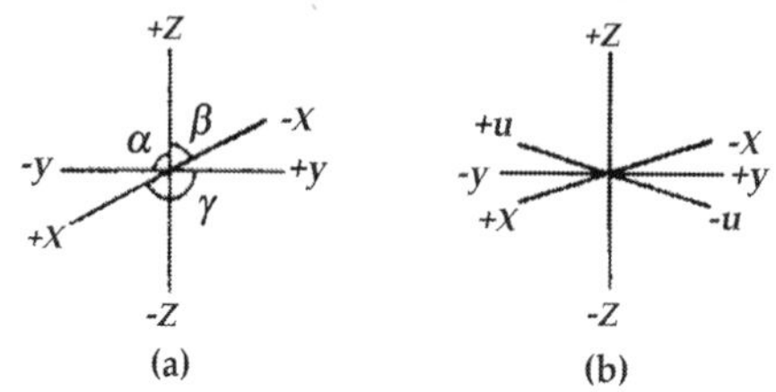

Figure 1.4: Crystallographic axes for describing the seven crystal systems: (a) three axes (b) four axes (hexagonal system) (Mullin, 2001).

Regarding the trigonal system, it differs from other systems in the sense that some workers favor to describe it with reference to its four axes. Descriptions of the seven crystal systems, accompanied by some of the other names occasionally employed, are mentioned in Table 1.1 and schematically presented in figure 1.5 (Mullin, 2001).

Table 1.1: The seven crystal systems (Mullin, 2001).

System	*Other names*	*Angles between axes*	*Length of axes*	*Examples*
Regular	Cubic Octahedal Isometric Tesseral	$\alpha = \beta = \gamma = 90°$	$x = y = z$	Sodium chloride Potassium chloride Alums Diamond
Tetragonal	Pyramidal Quadratic	$\alpha = \beta = \gamma = 90°$	$x = y \neq z$	Rutile Zircon Nickel sulphate. $7H_2O$
Orthorhombic	Rhombic Prismatic Isoclinic Trimetric	$\alpha = \beta = \gamma = 90°$	$x \neq y \neq z$	Potassium permanganate Silver nitrate Iodine α-Sulphur
Monoclinic	Monosymmetric Clinorhombic Oblique	$\alpha = \beta = 90° \neq \gamma$	$x \neq y \neq z$	Potassium chlorate Sucrose Oxalic acid β-Sulphur
Triclinic	Anorthic Asymmetric	$\alpha \neq \beta \neq \gamma \neq 90°$	$x \neq y \neq z$	Potassium dichromate Copper sulphate. $5H_2O$
Trigonal	Rhombohedral	$\alpha = \beta = \gamma \neq 90°$	$x = y = z$	Sodium nitrate Ruby Sapphire
Hexagonal	None	z axis is perpendicular to the x, y and u axes, which are inclined at 60°	$x = y = u \neq z$	Silver iodide Graphite Water (ice) Potassium nitrate

***Minimize the table above (table 1.1) to fit into the previous page

Regarding the regular, tetragonal and orthorhombic systems, they are characterized by having the three (3) axes x, *y* and *z* mutually perpendicular. However, the systems display some difference in the relative lengths of these axes: they are all equal in the regular system; they are all unequal in the orthorhombic system; and the tetragonal system has two equal axes and a different third axis. Concerning the monoclinic and triclinic systems, they have unequal three axes; in the former, two of the angles are 90° and the third angle is different, while in the latter, all three angles are unequal and none of them is equal to 90°. Occasionally the restriction ‹not equal to 30°, 60° or 90° is also applied to the triclinic system. In the trigonal system, three equal axes intersect at equal angles, but the angles are not 90°. The hexagonal system is described to have four axes. The axis of six-fold symmetry (hexad axis) is usually chosen as the *z* axis, and the other three equal-length axes, located in a plane at 90° to the *z* axis and intersect with one another at 60° (or 120°).

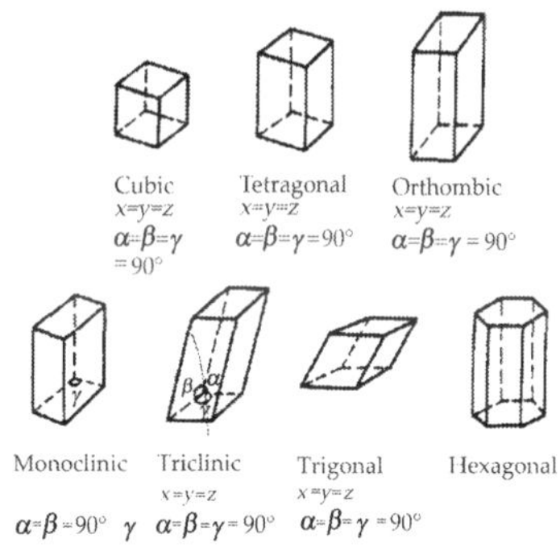

Figure 1.5: The seven crystal systems (Mullin, 2001).

Every crystal system comprises several classes that display only a partial symmetry; for illustration, only one-half or one-quarter of the determined number of faces allowed by the symmetry may have been developed. The maximum number of similar faces can be exhibited in the *holohedral* class, i.e. that class that owns the highest degree of symmetry. In the *hernihedral* class, only half the determined number of faces have been developed, and in the *tetrahedral* class only one-quarter of the determined number of faces have been developed. For example, the regular tetrahedron (4 faces) is the hemihedral form of the holohedral octahedron (8 faces) and the wedge-shaped sphenoid is the hemihedral form of the tetragonal bipyramid *(Figure 1.6).*

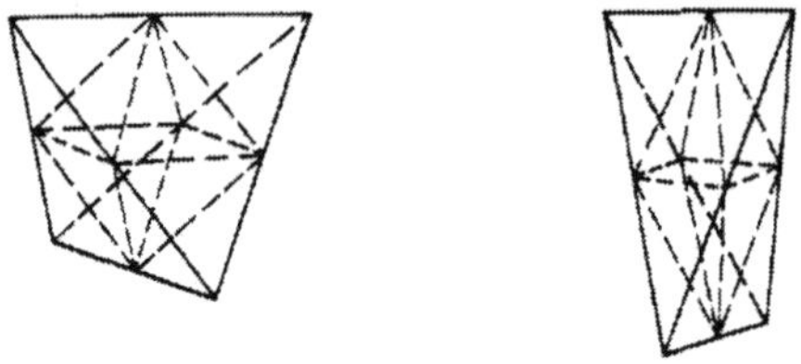

Figure 1.6: Hemihedral forms of the octahedron and tetragonal bipyramid (Mullin, 2001).

Miller indices

Numerical definition of the shape of crystals in relation to their faces is performed by using the Miller indices. Description and numbering of all the faces of a crystal can be performed in terms of their axial intercepts, which are usually three (3) though sometimes four are required. For example, if a decision has been made about three crystallographic axes, the standard or parametric plane is chosen on basis that it is inclined to all three axes. The intercepts X, Y, Z of this plane on the axes x, y, and z are named parameters a, b and c. The ratios of the parameters *a:b* and *b:c* are named the axial ratios. Conventional reduction of the values of the parameters can be done so that the value of b is unity.

The Miller indices, h, k and l of the crystal faces or planes are then defined by;

$$h = \frac{a}{X}, k = \frac{b}{Y}, l = \frac{c}{Z} \qquad (1.1)$$

For the parametric plane, the axial intercepts X, Y, and Z are the parameters a, b and c, so that indices, h, k and *l* are a/a, b/b and *c/c,* i.e. 1, 1 and 1, usually written (111). Calculation of the indices for the other faces of the crystal is conducted from the values of their respective intercepts X, Y, and Z, and these intercepts can constantly be symbolized by ma, nb and *pc,* where m, n and p were whole numbers.

Regarding the cube (see Figure 1.7), none can be chosen as the parametric plane, because no face is included (111). The intercepts Y and

Z of face A on the axes y and z are at infinity, so the Miller indices h, k and 1 for this face will be a/a, b/∞ and c/∞, or (100). Correspondingly, faces B and C are designated (010) and (001), respectively (Jones, 2002).

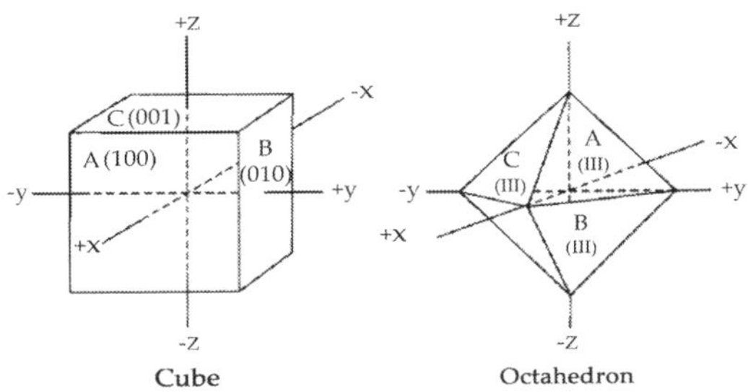

Figure 1.7: Miller indices: cube and octahedron (Jones, 2002).

THE LATTICE AND ITS PROPERTIES

All crystals always contain certain arrangement of a three-dimensional periodic arrangement of atoms, ions or molecules. This arrangement is particularly observable for the a-polonium crystal illustrated in Figure 1.8a. If every atom has a simple representation by its center of gravity, what remains is a *point* or *space lattice* (Figure 1.8b).

The description of a point or space lattice can be done as a three-dimensional periodic arrangement of points, and it is considered a pure mathematical concept. Establishment of the concept of a lattice will now be developed from a lattice point through the line lattice and the plane lattice, lastly to the space lattice.

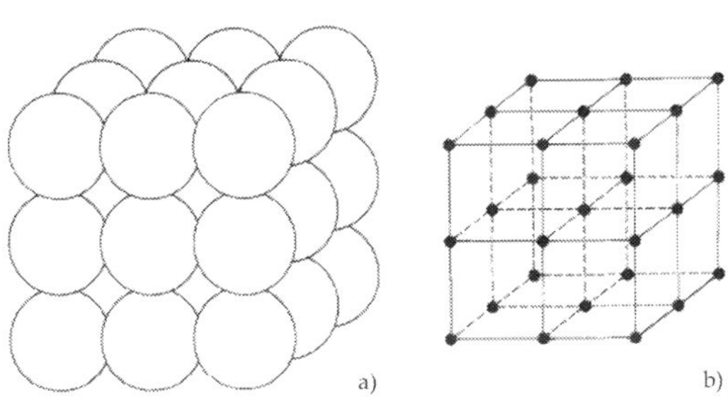

Figure 1.8: Three-dimensional periodic arrangements of the atoms in a crystal of a-polonium (a) and the space lattice of the crystal (b) (Borchardt-Ott , 2011).

Line Lattice

As showed in Figure 1.9a**,** we may consider moving from point 0 along the vector to point 1. By a similar movement of 2, we will reach point 2, etc. Through this movement, one point is brought into coincidence with another point, and this operation is repeated. Such operation is known as a *lattice translation*, and results in generating a *line lattice*. The points that may be brought into coincidence with one another by a lattice translation are termed *identical points or point's equivalent* by translation. is called the lattice parameter, and the one-dimensional lattice is entirely defined by this constant alone.

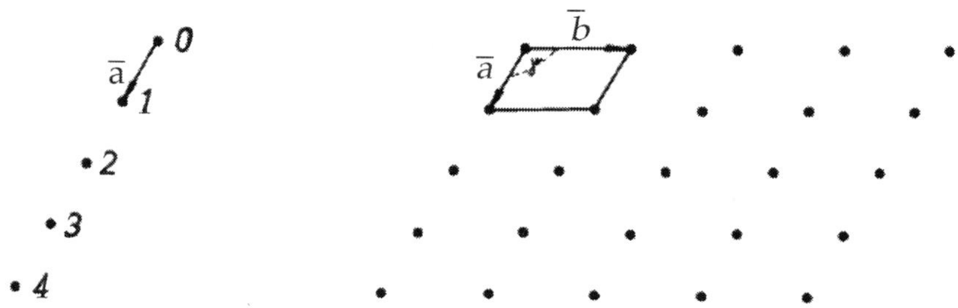

Figure 1.9: (a) Line lattice with its lattice parameter (b) Plane lattice with the unit mesh defined by the vectors and (Borchardt-Ott , 2011).

Plane Lattice

When a lattice translation is then permitted to operate on the line lattice in Figure 1.9a, the result is the plane lattice or plane net (see Figure 1.9b). The vectors and define a unit mesh. The whole plane lattice may now be constructed from the knowledge of three lattice parameters, , and γ, the included angle. If any random lattice translation caused any point to move, it will come into coincidence with another point. Therefore, a plane lattice has lattice translations not only parallel to and , but also to any number of combinations of them, i.e. an unlimited number of lattice translations.

Space Lattice

If another lattice translation is now introduced in a direction not coplanar with and , its action on the plane lattice in Figure 1.9b produces the space lattice presented in Figure 1.10. The operations of three-dimensional lattice translations can also produce this space lattice. Contrasting a finite crystal, a space lattice is infinite.

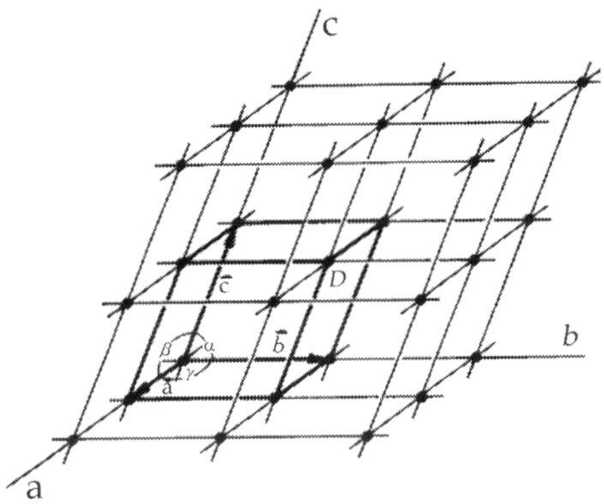

Figure 1.10: Space lattice with the unit cell defined by the vectors , and (Borchardt-Ott , 2011).

In relation to the arrangement of the vectors , and , introduction of an *axial system* with the *crystallographic axes a, b and c* may be done. The vectors , and and their respective crystallographic axes a, b and c are selected to be *right-handed.* So, if the right thumb points in the direction of (a) and the index finger is along (b), the middle finger will eventually point in the direction of (c). Consequently, a lattice, or a crystal described by it, may always be positioned so that (a) points toward the observer, (b) toward the right, and (c) upwards, as exhibited in Figure 1.10

The vectors , and outline a *unit cell,* which may alternatively be described by six lattice parameters:

length of lattice translation vectors	*interaxial lattice angles*
$\lvert\vec{a}\rvert = a_0$	$\vec{a} \wedge \vec{b} = \gamma$
$\lvert\vec{b}\rvert = b_0$	$\vec{a} \wedge \vec{c} = \beta$
$\lvert\vec{c}\rvert = c_0$	$\vec{b} \wedge \vec{c} = \alpha$

Additional application of lattice translations to the unit cell will produce the entire space lattice. Thus, the entire lattice is completely defined by the unit cell. Every unit cell is known to have eight vertices and six faces. An identical point exists at all vertices. Can these points be considered as a part of the unit cell? Figure 1.10 shows the lattice point D, which is not only part of the marked-out unit cell, but also part of all eight cells which meet at that point. In other words, only one eighth of it may be attributed to the marked unit cell, and since 8 x = 1, the unit cell contains only one lattice point. Such unit cells are termed simple or primitive, and are symbolized by P (Borchardt-Ott , 2011).

THE DESIGNATION OF POINTS, LINES AND PLANES IN A SPACE LATTICE

The Lattice Point uvw

Each lattice point is distinctively defined according to the origin of the lattice by the vector . The lengths of , and are simply the lattice parameters, hence only the coordinates u, v and w need to be specified. They are written as a "triple" uvw. In Figure 1.11, the vector describes the point 231 (which is read as two-three-one). The coordinates u, v and w normally are integers, but can also have values of other integer forms such as . When they possess integral values, the points uvw are the coordinates of the points of a P-lattice. The coordinates of the vertices of a unit cell are given in Figure 1.11.

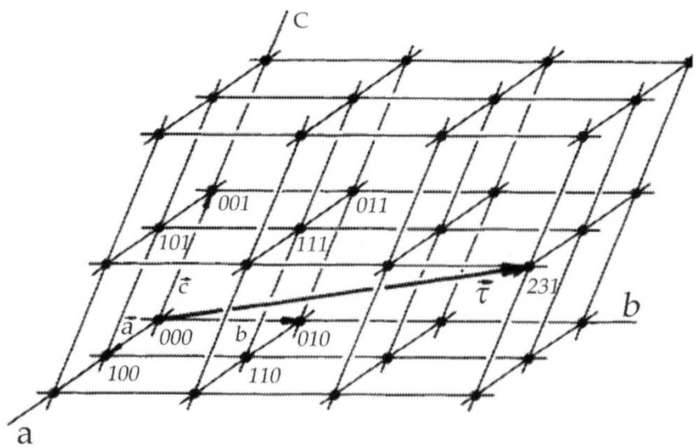

Figure 1.11: Designation of lattice points using the coordinates uvw that define the vector from the origin to the lattice point uvw, (Borchardt-Ott , 2011).

Lattice Lines [uvw]

Mathematical specification may be applied to a line in any coordinate system by two points. The lattice line I in Figure 1.12 contains the points 000 and 231. Since the lattice line usually passes through the origin, the other point on its own states to the direction of the line in the lattice, and the coordinates of this point define the line. Therefore, they are put in square brackets [231], or in general [uvw], to display that they characterize the direction of a line.

The lattice line II' passes through the points 100 and 212. Line II is parallel to this line, and passes through the origin along with the point 112 and subsequently both lines may be referred to by the representation [112].

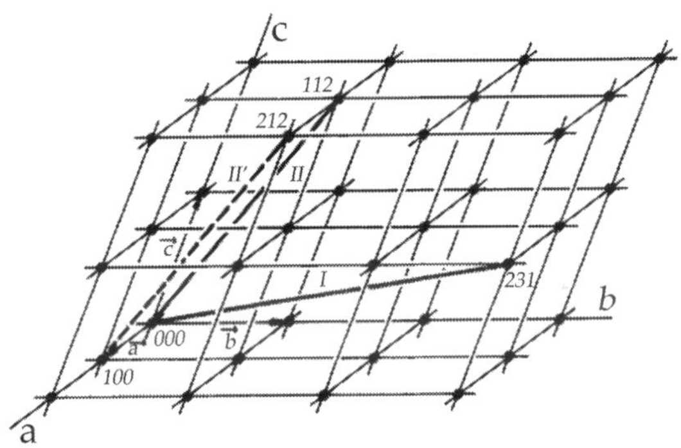

Figure 1.12: Designation of lattice lines using the coordinates [uvw] (in square brackets) that define the vector from the origin to the given point (I: [231], II: [112]) (Borchardt-Ott , 2011).

Lattice Planes (hkl)

If a plane in the lattice intersecting the axes a, b and c at the points m00, 0n0 and 00p, the position of a lattice plane is completely defined by the coordinates of the three intercepts (Figure 1.13). Though, usually the reciprocals of these coordinates are used instead of the coordinates themselves:

$$\text{a-axis: } h \sim \frac{1}{m}$$

$$\text{b-axis: } k \sim \frac{1}{n}$$

$$\text{c-axis: } l \sim \frac{1}{p}$$

For the reciprocal intercepts, the smallest integral values are chosen and they are then written as a triple (hkl) in round brackets.

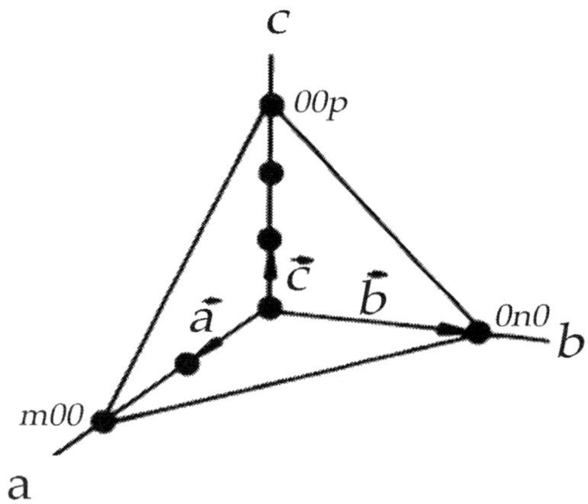

Figure 1.13: The intercepts on the axes of a lattice plane with the Miller indices (362) (Borchardt-Ott , 2011).

In the space lattice displayed in Figure 1.14, two lattice planes have been drawn in

	m	n	p	$\frac{1}{m}$	$\frac{1}{n}$	$\frac{1}{p}$	(hkl)
I	1	1	1	1	1	1	(111)
II	1	2	2	1	$\frac{1}{2}$	$\frac{1}{2}$	(211)

**rewrite the lattice above generating your own Table since this is not a figure… this should be in tabular form and you can label this one another table.

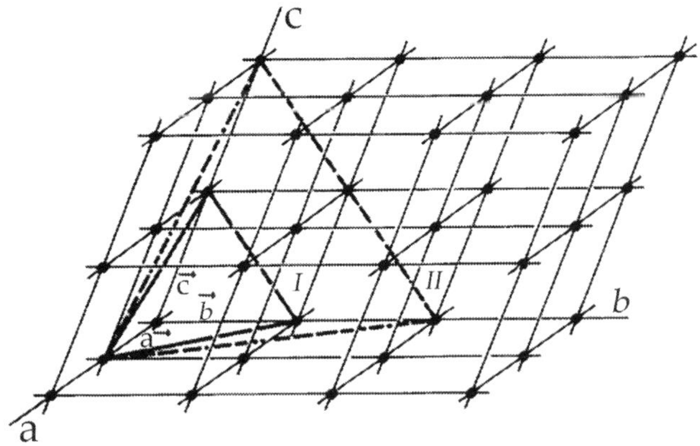

Figure 1.14: The indexing of lattice planes by using Miller indices, the smallest integral multiples of the reciprocals of the intercepts on the axes; I (111), II (211) (Borchardt-Ott , 2011).

In Figure 1.15, a projection of a lattice is presented in common with the lines demonstrating the traces of lattice planes perpendicular to the plane of the paper and parallel to the c-axis. These lattice planes are indexed as follows:

	m	n	p	$\frac{1}{m}$	$\frac{1}{n}$	$\frac{1}{p}$	(hkl)
A	2	4	∞	$\frac{1}{2}$	$\frac{1}{4}$	0	(210)
B	$\frac{3}{2}$	3	∞	$\frac{2}{3}$	$\frac{1}{3}$	0	(210)
C	1	2	∞	1	$\frac{1}{2}$	0	(210)
D	$\frac{1}{2}$	1	∞	2	1	0	(210)
E	–	–	–	–	–	–	
F	$\frac{\bar{1}}{2}$	$\bar{1}$	∞	$\bar{2}$	$\bar{1}$	0	(210)
G	$\bar{1}$	$\bar{2}$	∞	$\bar{1}$	$\frac{1}{2}$	0	(210)
H	3	$\bar{2}$	∞	$\frac{1}{3}$	$\frac{1}{2}$	0	(230)

**again rewrite this into a table…

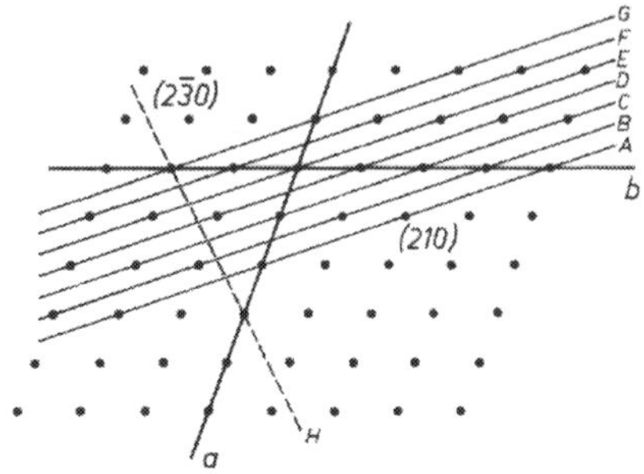

Figure 1.15: The projection of a space lattice along c onto the a, b-plane. The lines *A-G* are the traces of the lattice planes parallel to c with the Miller indices (210). The line *H* is the trace of a lattice plane (230) (Borchardt-Ott , 2011).

Considering the lattice plane A - G, they belong to a set of equally spaced, parallel planes bringing about the matching indices. Commonly, they are all defined by the triple (hkl), which symbolizes not only a single lattice plane, but an infinite set of parallel planes as well. The plane E intersects the origin, so it cannot be indexed in this position. It should be noticed that (210) and (0) define the same parallel set of planes. The projection of a lattice onto the a,b-plane is exhibited in Figure 1.16. This projection is characterized by having a square unit cell. The projections of six sets of lattice planes with several indices are presented as well. With raising of the indices, reduction of the spacing between the planes happens, as does the density of points on each plane. Planes which are parallel to b and c will thus only have intercepts with a and are indexed as (100). In the same way, (010) intersects only the b-axis, and (001) only c. The line H in Figure 1.15 is the trace of the plane (20).

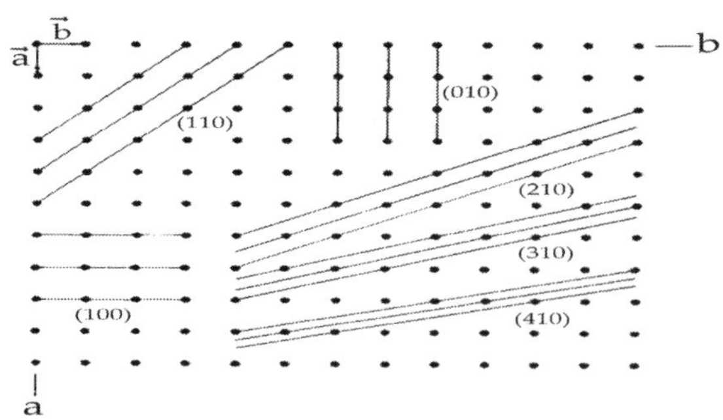

Figure 1.16: Projection of a lattice onto the a,b-plane, with the projections of six sets of lattice planes (Borchardt-Ott , 2011).

CRYSTAL DEFECTS

Crystalline solids always lack perfection at temperatures above absolute zero. All lattice sites are occupied in a wholly regular manner. Imperfections exist. Establishment of such sites is characterized by being endothermic; it requires a certain amount of energy to be visible. Point defects happen, of which the main two are a vacant lattice site, and an interstitial atom that occupies a volume between a group of atoms on normal sites that affects crystal purity.

Since a diversity of lattice defects with different density will be contained in crystals, these defects can have a role in determining the properties of the crystals that include mechanical strength or electric conductivity. Also, lattice defects play an essential role in the growth of the crystals, specifically, the screw dislocation.

Dimensionality, D is a characteristic of lattice defects; through which they extend into the lattice. This can be clarified by often encountered defects (Table 1.2).

Table 1.2: Lattice defects ordered by their dimensionality D (Jones, 2002).

Defect	D	Example
Point	0	Vacancy, interstitial atom
Line	1	Step dislocation, screw dislocation
Surface	2	Twin, small-angle grain/tilt boundary
Volume	3	Inclusion

** make your own table. This is a figure at its present state.

Figure 1.17 displays a two-dimensional lattice having two different defects, a vacancy and an interstitial atom. The two defects are restricted to a point and do not have an extension into the lattice, thus their dimensionality is D = 0. An example for line defects is screw dislocations, which can be produced by a shearing of the crystal (Figure 1.18). This defect extends in one dimension along the line shown, so that D = 1.

The most noticeable example of faults is twinning, with D = 2. The fault extends over a complete plane in the crystal (Figure 1.19). Concerning a face-centered cubic (fcc) lattice, twinning can arise by stacking faults; the order of ABCABC . . . is inverted at a specific point to ABCBAC . . . . Regarding the outer shape of the crystal, twinning will seem as an inflection, or forming of a mirror image, a twin. Figure 1.19 displays a twin of a silver halide crystal, in which the twin plane is clearly seen.

Lastly, the inclusion of mother liquid can form volume defects D =3 (Figure 1.20). Interestingly, the liquid inclusions can be kept by well-expressed low-indexed (negative) faces. Liquid inclusions, for instance, occur under high growth rates, such as in precipitations.

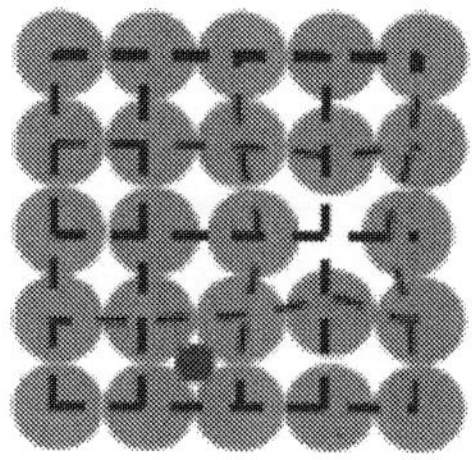

Figure 1.17: Two lattice defects: vacancy and interstitial atomfaces in a two-dimensional lattice. These lattice defects are limited to a point; thus, D = 0 (Beckmann, 2013).

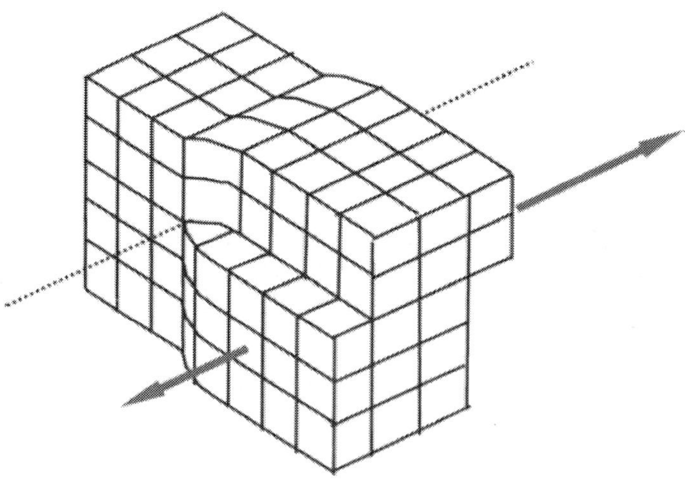

Figure 1.18: Screw dislocations as produced by a sheering of the lattice. This defect can be found along the dotted line, leading to a dimensionality of D = 1 (Beckmann, 2013).

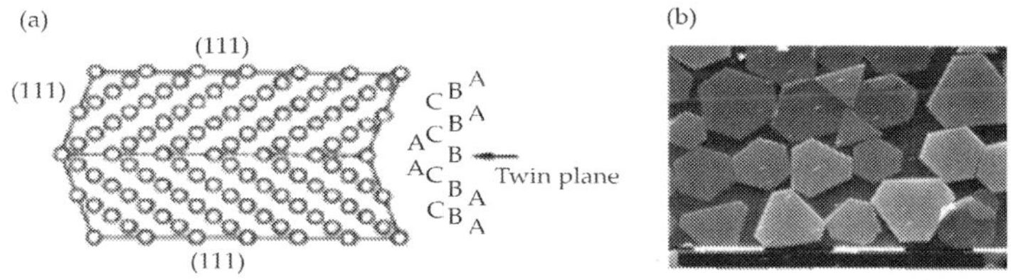

Figure 1.19: Establishment of a twin in a fcc lattice. (a) The stacking ABC is reversed at the twin plane to ABCABACBA. The lattice defect extends in the twinning plane, D = 2. (b) A twin of a silver halide crystal is presented and the twin plane can be obviously seen using (what method?... X-ray diffraction, SEM, plain X-ray… please indicate) Then emphasize by putting an arrow or any mark the silver halide crystal and its twin plane (Beckmann, 2013).

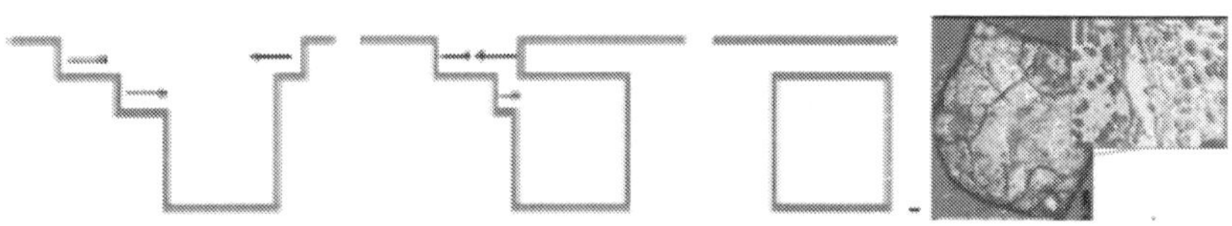

Figure 1.20: Creation of volume defects such as liquid inclusion via the overgrowth of mother liquid. Restriction of these liquid inclusions can be done by well-expressed low-indexed negative faces, D = 3 (Beckmann, 2013). ** label this figure a and b. Then in figure b..the xray image, indicate where is the defect thru arrows or circles.

PARTICLE CHARACTERISTICS

The relatively wide range of particle sizes encountered is considered a characteristic interesting feature of industrial crystallization systems. Particle sizes range over several orders of magnitude from the sub-micron (nanometers) to several millimeters or more, in other words, from "colloidal" to "coarse" particles. These involve a large part of the world on a human scale and an unlimited source of industrially generated wealth.

The purely physical laws mainly control the behavior of very large particles. Additional decrease of the particle size range causes a rapid increase of the specific surface area, more properly termed as surface area per unit mass. In addition, Chemical effects such as in the nucleation and growth of crystals, is another important phenomenon in the development of crystals. Consequently, much interest is being attributed in the study of particulate systems of different size range, within the scope of many fields such as chemical engineering, physical chemistry and materials science.

The distinct form that characterizes the solid particles greatly affect their appearance, product quality and processing behavior. Consequently, particulate solids need to be additionally characterized by particle size and shape, as well as chemical composition. Moreover, generation of particles can be done at any point inside the process. For instance, within a crystallization process, nucleation happens and large particles are broken down to abundant smaller ones in a comminution chamber or simply within the drier.

So, the formation of particle populations with a range of characteristics, is considered a common feature of crystallization process systems. Yet, real particles are seldom just the same form as each other and irregular particles possess several individual dimensions - they are actually "multi-dimensional", in common with several angles and faces giving rise to the analogous problem of mathematical dimensionality (Kaye, 1986). Thus, some simplification of the problem is employed to analyze such particulate populations to first, necessary to define what is meant by "particle size", "size distribution" and "particle shape" (or "habit" in crystallization parlance). Conventionally, this starts with a consideration of the simplest single particles and then trying to relate real particles to them (Jones, 2002).

Single particles

Particle size: the dimensions of particles are expressed in terms of *length;* with units: 1 m = 10^3 mm = 10^6 μm = 10^9 nm. The micrometer (micron) is probably the most common unit used in practice. The dimensions of individual particles of various shapes will now be considered.

Sphere

The sphere is unique. It is the only particle whose form can be totally described by a single dimension: its diameter, d. This dimension is primarily used to determine the spherical volume ; Volume = $\pi d^3/6$ as well as the area; Area = πd^2. Because of the uniqueness and simplicity of the sphere, the characteristics of non-spherical particles (which most real ones are) are often related back in some way to the size of an "equivalent" sphere which has some shared characteristic, such as the same volume or surface area.

Cube

The cube and other regular geometric shaped crystals, possess more than one dimension. In the cube, there is a choice for the characteristic dimension. Firstly, the side length, L could be selected whence the surface area = $6L^2$ and volume = L^3. Alternatively, the cube 'diameter' could be chosen, i.e. the distance between diagonally opposed corners.

Irregular particles

Irregular particles have a unique maximum and/or sometimes a mean characteristic chord length used as a characteristic dimension. Normally, however, irregular particles are characterized by some form of 'equivalent dimension'.

Equivalent dimensions

Any 'size' quoted for real particles normally depends on the method of measurement and is usually expressed in terms of an equivalent dimension of which there are two main types:

1. Equivalent diameter (d): This is frequently used and is usually related to the sphere.
2. Characteristic length (L): This is most frequently determined by sieving.

Particle shape

Qualitative description of particle shape can be performed using standard terms such as *granular, flaky, needle,* etc. Quantitative description, however, requires use of 'shape factors' that enables a particle's surface area and volume to be calculated by knowing its 'size'.

Shape factors

Shape factors can relate the characteristic dimension of a particle to its surface area or volume, respectively. Any characteristic dimension can be chosen, but the corresponding shape factor "is unique" to every characteristic dimension.

Surface shape factor (f_s)

The surface shape factor f_s relates the external surface area A of a particle to its characteristic dimension L:

(equation no???)

For a sphere with a characteristic diameter $L = d$, $f_s = \pi$. For all other particles $f_s > \pi$. For example, for a cube characteristic side length, L then $f_s = 6$.

Volume shape factor (f_v)

The volume shape factor, f_v relates the particle solid volume, V to its characteristic dimension, L.

(equation no???)

For a sphere: $f_v = \pi/6$. For all other particles $fv > \pi/6$. For example, for a cube of characteristic side length, L then $fv = 1.0$.

Specific surface shape factor (F)

The specific surface shape factor F is simply the ratio of surface to volume shape factors respectively considered above, i.e.

(equation no???)

For both a sphere of diameter d and cube of characteristic side length L, $F = 6$. Note the ambiguity of using F alone to characterize particle shape factors.

Sphericity (Ψ)

The sphericity Ψ quantifies the ‹deviation› of the particle from spherical

Ψ = surface area of sphere/surface area of particle (of equal volumes)

For the sphere $\Psi=1$, for all other particle shapes $\Psi < 1$.

Specific surface (S_p)

The specific surface S_p defines the surface area, A per unit volume, V (or mass) of particle

(Equation no???)

Therefore, the specific shape factor F is the ratio between surface shape factor, f_s and that of the volume shape factor, f_v, and is then directly proportional to the specific surface (S_p) and the cube characteristic length, L.

(Equation no???)

Solid state bonding

Crystalline solids may be specified according to the method of bonding in the solid state in four main types; ionic, covalent, molecular and metallic. There are materials intermediate between these classes, but most crystalline solids can be classified as predominantly one of the basic types.

The *ionic crystals* (e.g. sodium chloride) are composed of charged ions held in place in the lattice by electrostatic forces, and separated from the oppositely charged ions by regions of negligible electron density. In *covalent crystals* (e.g. diamond) the constituent atoms do not carry effective charges; they are connected by a framework of covalent bonds, the atoms sharing their outer electrons. *Molecular crystals* (e.g. organic compounds) are composed of discrete molecules held together by weak attractive forces (e.g. π-bonds or hydrogen bonds).

Metallic crystals (e.g. copper) is composed of ordered arrays of identical cations. The constituent atoms share their outer electrons, but these are so loosely held that they are free to move through the crystal lattice and confer 'metallic' properties on the solid. The free motion of the electrons is a metallic crystal produces a "sea of electrons" (imagine the ocean waves are to be made up electrons not water) which allows them to conduct heat and electricity. Ionic, covalent and molecular crystals are essentially non-conductors of electricity because the electrons are all locked into fixed quantum states. This means that metals are good conductors of heat and electricity because of the presence of mobile electrons.

Semiconducting crystals (e.g. germanium) are usually covalent solids with some ionic characteristics, although a few molecular solids (e.g. some polycyclic aromatic hydrocarbons such as anthracene) are known in which under certain conditions a small fraction of the valency electrons are free to move in the crystal. The electrical conductivity of semiconductors is electronic in nature, but it differs from that in metals. Metallic conductivity decreases when the temperature is raised, because thermal agitation exerts an impeding effect. On the other hand, the conductivity of a semiconductor increases with heating, because the number of "electron-hole pairs", the electricity carriers in semiconductors, increases greatly with temperature. Metals have electrical resistivity in the ranges 10^{-8} to 10^{-6} $S^{-1}m$. Insulators cover the range of 10^{8} to 10^{20} (diamond) and semiconductors in 10 to 10^{7} $S^{-1}m$.

The presence of impurities can affect the electrical conductivity of a semiconductor. For instance, if x silicon atoms in the lattice of a silicon crystal are replaced by x phosphorus atoms, the lattice will gain x electrons and a negative (n-type) semiconductor results. On the other hand, if x silicon atoms are replaced by x boron atoms, the lattice will lose x electrons and a positive (p-type) semiconductor is formed. The impurity atoms are called 'donors' or 'acceptors' according to whether they give or take electrons to or from the lattice (Mullin, 2001).

CHAPTER 2

PHYSICAL, CHEMICAL AND THERMAL PROPERTIES OF CRYSTALLINE SYSTEMS

INTRODUCTION

Generally, any product obtained from a chemical process must be characterized. If the product is a solid, an even larger set of characterization data is required than that required for a liquid or a gas. Concerning liquids and gases, they are homogeneous on a molecular level and therefore the properties of the bulk are identical to the properties of any volume element larger than a few molecules. On the other hand, a solid can be non-homogeneous so the properties of any volume element may be different from each other. If products from crystallization processes are considered, the product is considered homogeneous on a medium scale, however, it is still essential to determine both the intrinsic properties of the solid such as the particle shape, particle size, and particle size distribution as well as the mechanical properties of the solid such as powder flow properties (Beckmann, 2013).

Density

Solids

The density of solids is characterized by theoretical density (ρ_C^*.) of a crystal which is calculated from equation (2.1) as described below

$$\rho_C^* = \frac{nM}{VN} \qquad (2.1)$$

Where n is the number of formula units in the unit cell, M is the molar mass of the substance, V is the volume of the unit cell and N is the Avogadro number (6.023 x 10^{26} $kmol^{-1}$).

The density of solids has non-significant temperature dependence, but in some industrial crystallization purposes thermal expansion coefficient should be taken in consideration. For example, sodium chloride (salt) density is decreased by about 0.7 per cent when heated from 10 to 80°C (Mullin, 2001). In the actual, the density of a solid particles can assume different values depending on how the solid density is measured. Crystal density or true density only includes the solid fraction of material. The particle density includes the volume due to the intraparticulate pores and the bulk density or apparent density which further includes the interparticulate void volumes.

Liquids

In contrast to solids, liquid density has significant temperature dependence. Specific gravity of the liquid is defined as the ratio of the density of a given liquid at specific temperature to the density of a standard substance (which is water) at the same temperature. Liquid specific gravity may be also defined as the ratio between liquid density at specific temperature to the density of water at another temperature.

For example, specific grity for a certain liquid which is denoted by 20°C/4°C is mathematically equal to the liquid density at 20 °C expressed in g cm^{-3} compared to water density at 4°C as water density is maximum at this temperature (1 g cm^{-3}).

Liquid density is measured by hydrometer, which considered the simplest way to measure liquid density. A Hydrometer is composed of a float with a graduated stem. It is calibrated at a certain temperature marked on the body of the hydrometer, so liquid density should be measured at that temperature to get acceptable accuracy. There are other accurate methods to measure liquid density rather than hydrometer. The pyknometer and the Westphal balance are two other popular methods in doing so.

Bulk solids

The trapped air in solids results in the formation of void space in its solid-state structure. This makes the structure of solids porous. The number of void spaces contributes to the void volume of the solid and results in another type of density term called the bulk density or apparent density. So bulk density represents a non-fixed property for density which depends on void space in a particular solid. The relationship between the densities of the solid particles, ρ_s, and the bulk solids density, ρ_{BS}, is

$$\rho_{BS} = \rho_S (1-\varepsilon) \quad (2.2)$$

Where ε is the void space coefficient, the volume fraction of voids, which is significantly dependent on particle shape, particle size distribution and the packing of the particles. Also, ε. can vary significantly depending on the formation or processing method of particulate material. For example, when solids are being poured or transferred to another location or container, the increase in ventilation increases the bulk density while it is decreased by vibration, which happens after transportation in packaged form.

Viscosity

The viscosity of a fluid is a meure of its resistance to gradual deformation by shear stress or tensile stress. For liquids, it corresponds to the informal concept of "thickness"; for instance, sugar syrups have a much higher viscosity than water. However, heating the sugar syrup promotes its ability to flow easier, since the viscosity is lowered. As such, changes in the viscosity of a given material is expected depending on its temperature. Significantly, viscosity depends on temperature as it decreases with increasing temperature. This relationship between viscosity and temperature can be confirmed by equation 2.3.

$$\eta = A\exp(-B/T) \qquad (2.3)$$

Where, A and B are constants and T is the temperature expressed in Kelvins.

The absolute viscosity is measured by the poise (P) (1 g cm^{-1} s^{-1}) and its useful sub-multiple the centipoise (cP), have now been replaced by the SI unit (kg m^{-1} s^{-1}) which is generally written as Pa s and sometimes as N s m^{-2} (Mullin, 2001).

The relationships between these units as follows:

$$1cP = 0.01P = 1mPa{\cdot}s = 1mN{\cdot}s\ m^{-2} = 10^{-3}\,Kg\ m^{-1}s^{-1} = 2.42\ Ib.ft^{-1}h^{-1}$$

Similarly, the once common units of kinematic viscosity (= absolute viscosity/ density), the stokes (St) (1 cm^2 s^{-1}) and its sub-multiple the centistokes (cSt), have been replaced by the SI unit (m^2 s^{-1}). The following relationships hold:

$$1\ cSt = 0.01\ St = 10^{-6}\,m^2 s^{-1} = 0.0388\,ft^2 h^{-1}$$

As mentioned before, viscosity depends significantly on temperature. As a result, Plots of $log\eta$ versus T^{-1} or $log\eta$ versus log T commonly produce straight lines and this property is usually used for calculating viscosities at temperatures within the range covered. Generally, adding dissolved solids to water increases the viscosity, although this rule is known to have few exceptions. Usually, the viscosity has a considerable increase. For example, the system sucrose-water, in which the viscosity increases from around 2 to 60 mPa·s for a concentration increase from 20 to 60 g/100 g of solution at 20 °C (Mullin, 2001).

Figure 2.la illustrates the example of a solute that causes a decrease in the viscosity of the solvent; this system (KI-water) exhibits a minimum viscosity. Many other potassium and ammonium salts exhibit a similar behavior as well. Figure 2.lb displays the effect of concentration and temperature on the ethanol-water system that shows a maximum viscosity.

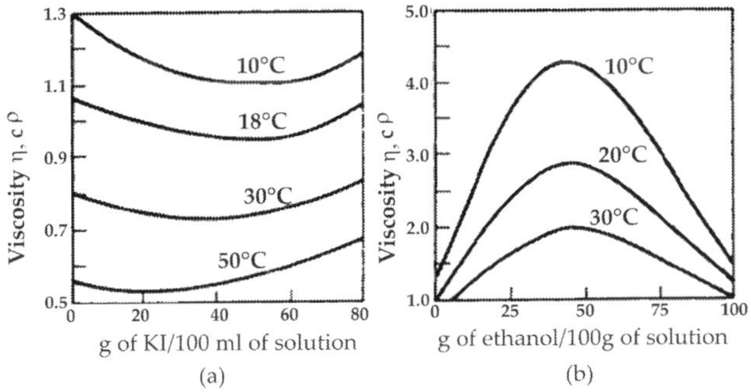

Figure 2.1: Aqueous solutions exhibiting (a) minimum, (b) maximum viscosities (Mullin, 2001).

The measurement of liquid viscosity has been performed through developing numerous instruments, the basis for many of these instruments depends on the flow of a fluid through a capillary tube and the application of Poiseuille's law in the form:

$$\eta = \frac{\pi r^4 \Delta P}{8lV} \qquad (2.4)$$

where ΔP = the pressure drop across the capillary of length, l and radius, r; and V = volume of fluid flowing in unit time. Figure 2.2 shows one simple type of U-tube viscometer. The liquid under test is suctioned (using a rubber tube connected to a compressor) into leg B to its mark (z). Then, the tube is held vertical, the temperature of the liquid is measured and maintained constant. The liquid is then sucked up into leg A to a point above x and the time t for the meniscus to fall from x to y is recorded. Calculation of the kinematic viscosity of the liquid v can be done from

$$v = kt \qquad (2.5)$$

Where, k is a constant parameter for the apparatus, determined by measurements on a liquid of well-known viscosity, e.g. water, and t is the time it takes for the meniscus to fall from x to y.

Another method known to measure viscosity is the falling-sphere method which is used particularly with solids of very high viscosity. Stokes' law could be applied in the form:

$$\eta = \frac{(\rho_S - \rho_l)d^2 g}{18u} \qquad (2.6)$$

Where d, ρ and u are the diameter, density and terminal velocity, respectively, of a solid sphere falling in the liquid of density ρ_l.

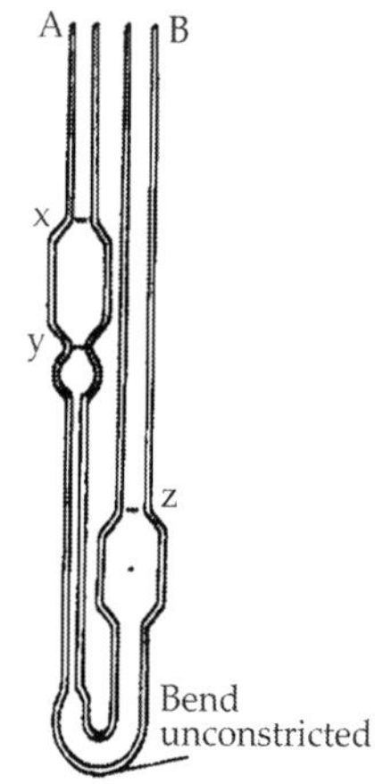

Figure 2.2: Simple U-tube viscometer (Mullin, 2001)

Figure 2.3 shows a simple falling-sphere viscometer. The tested liquid is confined in the inner tube of a diameter of 30 mm and a length of about 300 mm. Two reference marks a and b, 150 mm apart, are present in the central portion of the tube. The tube is kept vertical, and the temperature of the liquid is measured and maintained constant. A 1.5mm diameter steel ball, previously warmed to the test temperature, is inserted through a small guide tube and its time of passage between the two reference marks is measured. The mean of several measurements is taken. Calculation of the viscosity of the liquid can then be made using equation 2.6. Current instrumental versions of the falling-sphere technique are demanded to measure viscosities in the range 0.5-500 mPa with high accuracy at controlled temperatures via sample volumes as low as 0.5 mL (Mullin, 2001).

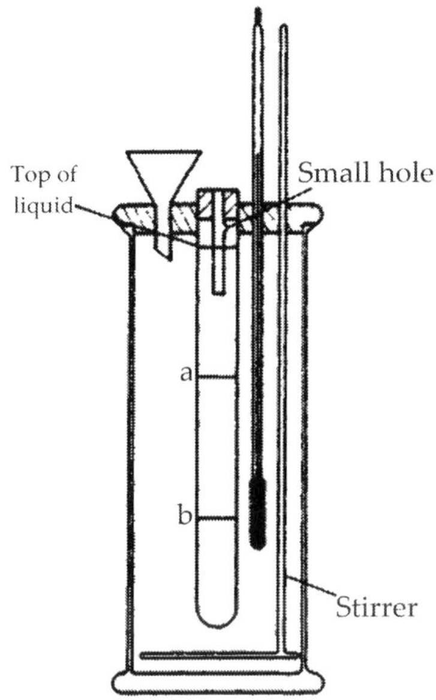

Figure 2.3: Falling-sphere viscometer (Mullin, 2001)

Solid-liquid systems

Considerable alteration of the viscosity characteristics of liquids can be made by the existence of finely dispersed solid particles, particularly of colloidal size. Expression of the viscosity of a suspension of rigid spherical particles in a liquid, once the distance between the spheres is much greater than their diameter, could be performed through the Einstein equation:

$$\eta_S = \eta_0 (1 + 2.5\varnothing) \qquad (2.7)$$

Where, η_S denotes the effective viscosity of the disperse system, η_0 denotes the viscosity of the pure dispersion medium, and 4 the ratio of the volume of the dispersed particles to the total volume of the disperse system. That means, $\varnothing = 1 - \varepsilon$, where ε is the voidage of the system. Equation 2.7 applies reasonably well to lyophobic sols and very dilute suspensions, but for moderately concentrated suspensions and lyophilic sols the Guth-Simha modification is preferred:

$$\eta_S = \eta_\circ \left(1 + 2.5\varnothing + 14.1\varnothing^2\right) \tag{2.8}$$

Surface tension

The capillary rise and ring techniques are considered the most useful methods available for measuring the surface tension of liquids, especially for general applications. Regarding the capillary rise method, determination of the surface tension, γ, of a liquid can be calculated from the height, h, of the liquid column in a capillary tube of radius r. If the liquid totally wets the tube (zero contact angle),

$$\gamma = \frac{1}{2} rh\Delta\rho g \tag{2.9}$$

The $\Delta\rho$ denotes the difference in density between the liquid and the gaseous atmosphere above it. Accurate measurement of the height, h, can be done with a cathetometer from the base of the liquid meniscus to the flat surface of the free liquid surface in a containing vessel. Yet, to reduce errors, this reference to a flat surface can be eliminated by measuring the difference in capillary rise in two tubes of different bore as presented in Figure 2.4.

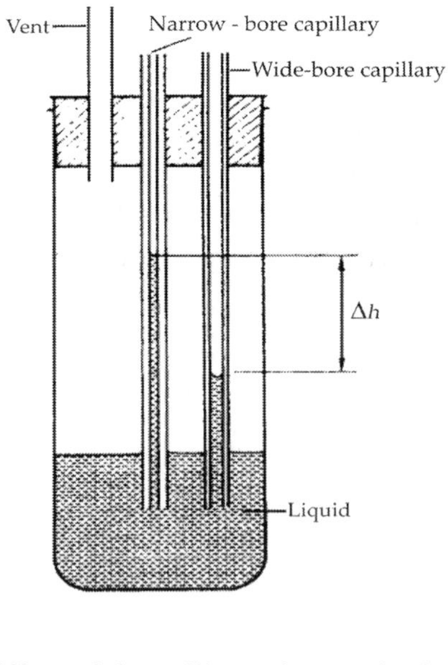

Figure 2.4: Simple differential capillary rise methods for measuring surface tension by the (Mullin, 2001)

$$\gamma = \frac{1}{2} r_1 h_1 \Delta\rho g = \frac{1}{2} r_2 h_2 \Delta\rho g \tag{2.10}$$

Then the following equation could be driven to calculate surface tension:

$$\gamma = \frac{\Delta h \Delta\rho r_1 r_2 g}{2(r_1 - r_2)} \tag{2.11}$$

The differential height Δh is calculated from differential capillary rise method as illustrated in Figure 2.4.

On the other hand, the ring technique is broadly used in industrial applications. There are many kinds of commercial instrument which based on ring technique and incorporating a torsion balance is available under the name du Nouy tensometer. The method has many advantages including its simplicity, rapid data generation, and capability of measuring the surface tension of a pure liquid to a precision of 0.3% or better.

Surface tension in ring technique is defined by the force necessary to pull a ring (fabricated from platinum or platinum-iridium wire) from the surface of the liquid. The surface tension is calculated from the pull and the dimensions of the ring after the appropriate correction factors have been applied.

It is often possible to predict the surface tension of a non-aqueous mixture of solvents by assuming a linear dependence with the mole fraction. Aqueous solutions, however, generally show a pronounced non-linear behavior and prediction is not recommended. The surface tension of a liquid decreases with an increase in temperature, but the decrease is not always linear.

The addition of an electrolyte to water generally increases the surface tension very slightly, although an initial decrease is usually observed at very low concentrations (< 0.002 mol L^{-1}). Non-electrolytes generally decrease the surface tension of water. For example, saturated aqueous solutions of a-naphthol, adipic acid and benzoic acid at 22°C is at 48, 55 and 60 mNm^{-1}, respectively, whereas a saturated solution of potassium sulphate at the same temperature has a surface tension of 73 mN m^{-1} (Mullin, 2001).

Interfacial tension

The surface tension of a liquid, as normally measured, is the interfacial tension between a liquid surface and air saturated with the relevant vapor. The interfacial tension of a crystalline solid in contact with a solution of the dissolved solid is a quantity of considerable importance in crystal nucleation and growth processes. It is also sometimes referred to as the 'surface energy'.

Diffusivity

Two examples of a theoretical approach to the problem of the prediction of diffusion coefficients in fluid media are the equations postulated in 1905 by Einstein and in 1936 by Eyring. The former is based on kinetic theory and a modification of Stokes' law for the movement of a particle in a fluid, and is most conveniently expressed in the form:

$$D = \frac{kT}{\varnothing r\eta} \tag{2.12}$$

where D = diffusivity ($m^2\ s^{-1}$), T = absolute temperature (K), = viscosity ($kg\ s^{-1}\ m^{-1}$), r = molecular radius (m), k = Boltzmann's constant and the dimensionless factor ϕ has a numerical value between 4π and 6π depending on the solute-to-solvent molecular size ratio. Eyring's approach, based on reaction rate theory, treats a liquid as a disordered lattice structure with vacant sites into which molecules move, i.e. diffuse. For low solute concentration Eyring's equation may be expressed in the form identical with that of equation 2.12 but with a different value of ϕ.

The usefulness of these equations, however, is strictly limited because they both contain a term, r, which denotes the radius of the solute molecule. Values of this quantity are difficult to obtain. Consequently, the most directly useful relationship that emerges is:

$$D\eta / T = constant \tag{2.13}$$

This equation is of considerable value in predicting, for a given system, the effect of temperature and viscosity on the diffusion coefficient. This simple relation is often referred to as the Stokes-Einstein equation. The very limited success of the theoretical equations has led to the development of many empirical and semi-empirical relationships for the prediction of diffusion coefficient (Mullin, 2001).

Refractive index

Refractometric measurements can often be used for the rapid measurement of solution concentration. Several standard instruments are available commercially. A sodium lamp source is most usually used for illumination, and an instrument reading to the fourth decimal place is normally adequate for crystallization work. It is advisable that calibration curves be measured, in terms of temperature and concentration, prior to the study with the actual system.

If a dipping-type refractometer is used, a semi-continuous measurement may be made of the change in concentration as the system crystallizes. However, if nucleation is heavy or if large numbers of crystals are present, it may be difficult to provide sufficient illumination for the prism because of the light scattering. One solution to this problem is to use a fiber optic (a light wire) fitted into a collar around the prism illuminated from an external source (Figure 2.5). In this way, undue heating of the solution is also avoided.

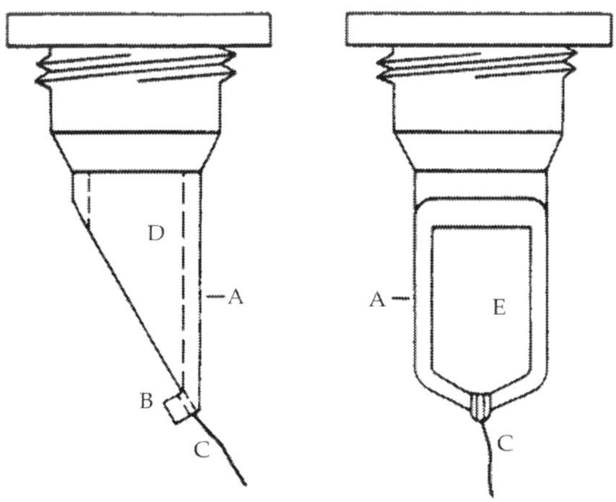

Figure 2.5: A technique for illuminating the prism of a dipping-type refractometer in an opaque solution: A-Perspex collar; B-fiber optic holder; C-fiber optic; D-refractometer prism; E-polished face of prism (Mullin, 2001)

Crystal hardness

Crystals vary in hardness not only from substance to substance but also from face to face on a given crystal. One of the standard tests for hardness in non-metallic compounds and minerals is the scratch test, which gave rise to the Mohs scale. Ten 'degrees' of hardness are designated by common minerals in such an order that a given mineral will scratch the surface of any of the preceding members of the scale (see Table 2.1).

The hardness of metals is generally expressed in terms of their resistance to indentation. A hard indenter is pressed into the surface under the influence of a known load and the size of the resulting indentation is measured. A widely used instrument is the Vickers indenter, which gives a pyramidal indentation, and the results are expressed as a Vickers hardness number (kgf mm^{-2}). Other tests include the Rockwell, which uses a conical indenter and the Brinell, which uses a hard steel ball (Mullin, 2001). The relation between Mohs hardness, M, and the Vickers hardness, V, is not a clear one. However, if diamond (M = 10) is omitted from the Mohs scale, the relationship below (equation 2.14) may be used for rough approximation purposes for values of M < 9.

$$log\, v = 0.2M + 1.5 \tag{2.14}$$

Table 2.1: Mohs scale of hardness (Mullin, 2001).

Mohs hardness number M	*Reference substance*	*Formula*	*Vickers hardness number V*
1	talc	$3Mgo \cdot 4SiO_2 \cdot H_2O$	50
2	gypsum	$CaSO_4 \cdot 2H_2O$	80
3	calcite	$CaCO_3$	130
4	fluorite	CaF_2	200
5	apatite	$CaF_2 \cdot 3Ca_3(PO4)_2$	320
6	orthoclase	$K_2O \cdot Al_2O_3 \cdot 6SiO_2$	500
7	quartz	SiO_2	800
8	topaz	$(AlF)_2 \cdot SiO_4$	1300
9	corundum	Al_2O_3	2000
10	diamond	C	10000

**rewrite this table 2.1. It is actually, a figure at its present state.

Table ___ : Typical values of Mohs' surface hardness of some common substances

Sodium	0.5	Aluminum	2-3
Potassium	0.5	Gold	2.5-3
Lead	1.5	Brass	3-4
Magnesium	2	glass	3-4

** label this as another table: Column 1-Material and Column2-Mohs' Hardness

The scratch test is not suitable for specifying the hardness of substances commonly crystallized from aqueous solutions, because their Mohs values lie in a very short range-frequently between 1 and 3 for inorganic salts and below 1 for organic substances. For a reliable measurement of hardness of these soft crystals the indentation test is preferred (Mullin, 2001).

Crystal Structure

Several methods are suitable to characterize the crystal structure of a solid, such as single-crystal X-ray diffraction, X-ray powder diffraction (XRPD), Raman, infrared (IR), near infrared (NIR), terahertz (THz), and solid-state NMR (ssNMR) spectroscopy. Each method has its advantages and disadvantages. The most widely used ones will be described in some more detail.

While single-crystal X-ray diffraction provides the ultimate characterization of the crystal structure, its practical usefulness, particularly as a method for quality control, is limited by the fact that single crystals are needed for this technique, and that it provides information only on that single crystal that might not be representative of the whole sample, and that it is time-consuming. All these restrictions do not apply for X-ray powder diffraction, where a certain amount of powder is irradiated giving an averaged pattern. This makes it one of the most widely used methods for polymorph characterization (Beckmann, 2013).

X-Ray Powder Diffraction (XRPD)

The underlying principle of both single-crystal X-ray diffraction and

XRPD is Bragg's law (Equation 2.15), which describes the relationship between the angle for constructive interference (Θ), layer spacing (d), wavelength (l), and order of the interference (n):

$$n\lambda = 2d\sin(\Theta) \tag{2.15}$$

In XRPD, the sample exposed to the X-ray beam is a powder and if the particles are randomly oriented, all crystal planes in the crystal lattice will be suitably oriented with respect to the X-ray beam and give rise to constructive interference at the respective angle (Figure 2.6). Therefore, a pattern of the form depicted in Figure 2.7 will be obtained. Different polymorphs will have different crystal plane spacing and will therefore lead to a different XRPD. Rare exceptions exist, where the differences are so small that they may not be discernible using standard instrumentation.

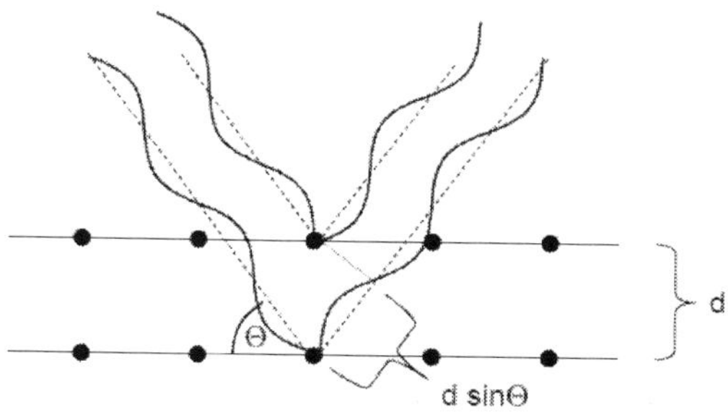

Figure 2.6: Principle of Bragg's law: constructive interference when the path difference is λ or multiples of λ (Beckmann, 2013).

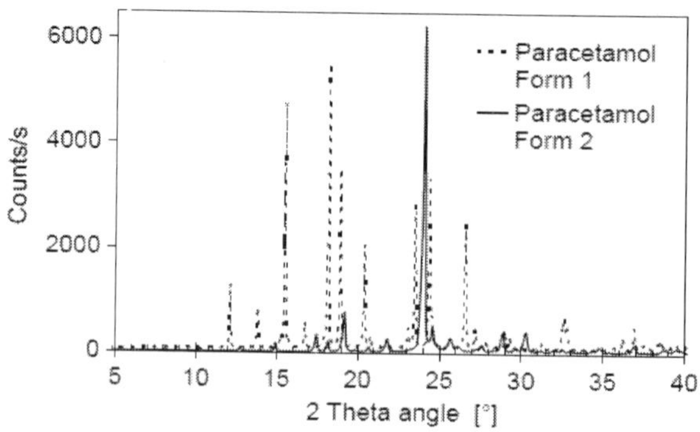

Figure 2.7: XRPD of paracetamol form 1 (broken line) and form 2 (solid line) (Beckmann, 2013).

A practical problem with XRPD can be the so-called preferential orientation. If the particles are non-spherical (e.g., plates or needles) and are not randomly oriented, some lines in the XRPD may become greatly attenuated or even disappear. When comparing XRPDs of various samples, good sample preparation is therefore essential. Orientation effects, for example, can be reduced by gentle grinding of the substance. Investigating the sample by optical microscopy may also be very helpful to see if the particles are preferentially oriented. When a sample is ground, care must be taken that the solid form is not changed due to the grinding. The absence of such changes can be confirmed if the positions of the lines in the XRPD diagram are not shifted by the grinding if the intensity of the grinding is varied.

Due to the principle of XRPD, the method will lead to strong constructive interference if the spacing of crystal planes is constant over an extended distance. It is therefore sensitive to long-range order. In contrast to XRPD, vibrational and NMR spectroscopy is essentially sensitive to the immediate environment of the molecule (Beckmann, 2013).

Vibrational Spectroscopy (Raman, IR, NIR, THz)

Vibrational spectra are mainly governed by vibrations within the molecule. But since these vibrations are influenced by the environment of the molecule and hence the crystal packing, they will also differ for different polymorphs of the same molecule. The differences will in general not be as obvious as for XRPD, but nevertheless the specificity of spectroscopic and diffraction methods with respect to the differentiation of polymorphs are similar. Figure 2.8a shows the Raman spectra of the same two polymorphs of paracetamol, as given in Figure 2.7. While the spectra look essentially similar, clear line shifts, for example, can be observed between 1700 and 1200 cm^{-1} (Figure 2.8b). In some cases, such as isomorphous solvates (solvates which are identical in structure, except that different solvents occupy a certain lattice space), the specificity of IR and Raman is often superior to XRPD (Beckmann, 2013).

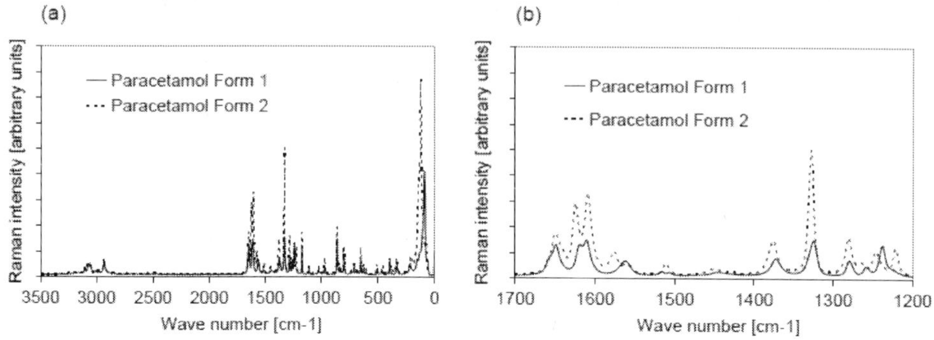

Figure 2.8: Raman spectrum of paracetamol form 1 (solid line) and form 2 (broken line). (a) Whole spectrum. (b) Section of Raman spectrum where differences between form 1 and form 2 are most obvious (Beckmann, 2013).

Raman and IR are complementary methods. They differ in their selection rules (change of dipole moment or polarizability during the vibration) and in practical aspects. The main advantages of Raman compared to IR are as follows:

No sample preparation necessary (IR often uses KBr pellets; modern ATR-IR also eliminates sample preparation).

- In general, narrower bands.

Raman microscopy needs less sample preparation and has a better resolution than IR microscopy.

Raman spectroscopy can easily be applied for online monitoring

Low frequencies more readily attainable; this is the range of lattice vibrations, where the difference between polymorphs is often very pronounced.

Can measure through glass/quartz containers.

- Little interference from water (weak Raman scatterer).
- Hydrogen bonding shows up weakly.
- The main advantages of IR are as follows:
- More widely available.
- Cheaper instruments
- Hydrogen bonding shows up strongly.
- No problems with samples that exhibit fluorescence.

NIR spectra are due to overtone and combination bands, mostly of OH, NH, and CH stretching vibrations. They are generally less specific with respect to polymorphs than IR or Raman spectra, but may offer practical advantages due to sample preparation and wide availability for online monitoring.

Terahertz spectroscopy accesses the wave number region approximately between 2 and 100 cm^{-1}, that is, even lower wave numbers than Raman spectroscopy. This is a region where differences between polymorphs may be very large, which makes it a powerful method for differentiating polymorphs. The technique however is still expensive and not widely available (Beckmann, 2013).

Solid-State NMR (ssNMR)

Nuclear magnetic resonance (NMR) spectroscopy is one of the most important methods for determining the structure of molecules in solution. Recording NMR spectra of solids is technically more challenging and requires specialized equipment and techniques. One such technique is known as "magic angle spinning" and involves the very fast rotation (often at rates of tens of kilohertz-that is, millions of revolutions per minute) of the sample about a specified angle (54.7°). Two other fundamental techniques involve the application of series of precisely timed, high-power radiofrequency pulses for purposes of spectral simplification (in the case of "dipolar decoupling") or sensitivity enhancement (in the case of "cross polarization"). These techniques can be combined in a variety of ways so that the spectral resolution of modern solid-state NMR (ssNMR) spectra currently approaches that of solution-state spectra. The wide range of both one- and two- dimensional NMR spectroscopic techniques known to the organic chemist can now, with appropriate modifications, be applied to solids (Beckmann, 2013).

Like IR and Raman spectra, solid-state NMR spectra are sensitive to local order and local properties, i.e. "chemical information", whereas X-ray diffraction is sensitive to long-range order. As different polymorphs will also differ in their local crystal lattice packing and/or molecular conformation, they can generally be distinguished by ssNMR. Figure 2.9 shows a section of the ssNMR spectra of two different polymorphs of an active pharmaceutical ingredient.

Solid-state NMR offers many advantages compared to other techniques, such that it is/has:

Chemical-specific (^{13}C, ^{19}F, ^{1}H, ^{31}P, etc.) structural information.

Narrow, well-resolved bands (especially compared to NIR).

Possibility of rigorous peak assignment (spectral editing, 2D methods) in many cases.

Nondestructive sampling process.

Insensitivity to particle size effects.

Applicability to both drug substances and formulated drug products.

Quantitative (in principle) with limit of quantification (LOQ) as low as ~1%.

Suitable for studying solvation and desolvation, with both organic solvent "guest molecules" and solid "host molecules" being directly observable.

Dynamical information on the nanosecond to second timescale can be derived, which may be useful to detect fast isomerization.

However, the method also has the following disadvantages:

Large sample sizes (up to ~300 mg) required.

Time-consuming (more than 2 days required to obtain some spectra).

Possible sample transformation under certain measurement conditions (fast sample spinning speeds, high-power radiofrequency pulses).

Extensive calibration studies generally required for quantitation.

Need for isotopic labeling to apply some methods.

Lack of spatial resolution.

Expensive.

Requires specialized knowledge to both implement and evaluate experiments.

Solid-state NMR, therefore, is a very powerful tool for research and development, but is not particularly suitable as a routine technique for quality control in the industrial setting (Beckmann, 2013).

Thermodynamic Properties

Thermal methods play a key role in both the identification and the characterization of solid forms. A thermal method can be defined as a method where the heat flow is measured as a function of an externally varied parameter. If this parameter is, for example, time and all other parameters are held constant, the method is called isothermal calorimetry. In solution calorimetry, a solid is dissolved in a solvent and the associated heat of dissolution is measured. Probably the most frequently used thermal technique is differential scanning calorimetry (DSC), where temperature is varied and the associated heat flow is measured (Beckmann, 2013).

Differential Scanning Calorimetry (DSC)

Two major types of DSC instruments are available on the market. The first one is the power compensation mode used in Perkin Elmer instruments. In this mode, sample and reference are kept at the same (changing) temperature and the power input that is needed to achieve this temperature is measured for both sample and reference. The second mode, which is used by the majority of DSC instrument providers, is called heat flux DSC. Here, the temperature difference between sample and reference is measured as a function of temperature, while both are heated or cooled in two compartments of one oven. The heat flow difference between sample and reference is then calculated from this temperature difference and the independently determined thermal resistance between sample and reference. Both methods yield essentially the same data since the change is measured in terms of:

$$dQ/dt = m \times C_P \times (dT/dt) \tag{2.16}$$

The heat flow dQ/dt is measured as a function of temperature when this temperature is changed at a certain rate dT/dt. Equation 2.16 relates the two quantities to C_P, the heat capacity at constant pressure, and m, the sample mass. In the absence of phase transitions, the DSC signal is proportional to the heat capacity of the sample. If the sample undergoes a first-order phase transition (e.g., melting), the effective heat capacity of the system goes to infinity at the transition temperature due to the latent heat associated with that event. This leads to a peak in the DSC curve. If melting is the only event observed in a DSC scan, the melting point of that solid can be determined accurately by applying a tangent to the melting peak (Figure 2.9). The melting enthalpy can be determined from the area under the peak. Melting point and enthalpy are essential parameters to describe a solid.

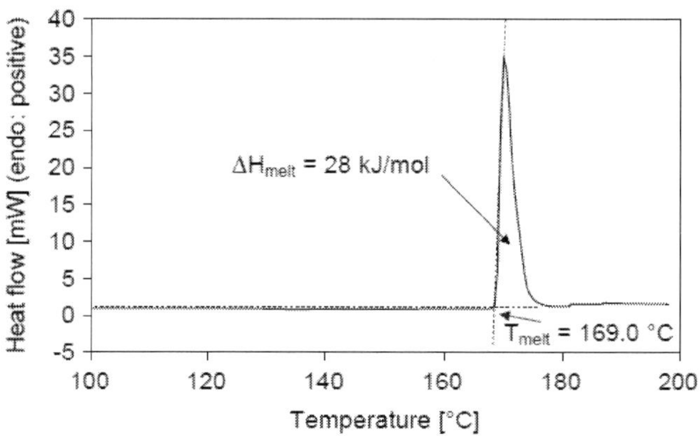

Figure 2.9: DSC thermo-gram of paracetamol form 1. Form 1 is the thermodynamically stable form (monotropic) of paracetamol. By applying a tangent to the melting peak, the melting point is determined to be 169.0 °C. From the area under the melting peak, a melting enthalpy of 28 kJ/mol is determined (Beckmann, 2013).

If a substance can exist in various polymorphs and the polymorph that is not thermodynamically stable at the melting temperature, and is subjected to a DSC scan, then either of the following occurs:

The melting point and melting temperature of that polymorph may be observed.

The start of melting is immediately followed by recrystallization into the high temperature form, which is then followed by melting of the high-temperature form.

A solid–solid transition from the metastable to the stable polymorph and subsequent melting of the stable polymorph may be observed.

Figure 2.10a and b shows cases (II) and (III) for two polymorphs of carbamazepine. Whether a solid–solid transition as described in case (III) above is observed or not, depends on the moiety under investigation and on properties of the sample (such as particle size, crystal imperfections, small amounts of the other form, or residual solvent). In addition, it may also depend on instrument parameters: fast scanning rates favor melting/recrystallization, while solid–solid transitions are preferentially seen for slower scanning rates. In all three cases, DSC provides important information on the properties of the two forms. If melting temperatures and heats of fusion of both forms can be determined, it can be deduced from Burger–Ramberger rule 2 whether the forms are monotropically or enantiotropically related and if a solid–solid transformation might be observed.

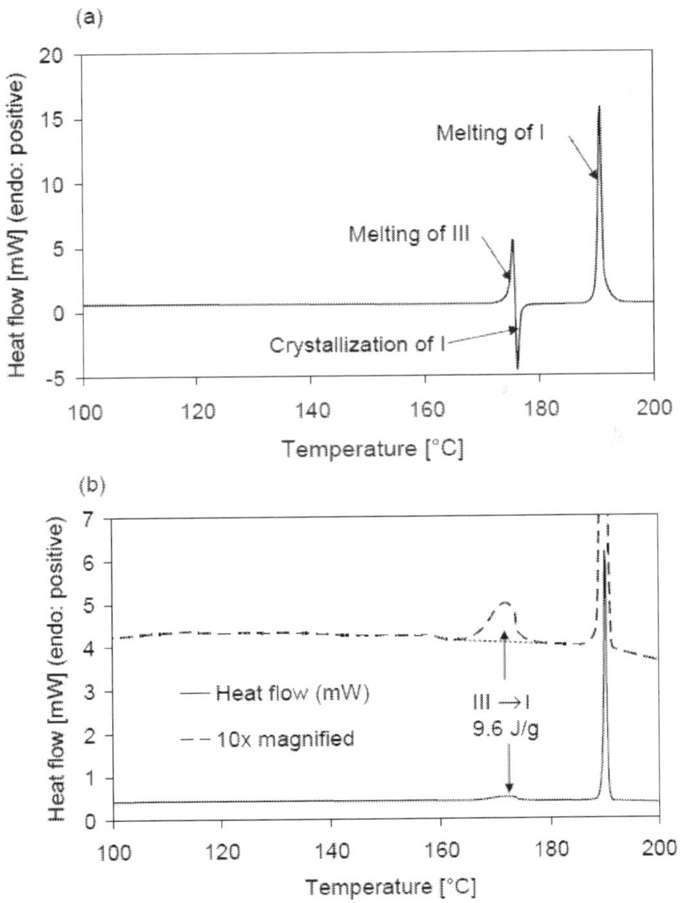

Figure 2.10: (a) DSC thermogram of carbamazepine form III. Carbamazepine form III is the stable form at room temperature. At 175 °C, melting of form III is observed, which is immediately followed by (exothermic) recrystallization into form I. Form I is the thermodynamically stable form at temperatures above 118 °C. Form I and form III are therefore enantiotropically related. At 190°C,

melting of form I is observed. Scanning rate was 10 K/min. (b) DSC thermogram of carbamazepine form III at a lower scanning rate of 2 K/min. Now an endothermic solid–solid transition of form III to form I at ~170 °C is observed in accordance with the Burger–Ramberger rules. Subsequently, form I melts at 190 °C (Beckmann, 2013).

Rule 1 provides the same answer in the other two cases. The corresponding temperature of a solid–solid transformation can vary from experiment to experiment according to the nature of the sample and in general will not correspond to the thermodynamic transition temperature of the two forms. For enantiotropic systems, it will be higher than the thermodynamic transition temperature; Figure 2.10b shows that the transition from form III to form I occurs at ~170 °C, while the thermodynamic transition temperature is 118°C. For monotropic systems, there is no thermodynamic transition temperature at all.

Finally, DSC is also well suited for the characterization of amorphous forms. At the glass transition temperature, where the amorphous form converts from the glassy to the rubbery state, a jump in the heat capacity will be observed. It may also happen that the amorphous substance crystallizes when it becomes rubbery. This will manifest itself in the DSC by an exothermic heat of crystallization and subsequently the crystalline solid will melt (Figure 2.11).

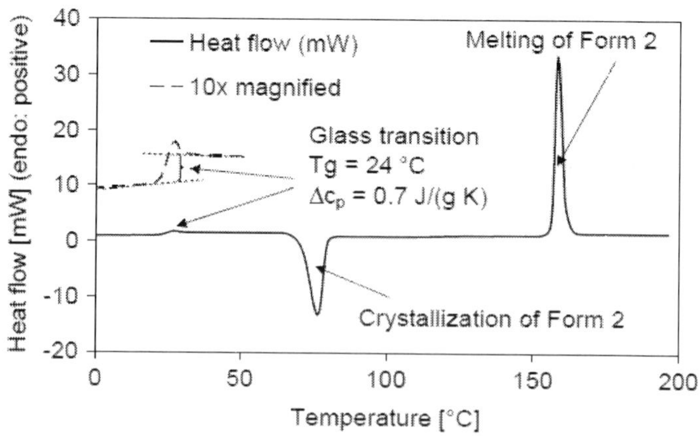

Figure 2.11: DSC thermogram of amorphous paracetamol. At 24 °C, the glass transition from the glassy to the rubbery state is observed by a change of the heat capacity of 0.7 J/g/K. At ~76 °C, the rubbery paracetamol crystallizes into the metastable form 2, which subsequently melts at 156 °C (Beckmann, 2013).

Isothermal Microcalorimetry

Isothermal microcalorimetry is a versatile method that can be used for diverse applications such as assessing chemical stability, heats of dissolution by solution calorimetry, and binding affinities by titration calorimetry. An important application of isothermal micro-calorimetry related to characterization of solids is the determination of the amorphous content. Typically, it is desired that the product of a crystallization process is 100% crystalline. If the crystallization process is carried out in a non-optimized way, the resulting solid may contain amorphous fractions. One possible cause for this could be that the crystallization is carried out too fast. Also, processes following the crystallization process, such as drying or milling, may induce formation of amorphous fractions. In most cases, such amorphous fractions are undesired, since they may reduce the chemical stability and change the dissolution characteristics of the product (Beckmann, 2013).

For a lot of compounds, it is possible to induce crystallization of amorphous fractions by subjecting the sample to humidity or solvent vapors, since adsorption of water or solvent in amorphous sections will reduce the glass transition temperature. By exposing samples to a controlled atmosphere in an isothermal microcalorimeter, the heat of crystallization can be measured and the amorphous content may be calculated after having performed a calibration of the system (Figure 2.12).

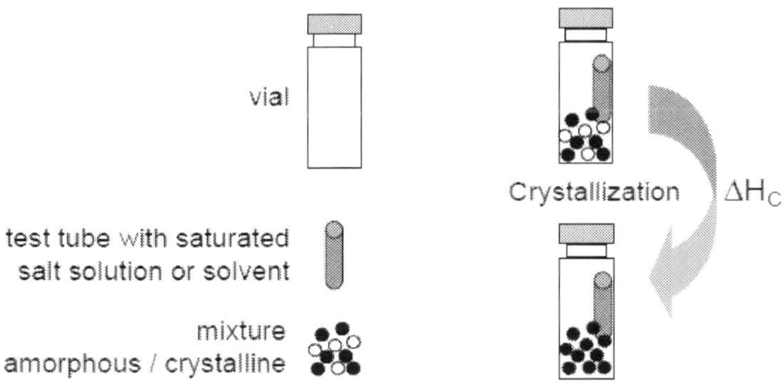

Figure 2.12: Amorphous content by isothermal microcalorimetry. After having identified a suitable solvent, the substance is placed in the vial together with a small test tube filled with a salt solution or solvent to provide the appropriate

humidity or solvent atmosphere. Crystallization of the amorphous fraction will then occur in the isothermal microcalorimeter (Beckmann, 2013).

In addition to finding conditions that are suitable to induce crystallization of the amorphous fractions, it is necessary to optimize the conditions in such a way that the kinetics of the crystallization is suitable. The crystallization event should only start after about an hour so that the vial has sufficient time to equilibrate after it has been placed in the isothermal microcalorimeter, and the crystallization event should be completed after 5–10 h to be easily discernible from the background. Figure 2.13a shows the thermal events observed in an isothermal microcalorimeter for a sample with a fraction of 0% and 0.5% amorphous material.

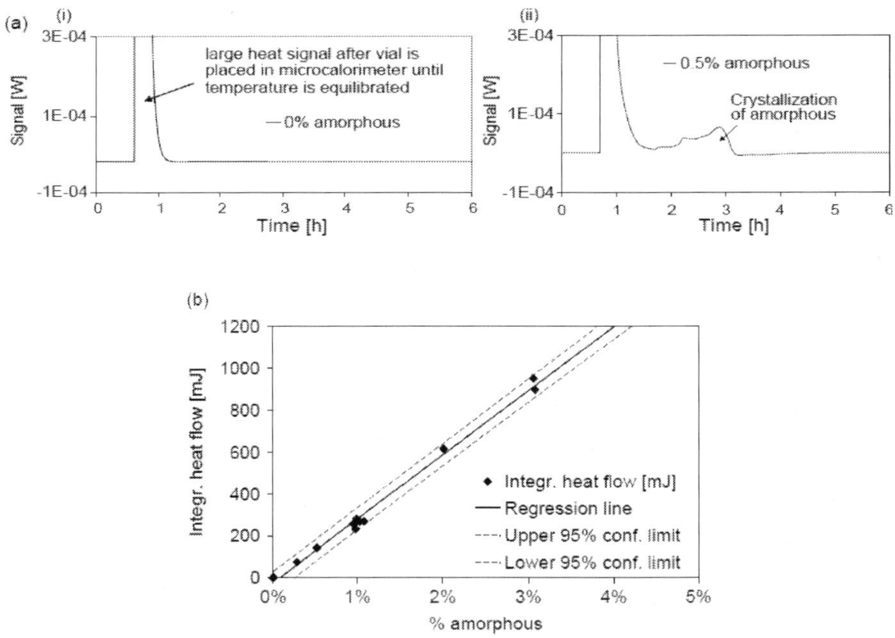

Figure 2.13: (a) Signal in isothermal microcalorimeter of 100% crystalline (i) and 99.5% crystalline (ii) substance. (b) Regression line with 95% confidence limits to the measured heat flow values of mixtures with varying amorphous content. From the linear regression analysis, it can be calculated that the limit of detection is 0.3% and the limit of quantification is 0.8% (Beckmann, 2013).

Figure 2.13b shows the respective calibration curve indicating the limit of detection for amorphous fractions in the solid. Accurate determination of the amorphous content will allow optimization of process parameters with a minimized amorphous content as the primary criteria.

Size classification of crystals

The most widely employed physical test applied to a crystalline product is the one by means of which an estimate may be made of the particle size distribution. Product specifications invariably incorporate a clause that defines, often quite stringently, the degree of fineness or coarseness of the material. For many industrial purposes, the demand is for a narrow range of particle size; regularity results in the crystals having good storage and transportation properties, a free-flowing nature and, above all, a pleasant appearance. Terms such as 'fine' and 'coarse' are frequently used, although usually without definition, to describe crystalline and powdered materials (Mullin, 2001).

For pharmaceutical products, however, some guidance is available from recommendations in the British Pharmacopoeia (Commission, 2000), based on sieve grading.

**label the table below as Table ___. Classification of Particle Size based on Sieve Grading (British Pharmacopeia, 2000)

Coarse	All passes 1700µm	$\not> 40\%\ passes\ 355\,\mu m$
Moderately coarse	All passes 710µm	$\not> 40\%\ passes\ 250\,\mu m$
Moderately fine	All passes 355µm	$\not> 40\%\ passes\ 180\,\mu m$.
Fine	All passes 180µm	$\not> 40\%\ passes\ 125\,\mu m$
Very fine	All passes 125µm	$\not> 40\%\ passes\ 45\,\mu m$
Microfine	$\not< 90\%\ passes\ 45\,\mu m$	

Particle size and surface area

Many methods are available to determine particle size distributions. Table 2.2 presents the different methods in determining particle size based on the desired product size range.

Table 2.2 Commomethods to measure particle size (Beckmann, 2013).

Method	Size range (μm)
Sieving	>20
Optical microscopy	0.5–150
Electron microscopy	0.001–5
Sedimentation	0.01–50
Colter counter	1–200
Laser light diffraction (Fraunhofer diffraction)	1–1000
Quasi-elastic laser light scattering (photon correlation spectroscopy)	0.001–1

Sieving

The most easily understood method of particle size determination is sieve analysis, where the powder is separated through sieves of different sizes. Accordingly, the PSD is defined in terms of discrete size ranges, for example, fraction of sample mass between 60 and 70 mm. The sieves with the smallest holes are around 20 mm; therefore, sieve analysis accesses the size range larger than 20 mm. It continues to be used for many measurements because of its simplicity, cheapness, and ease of interpretation. The simplest form of performing sieve analysis is shaking the sample on a stack of sieves of progressively smaller mesh size or aperture size until the weight stays constant in each sieve. Of course, the intensity with which the sieve is shaken may affect the measured particle size distribution as it may lead to the breaking up of aggregates. A reproducible mechanical shaker is used to perform sieving. (Beckmann, 2013).

For rod-like particles, sieving essentially determines the diameter of the rod, whereas laser light diffraction determines the diameter of a sphere with same volume as the rod. Therefore, the result obtained by sieving will be smaller than that obtained by laser light diffraction. The opposite applies for plate-like particles, where sieving measures the diameter of the plate. Sieving is basically a very simple and justifiably popular particle sizing technique, but the precautions necessary to produce reliable data do not appear to be widely appreciated. Some of the more important points to note about the use of standard test sieves for particle size analysis are as follows.

Particles must not be forced through the sieve apertures.

Sieving should be continued to an end-point, i.e. until the amount of material passing through ceases to affect the result significantly. When using a mechanical shaker, it is recommended that each sieve removed from the stack should be given a brief brisk tapping and shaking by hand to ensure that the end-point has been reached. If it has not, sieving must be continued.

It should be clearly understood that the aperture size marked on the sieve is only a nominal size. The actual value can vary from this value, within specified tolerances. For accurate work, it is advisable to calibrate the sieve, e.g. by sieving a reference sample of known size distribution (Mullin, 2001).

Microscopy

Microscopy provides the unique advantage that it can measure both particle shape and particle size. Its disadvantage is that compared to other methods, only a small number of particles are counted and that the result may therefore be accompanied by a large statistical error. Automatic image processing reduces this problem to some extent.

Figure 2.14 shows an example of particle size distribution determined by microscopy together with the determined differential and cumulative number and volume distributions. The diameter that is normally used in particle sizes determined by microscopy is the ECD (equivalent circle diameter) value, which is defined as the diameter of a circle that has the same area as the particle. Given the small dynamic range of particle size determinations by microscopy, the diameter is normally displayed on a linear rather than on a logarithmic scale. It can be seen again that q3 is significantly larger than q0 (Beckmann, 2013).

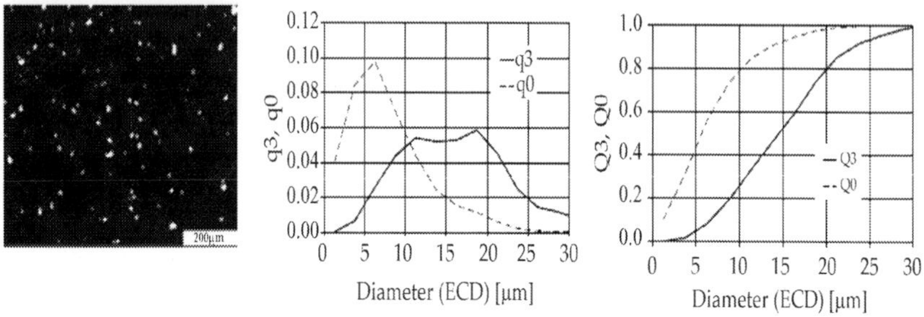

Figure 2.14: Particle size distributions by microscopy (Beckmann, 2013).

Sedimentation

A simple sedimentation technique, which readily lends itself to the determination of crystal size distribution in the range 1-50 µm, is the Andreasen pipette method. Although it is generally better to prepare a fresh suspension of the crystals under test in a suitable inert liquid, it is possible to classify crystals suspended in their own mother liquor. If the difference in density between the particles and suspending liquid is < 0.5 g cm^{-3} special care must be taken to avoid convection currents (Mullin, 2001).

The method, briefly, is as follows: a homogeneous suspension of the crystalline material in a suitable liquid is prepared in the graduated sedimentation cylinder of capacity ~600 cm^3 (Figure 2.15). Small samples (e.g. 10 cm^3) of the suspension are withdrawn through the fixed pipette, at a known depth, h, below the liquid level, at chosen time intervals. The samples, including the one taken at zero time, are analyzed for total suspended solids content by a suitable method. Ideally the suspension should be diluted (< 3 per cent) and a dispersion agent may be needed to prevent agglomeration: for particles in insoluble water, a 0.1 per cent solution of sodium pyrophosphate is generally suitable (Mullin, 2001).

Electrical sensing zone (Coulter) methods

In the Coulter technique, particles must be suspended in an electrolyte solution and then induced to pass through a small orifice, with surrounding electrodes, located in the measurement cell. Changes in electrical impedance in the orifice channel for each particle passage are measured and counted. The result is a number-size (volume) distribution of particles.

The method has found applicability in a wide range of industries. For application to crystallizing systems, however, it is important to choose a unit in which the voltage between the inner and the outer electrode is automatically adjusted to maintain a constant current. This renders the calibration, and hence the actual counting, insensitive to the type of electrolyte used as well as to concentration and temperature changes within the electrolyte. This precaution is of paramount importance for crystallization studies where changes in electrolyte properties due to phase transitions are inevitable (Mullin, 2001).

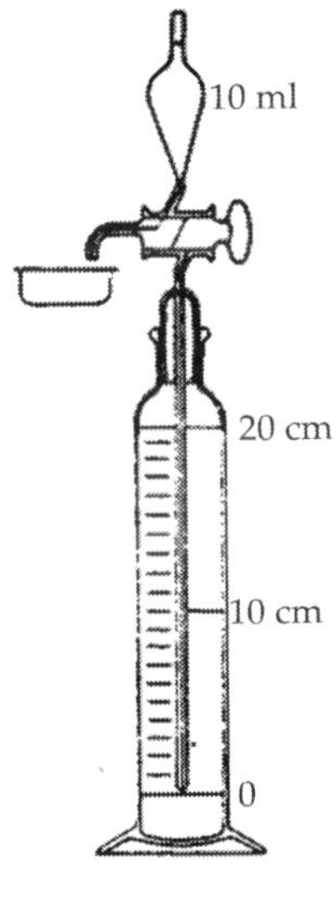

Figure 2.15: Andreasen fixed pipette method (Mullin, 2001).

Laser Light Diffraction

The most frequently used method for particle size determination is laser light (Fraunhofer) diffraction. It has the advantages of a very large dynamic range in a size domain where most crystallization products exist, ease of use, excellent statistics, and speed. Modern instruments provide a dynamic range of particle size diameters between 0.2 and 2000 μm when in the size range below 20 mm corrections due to Mie scattering are applied.

The method can be used for both dry powders and suspensions. The technique of laser light diffraction is based on the principle that particles that are illuminated by coherent light will scatter light at an angle that is directly related to their size. As the particle size decreases, the observed scattering angle increases. The principal set up of a Fraunhofer diffraction instrument is shown in Figure 2.15. The sample in the measuring cell is illuminated by an expanded parallel coherent and monochromatic laser beam. The light scattered by the particles in the cell is focused by a lens system on a position-sensitive detector. From the measured intensity distribution on the detector, the particle size distribution can be calculated. The result obtained is the size distribution of equivalent spherical particles, that is, the result refers to a distribution of spherical particles that interact with the laser light in the same way as the sample consisting of non-spherical particles. Therefore, differences can occur in comparison to sieving, sedimentation, or other particle size distribution techniques. Typically, volume distributions (q_3 and Q_3) are given as results (Figure 2.16).

As with any physical measurement, sample preparation is of utmost importance. This is particularly true for particle size determinations. If the measurement is performed in suspension, dispersing aids may have to be added and ultrasound may have to be applied to ensure homogeneous dispersion. This may disrupt aggregates, which can be desired or undesired. In any case, it is highly advisable to observe the effects of sample preparation by additional methods such as microscopy. Guidelines for method development, instrument calibration, and acceptance criteria can be found in regulatory documents.

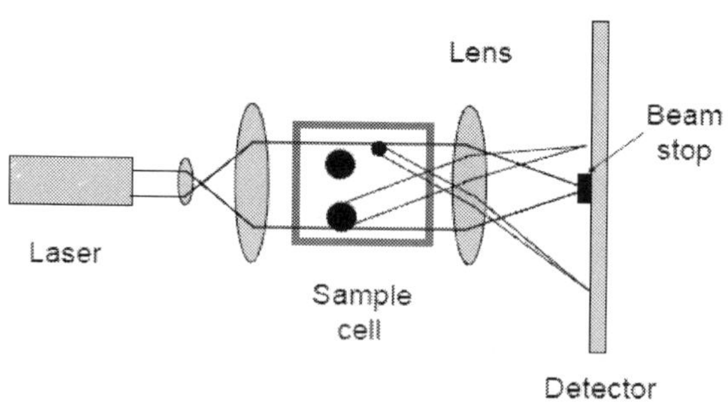

Figure 2.15: Principle of Fraunhofer diffraction instrument (Beckmann, 2013)

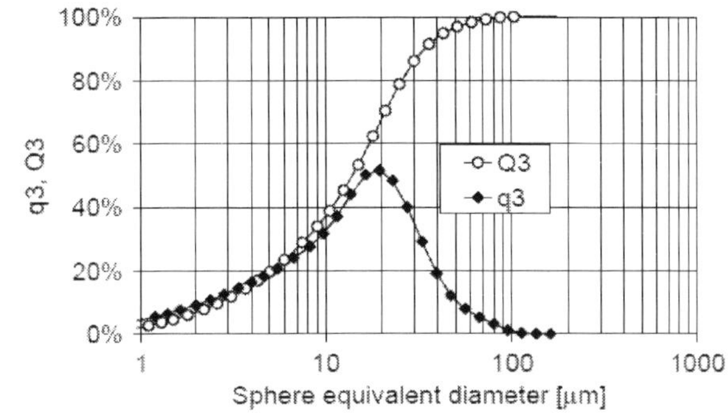

Figure 2.16: Typical result of a particle size distribution obtained by laser light diffraction (Beckmann, 2013).

Powder Flow Properties

In addition to the microscopic properties of a solid product, its bulk properties are also of great practical importance. These properties include bulk density, tapped density, and flow properties. The aim is to predict the behavior of a powder during various manufacturing processes from powder flow properties determined in a standardized setup. Naturally, the predictive power of such a measurement can never be perfect, but it may provide a useful guidance. The practical value of a method can be gauged based on its predictability, reproducibility, sensitivity, and ease of use. The following are the most commonly used methods to characterize the powder flow.

Hausner ratio or compressibility index,

- Angle of repose,
- Flow rate through an orifice, and
- Powder rheometer/shear cell.

The Hausner ratio and compressibility index are calculated from the measured "bulk volume" or "poured volume" (V_0) and "tapped volume" or "settled volume" (V_t) and are defined by Equations 2.17a and 2.17b, respectively. The bulk volume can be determined by filling a known amount of powder in a graduated cylinder, while the tapped volume is obtained by mechanically tapping a measuring cylinder containing a powder sample in a controlled way. A Hausner ratio close to 1 or a low compressibility index indicates good flow properties, while a Hausner index >1.35 or a compressibility index >26% would indicate poor flow properties.

$$Hausner\ ratio = \frac{V_{\circ}}{V_t} \tag{2.17a}$$

$$Compressibility\ index = 100\% \frac{V_{\circ} - V_t}{V_{\circ}} \tag{2.17b}$$

Various methods for the measurement of the angle of repose are used and both static and dynamic angles of repose can be determined. One way of evaluating the static angle of repose is letting the powder flow through a funnel on a flat surface allowing a cone of powder to be produced on the surface. During the formation of the cone, the funnel is moved in such a way that the distance between the top of the cone and the base of the funnel stays constant. The angle of repose (a) is defined by Equation 2.18. An angle of repose <30° indicates an excellent powder flow properties, while an angle of >45° would indicate poor flow properties:

$$\tan\alpha = \frac{2 \times height\ of\ cone}{diameter\ of\ base\ cone} \tag{2.18}$$

Measuring the flowrate through an orifice

maybe directly linked to the behavior during a manufacturing step, for example, the filling of a die with powder through a funnel, but the measured flow rates through orifices depend very much on the setup used (diameter and shape of orifice, type of container material, diameter and height of powder bed, etc.) and therefore nso universal scale is available for that property. If a defined setup is used, the measured flow rates of a series of different samples may correlate well with the behavior of the corresponding samples in a processing step on large scale.

The most fundamental powder properties can be determined in a powder rheometer. Its principle of operation is like that of a rheometer for liquids, where the sample is placed in a shear cell and stress/strain relationships are measured. Design of a shear cell for a powder sample is more demanding than for a liquid sample, as an adequate contact between cell surface and powder need to be ensured. Accordingly, many shear cell configurations exist. While a powder rheometer provides very detailed information on powder properties, the interpretation of the data is also very demanding (Beckmann, 2013).

CHAPTER 3

PHASE EQUILIBRIUM

INTRODUCTION

A phase is a homogeneous part of a system which has uniform chemical composition and physical state. Phases can be of gaseous, liquid, and solid states and can be a pure substance (pure phase) or a mixture of different substances (mixed phase). For example, oxygen in gaseous state is a pure phase, but a mixture of oxygen, nitrogen, argon, and carbon dioxide as constituents of air is a mixed phase.

One-phase system also could be any mixture of gases, vapors or mixtures of two or more completely miscible liquids or solids. Mixtures of non-miscible liquids or solids are heterogeneous systems of two or more phases. For example, a solution of sodium chloride in water is single liquid phase when it is under-saturated, but this solution becomes heterogeneous system when adding sodium chloride to this solution until the solubility at the given temperature is exceeded and another phase, a solid phase appears. In such a heterogeneous system, the macroscopic properties of the system jump-changes unsteadily, at the boundaries between the phases (Beckmann, 2013).

PHASE RULE

For a system in thermodynamic equilibrium, J.W. Gibbs in 1875 derived equation (3.1) which represents the fundamental phase rule.

$$P + f = N + 2 \qquad (3.1)$$

This equation describes the relation between number of independent components, N, the number of co-existing phases, P, and the number of degrees of freedom, F, of the system.

The number of independent components, N, is the minimum number of chemical species necessary to depict the existing phases completely. It is mathematically given as the difference of the sum of chemical individuals in the system and the sum of restrictive conditions such as chemical equation, electro-neutrality, and stoichiometry. Examples with focus on solution equilibrium are compiled in Table 3.1. Depending on the number of independent components, a system is defined as a single, binary, ternary, quaternary, or quinary system with N=1,2, . . . ,5. For example, the last case given in Table 3.1 characterizes the quinary system of the oceanic salts.

The number of degrees of freedom, F, characterizes the variance of the system in the given state of equilibrium. It is defined as the number of independent variables (temperature T, pressure p, and composition x) that can arbitrarily be verified without changing the number of phases present. The "2" found as the second term at the right hand side of Equation 3.1 relates to the physical variables pressure and temperature.

Table 3.1: Number of independent components, N, in different systems (Beckmann, 2013).

System	*N*
H_2O	1
H_2O + ethanol	2
$H_2O + NaCl$	2 (3 − 1)
D-threonine + L-threonine + H_2O	3
$H_2O + NaCl + KCl$	3 (4 − 1)
$H_2O + NaCl + NH_4HCO_3$	4 (5 − 1)
$Na^+, K^+, Mg^{2+}/Cl^-, SO_4^{2-}/H_2O$	5 (6 − 1)

**rewrite this as a table, this is actually a figure at its present state.

The phase rule is universally valid for systems in thermodynamic equilibrium and establishes the theoretical basis for all phase diagrams.

A Phase diagram is a two or three dimensions graph represents the equilibrium between various phases of a system at wide range of temperature, pressure, and concentration of various phases. It specifies the equilibrium conditions (T, p, and x) and the corresponding phases present at this state. Thus, in case of (solid liquid equilibrium) SLE, the phase diagram also tells about the solid phases occurring in a system, such as polymorphs, solvates, or intermediate compounds.

ONE-COMPONENT SYSTEM PHASE EQUILIBRIUM

In these systems, phase equilibrium in a one-component or a single-component system is affected by two variables; temperature and pressure. So, the phase diagram for one-component system is a temperature-pressure equilibrium diagram.

Figure 3.1 illustrates the equilibrium between the vapor, liquid and solid phases of water. Curve AB represent sublimation curve which traces the effect of temperature on the vapor pressure of ice. Curve BC represent vapor pressure curve for liquid water, and line BD represent the significant effect of pressure on the melting point of ice, i.e. the freezing point of water. Water is an unusual substance in that it expands on freezing, indicated by the slope of line BD towards the left of the diagram, i.e. pressure decreases its melting point (Mullin, 2001).

Greater majority of other substances behave in the opposite manner. The three curves meet at the triple point *B* (0.01 "C, 6.1 mbar (610 Pa)) where ice, liquid water and water vapor coexist in equilibrium. At the critical point *C* (374 "C, 220 bar (22 MPa)) liquid and vapor phases become indistinguishable. Above the critical point water is referred to as a supercritical fluid.

Regarding the solvent properties related to the supercritical fluids, they are mostly interesting. For example, liquid water is characterized by having a dielectric constant of around 80, while supercritical water has a value of around 2. Because of this low value, it does not act as a polar solvent any more. Therefore, various organic compounds can be dissolved in and crystallized from it (Mullin, 2001).

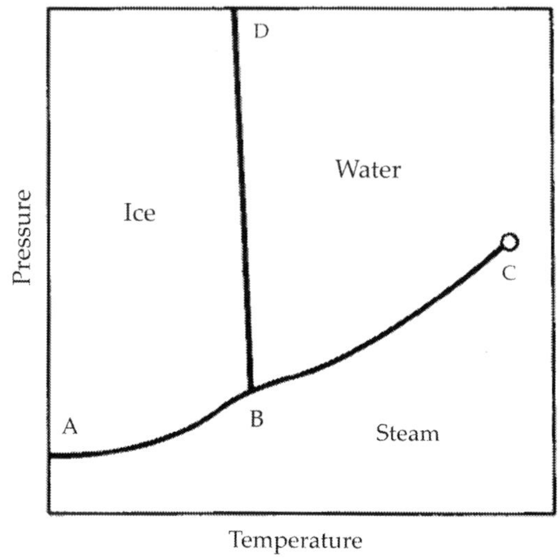

Figure 3.1: Phase diagram for water (not to scale) (Mullin, 2001).

Polymorphs

A system that displays two crystalline polymorphs, the sulfur, is illustrated in Figure 3.2. The area existing above the curve ABEF relates to the region where orthorhombic sulfur is in the stable solid form. The areas bounded by curves ABCD and FECD point to the presence of vapor and liquid sulfur, correspondingly. The region where monoclinic sulfur is in the stable solid form is represented by the 'triangular' area BEC. Curves AB and BC are the vapor pressure curves for orthorhombic and monoclinic sulfur, respectively, and these curves intersect at the transition point B.

The curve *BE* refers to the effect of pressure on the transition temperature for orthorhombic **S** $\rightleftharpoons$ monoclinic **S.** Point *B,* hence, is a triple point demonstrating the temperature and pressure (95.5 °C and 0.51 Nm^{-2} at which both orthorhombic sulfur and sulfur vapor can coexist in stable equilibrium. The curve *EF* refers to the effect of pressure on the melting point of orthorhombic sulfur;

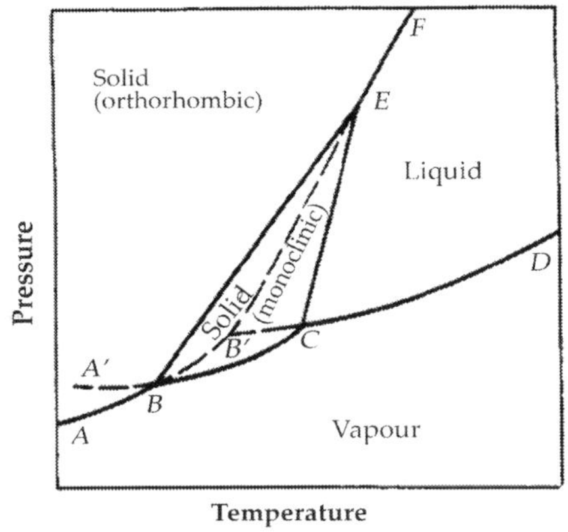

Figure 3.2: Phase diagram for sulfur (not to scale) (Mullin, 2001).

Stable equilibrium can occur between orthorhombic and monoclinic sulphur and liquid sulphur at point *E,* which is a triple point representing the temperature and pressure (151 °C and 1.31 x 10^8Nm^{-2}). The curve ***CD*** is representing the vapor pressure curve for liquid sulphur, and the curve *CE* refers to the effect of pressure on the melting point of monoclinic sulphur. So, point C is another triple point (115 °C and 2.4N m^{-2}) that indicates the equilibrium between monoclinic and liquid sulphur and sulphur vapor.

Figure 3.2 shows the broken lines, which illustrates the metastable conditions. When orthorhombic sulphur undergoes rapid heating beyond 95.5 °C, the conversion to the monoclinic form does not occur till a specific time has passed; curve BB', a continuation of curve AB, is the vapor pressure curve for metastable orthorhombic sulphur above the transition point. Similarly, if monoclinic sulphur is cooled rapidly below 95.5°C, the change to the orthorhombic form does not take place immediately, and curve BA' is the vapor pressure curve for metastable monoclinic sulphur below the transition point. Likewise, curve CB' is the vapor pressure curve for metastable liquid sulphur below the 115°C transition point, and curve B'E the melting point curve for metastable orthorhombic sulphur. Thus, point B', is considered a fourth triple point (110°C and 1.7 N m^{-2}) of the system. So, among the four possible phases: orthorhombic (solid), monoclinic (solid), liquid and vapor, only three of them can coexist in stable equilibrium at any one time, and then only at one of the three 'stable' triple points (Mullin, 2001).

Transformations

The process of transformation from one polymorph to another can be either reversible or irreversible; regarding the former case the two crystalline forms are known to be enantiotropic; in the latter, they are known to be monotropic. Demonstration of these phenomena can be done with guidance of the pressure-temperature phase diagram.

The phase reactions displayed by two enantiotropic solids, α and β are shown in Figure 3.3a. AB curve is the vapor pressure curve for the α form, BC curve is that for the β form, and CD curve is that for the liquid. The transition point is referred to point B, where the vapor pressure curves of the two solids intersect; that means the two forms can coexist in equilibrium under these specific conditions of temperature and pressure. Point C is a triple point at which vapor, liquid and β solid can coexist. This point can be considered as the melting point of the β form.

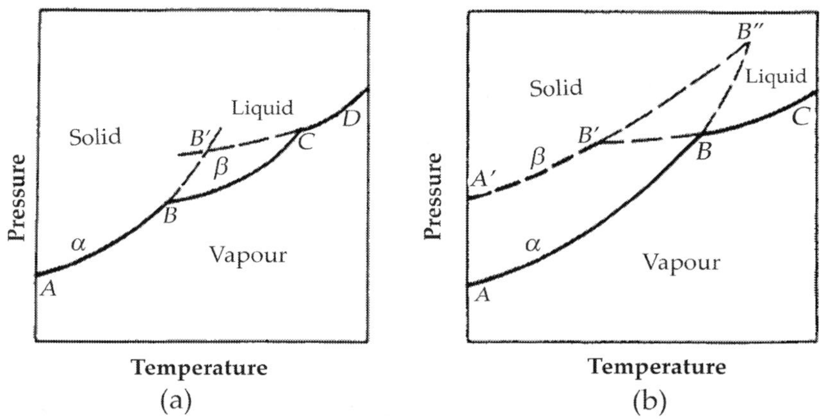

Figure 3.3: Pressure-temperature diagrams for dimorphous substances: (a) enantiotropy; (b) monotropy (Mullin, 2001).

As shown in figure 3.3a, heating *α* solid slowly transforms it to the *β* solid and further heating makes it melt and vapor pressure curve ABC is followed. On the other hand, when cooling the liquid slowly, the *β* form crystallizes out first and then transforms into *α* form. However, different behavior happened with rapid heating or cooling as shown also in figure 3.3a. The vapor pressure of *α* form can increase along curve *BB'*, an extension of *AB, α* form now being metastable. Also, the liquid vapor pressure can fall along curve *CB'*, an extension of ***DC,*** the liquid being metastable. Point *B'*, therefore, is a metastable triple point at which the liquid, vapor and *α* solid can coexist in metastable equilibrium (Mullin, 2001).

Phase equilibrium of sulphur (as shown in Figure 3.2) is an example of metastable triple equilibrium. Where the orthorhombic and monoclinic forms are enantiotropic; the transition point occurs at a lower temperature than does the triple point.

Figure 3.3b illustrates the pressure-temperature curves for a monotropic substance. AB and BC are the vapor pressure curves for α solid and liquid and A'B' is pressure-temperature curve for β solid. In this case the vapor pressure curves of α and β forms do not intersect, so there is no transition point within this range of temperature and pressure. The solid form with the higher vapor pressure at any given temperature (β in this case) is the metastable form. Curves BB' and BB" are the vapor pressure curves for the liquid and metastable α solid, so B' is a metastable triple point. If this system did exhibit a true transition point, it would lie at point B"; but as this denotes a temperature higher than the melting point of the solid, it cannot exist (Mullin, 2001). **The red highlighted portion in this paragraph… try to use a uniform symbol for "prime"… will it be like the AprimeBprime symbol of the BBprime, BBdoubleprime symbol—do this to all sections that apply.

A typical case of monotropy is the change from white to red phosphorus. Benzophenone is another example of a monotropic substance: the stable melting point is 49°C, whereas the metastable form melts at 29°C.

TWO-COMPONENT SYSTEM PHASE EQUILIBRIUM

Binary system phase equilibrium is affected by three main variables; temperature, pressure and concentration. As a result, the behavior of binary system should be represented by a 3D model with three mutually perpendicular axes of pressure, temperature and concentration. Otherwise, for simplicity three 2D diagrams with pressure temperature, pressure-concentration and temperature concentration axes, respectively, can be used as replacement to the 3D model. However, in most crystallization processes the main interest lies in the liquid and solid phases of a system; a knowledge of the behavior of the vapor phase is only required when considering sublimation processes. Because pressure has little effect on the equilibrium between liquids and solids, the phase changes can be represented on a temperature-concentration diagram; the pressure, usually atmospheric, is ignored. Such a system is said to be 'condensed', and a 'reduced' phase rule can be formulated excluding the pressure variable:

$\boldsymbol{P} + F' = C + 1$ (Equation no???)

Where F' is the number of degrees of freedom, not including pressure. Four different types of two-component system will now be considered. Detailed attention is paid to the first type solely to illustrate the information that can be deduced from a phase diagram. It will be noted that the concentration of a solution on a phase diagram is normally given as a mass fraction or mass percentage and not as 'mass of solute per unit mass of solvent', as recommended for the solubility diagram. Mole fractions and mole percentages are also suitable concentration units for use in phase diagrams (Mullin, 2001).

Let us now consider solid–liquid equilibrium in a binary system. Contrary to phase equilibrium between vapor and condensed phases, in the systems comprising solid and liquid phases the pressure as variable can usually be neglected. The behavior of condensed systems does not mostly depend on pressure and thus the appropriate phase diagrams are valid for arbitrary (but not too high) pressures. Thus, phase changes in a binary system can be represented on a temperature–composition diagram as shown in Figure 3.4 for two arbitrary components A and B.

With constant pressure and N=2 in a binary system, the phase rule follows: P+F=3, which mean the following:

- The maximum number of phases that can coexist in equilibrium is three, leaving no degree of freedom (F=0), that is, the system is in an invariant state. Temperature and composition are fixed as long as the three phases coexist. This applies to the eutectic point in the phase diagram in Figure 3.4. On heating, a mixture of A and B of eutectic composition melts at the eutectic temperature T_{eu}. Three phases coexist at that temperature, solid A, solid B, and the liquid melt. The temperature remains constant if all three phases are present and the mixture melts in eutectic composition. After melting, just one phase remains and the temperature can be varied by heating. On cooling, at T_{eu}, a melt of eutectic composition splits into two solid phases A and B. The eutectic invariant in the phase diagram is represented by a horizontal line connecting the three coexisting phases.

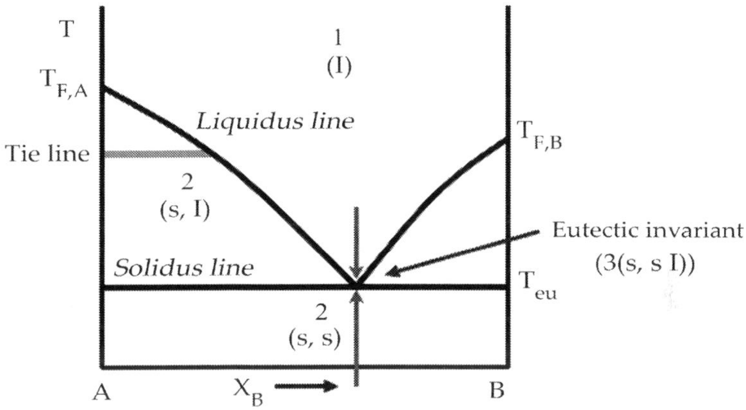

Figure 3.4: Binary phase diagram of components A and B with temperature, T, mole fraction of component B, X_B. Numbers refer to the number of phases present and s, l to their state of aggregation (Beckmann, 2013).

- When two phases are present in equilibrium, one degree of freedom (F=1) is left, that is, the system is in monovariant state. By specifying, for example, temperature, the composition is fixed. Such a situation is illustrated in Figure 3.4 in the two-phase region below the liquidus line of A. A tie line at a certain temperature connects the two coexisting phases, which are a solid and a liquid phase, in fact solid A and a liquid of corresponding composition on the liquidus line. If another temperature is chosen, again the composition of the phases is fixed.
- In case of just one phase, the degree of freedom is two (F=2, bivariant state). Temperature and composition can independently be varied to hold the present phase. The one-phase region in Figure 3.4 is the region above the liquidus lines of A and B, being a liquid phase.

For quantification in the phase diagram, the so-called lever rule can be applied: If two phases are in equilibrium, their amount behaves reciprocally to the appropriate part of the tie line (Beckmann, 2013).

Simple eutectic

A typical example of a system in which the components do not combine to form a chemical compound is shown in Figure 3.5. Curve AB and BC represent the temperatures at which homogeneous liquid solutions of naphthalene in benzene begin to freeze or to crystallize. The curves also represent, therefore, the temperatures above which mixtures of these two components are completely liquid. The name 'liquidus' is generally given to this type of curve. In aqueous systems of this type one liquidus is the freezing point curve, the other the normal solubility curve. Line DBE represents the temperature at which solid mixtures of benzene and naphthalene begin to melt, or the temperature below which mixtures of these two components are completely solid. The name 'solidus' is generally given to this type of line. The melting or freezing points of pure benzene and naphthalene are given by points *A* (5.5°C) and C (80.2°C), respectively. The upper area enclosed by the liquidus, *ABC,* represents the homogeneous liquid phase, i.e. a solution of naphthalene in benzene; that enclosed by the solidus, *DBE,* indicates solid mixtures of benzene and naphthalene. The small and large 'triangular' areas *ABD* and *BCE* represent mixtures of solid benzene and solid naphthalene, respectively, and benzene-naphthalene solution.

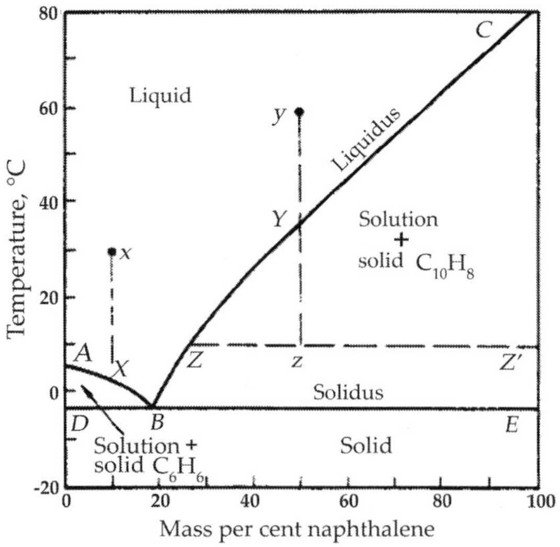

Figure 3.5: Phase diagram for the simple eutectic system naphthalene-benzene (Mullin, 2001).

If a solution represented by point x is cooled, pure solid benzene is deposited when the temperature of the solution reaches point *X* on curve *AB*. As solid benzene separates out, the solution becomes more concentrated in naphthalene and the equilibrium temperature of the system falls, following curve *AB*.

If a solution represented by point y is cooled, pure solid naphthalene is deposited when the temperature reaches point *Y* on the solubility curve; the solution becomes more concentrated in benzene and the equilibrium temperature follows curve *CB*. Point *B*, common to both curves, is the eutectic point (-3.5°C and 0.189 mass fraction of naphthalene), and this is the lowest freezing point in the whole system. At this point a completely solidified mixture of benzene and naphthalene of fixed composition is formed: it is important to note that the eutectic is a physical mixture, not a chemical compound. Below the eutectic temperature all mixtures are solid.

If the solution *y* is cooled below the temperature represented by point Yon curve *BC* to some temperature represented by point *z*, the composition of the system remains unchanged. The physical state of the system has been altered, however; it now consists of a solution of benzene and naphthalene containing solid naphthalene. The composition of the solution, or mother liquid, is given by point *z* on the solubility curve, and the proportions of solid naphthalene and solution are given, by the so-called 'mixture rule', by the ratio of the lengths *ZZ* and *zZ'*, i.e.

$$\frac{mass\ of\ solid\ C_{10}H_8}{mass\ of\ solution} = \frac{zZ}{zZ'}$$ **rewrite this using equation editor.

A process involving both cooling and evaporation can be analyzed in two steps. The first is as described above, i.e. the location of points *z*, *Z* and *Z'*; this represents the cooling operation. If benzene is evaporated from the system, *z* no longer represents the composition; thus, the new composition point *z'* (not shown in the diagram) is located along line *ZZ'* between points *z* and *2*. Then the ratio *z'Z/z'Z'* gives the proportions of solid and solution.

The systems KCl-H_2O and $(NH_4)_2S0_4$-H_2O are good examples of aqueous salt solutions that exhibit simple eutectic formation. In aqueous systems, the eutectic mixture is sometimes referred to as a cryohydrate, and the eutectic point a 'cryohydric point'.

The term 'pure' when commonly used does not mean absolute 100% purity. In industrial crystallization practice, this is neither necessary nor indeed achievable, and for many bulk produced chemicals a purity of >95% is often accepted as justifying the designation 'pure'. In any case, a single crystallization step cannot produce 100% pure crystals for a variety of reasons; first is, they can be contaminated with residual solvent or other impurities that have not been removed by washing, or have been incorporated into the crystal interstitially or as liquid inclusions, and so on. Furthermore, contamination commonly results from the existence of terminal solid solutions, which inevitably accompany both eutectic and chemical compound systems (Mullin, 2001).

Compound formation

The solute and solvent of a binary system frequently combine to form one or more different compounds. In aqueous solutions, these compounds are called 'hydrates'; for non-aqueous systems, the term 'solvate' is used. Two types of compound can be considered: one can coexist in stable equilibrium with a liquid of the same composition, and the other cannot behave in this manner. In the former case, the compound is said to have a congruent melting point; in the latter, an incongruent melting point.

Figure 3.6 illustrates the phase reactions in the manganese nitrate-water system. Curve *AB* is the freezing point curve. The solubility curve *BCDEFG* for $Mn(NO_3)_2$ in water is not continuous owing to the formation of several different hydrates. The area above curve *ABCDEFG* represents homogeneous liquid solutions. Mixtures of the hexahydrate and solution exist in areas *BCH* and *ICD*. The tetrahydrate is the stable phase in region *DEJ* and the dehydrate in *EKF*. The rectangular areas under *FH, ZJ* and *KL* represent completely solidified systems (ice and hexahydrate, hexa- and tetrahydrates, and tetra and dihydrates, respectively). Point *B* is a eutectic or cryohydric point with the co-ordinates -36 °C and 0.405 mass fraction of $Mn(NO_3)_2$.

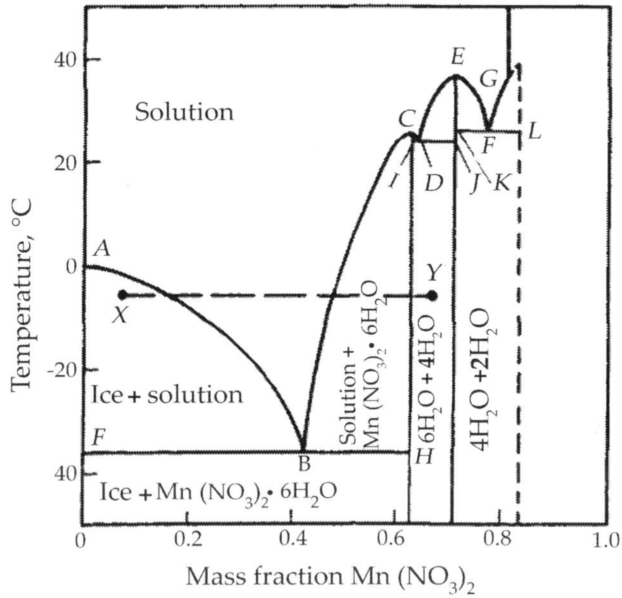

Figure 3.6: Phase diagram for the system $Mn(N0_3)_2$-H_2O (Mullin, 2001).

Point *C* in Figure 3.6 indicates the melting point (25.8°C) and composition (0.624 mass fraction) of the hexahydrate. Thus, when a solution of this composition is cooled to 25.8°C it solidifies to its hexahydrate form, i.e. no change in composition occurs. Point *C*, therefore, is a congruent point. Similarly, point *E* is the congruent point for the tetrahydrate (melting point 37.1°C, composition 0.713). Points ***D*** and F are the other two eutectic points of the system. Point *G* is the transition point at which the dihydrates decomposes into the monohydrate and water, i.e. it is the incongruent melting point of the dihydrate. The vertical broken line at 0.834 mass fraction represents the composition of the dihydrate.

The behavior of manganese nitrate solutions on cooling can be traced in the same manner as that described above for simple eutectic systems. The solution concentrations and the proportions of solid and solution can similarly be deduced graphically. The process of isothermal evaporation in congruent melting systems presents an interesting phenomenon. For example, the mixture represented by point *X* in Figure 3.6 represents slurry of ice and solution; but when sufficient water is removed to bring the system composition into the region to the right of curve *AB,* it becomes a homogeneous liquid solution. When more water is removed, so that region *BCG* is entered, the system partially solidifies again, depositing crystals of the hexahydrate. On further evaporation, once the composition exceeds 62.4 percent of $Mn(NO_3)_2$, e.,g . at point *Y,* the system solidifies completely to a mixture of the hexa- and tetrahydrates. The reverse order of behavior occurs on isothermal hydration.

The formation of eutectics and solvates with congruent points is observed in many organic, aqueous inorganic and metallic systems. The case illustrated above is a rather simple example. Some systems form many solvates and their phase diagrams can become rather complex. Ferric chloride, for example, forms four hydrates, and the $FeCl_3$-H_2O phase diagram exhibits five-cryohydric points and four congruent points.

A solvate that is unstable in the presence of a liquid of the same composition is said to have an incongruent melting point. Such a solvate melts to form a solution and another compound, which may or may not be a solvate. For instance, the hydrate $Na_2SO_4 \cdot 10H_2O$ melts at 32.4°C to give a saturated solution of sodium sulfate containing a suspension of the anhydrous salt; hence, this temperature is the incongruent melting point of the decahydrate. The terms 'meritectic point' and 'transition point' are also used instead of the expression 'incongruent melting point'.

Figure 3.7 illustrates the behavior of the system sodium chloride-water. The various areas are marked on the diagram. *AB* is the freezing point curve and *BC* is the solubility curve for the dihydrate. Point *B* (-21˚C) is a eutectic or cryohydric point at which a solid mixture of ice and $NaCl \cdot 2H_2O$ of fixed composition (0.29 mass fraction of NaCl) is deposited. At point *C* (0.15˚C) the dihydrate decomposes into the anhydrous salt and water; this is, therefore, the incongruent melting point, or transition point, of $NaCl \cdot 2H_2O$. The vertical line commencing at 0.619 mass fraction of NaCl represents the composition of the dihydrate. If this system had a congruent melting point, which it does not have, this line would meet the peak of the extension of curve *BC* (see Figure 3.6) (Mullin, 2001)

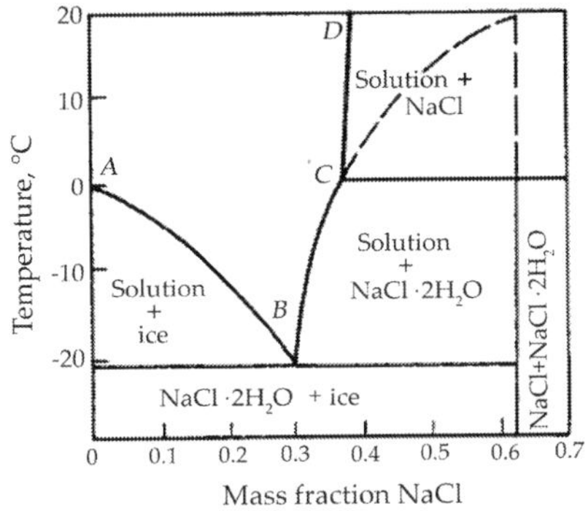

Figure 3.7: Phase diagram for the system NaCl·$2H_2O$ (Mullin, 2001).

Solid Solutions

Many binary systems when submitted to a cooling operation do not at any stage deposit one of the components in the pure state: both components are deposited simultaneously. The deposited solid phase is, in fact, a solid solution.

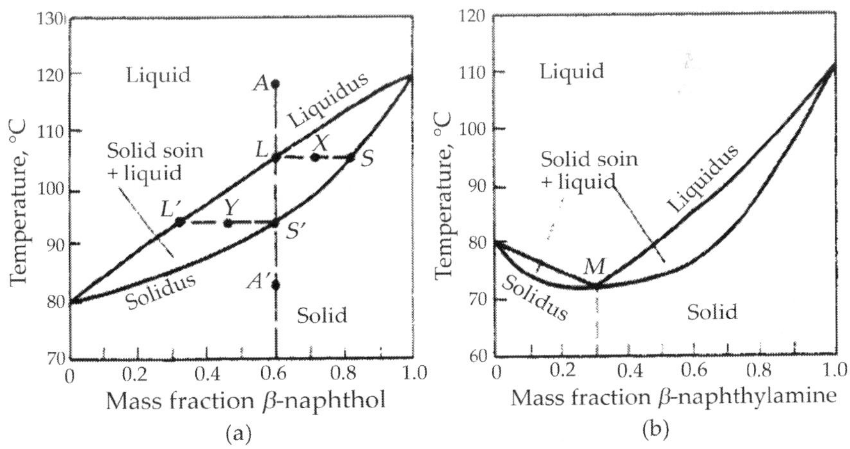

Figure 3.8: Solid solutions: (a) continuous series (naphthalene-β-naphthol; (b) minimum melting point (naphthalene-β-naphthylamine) (Mullin, 2001).

Only two phases can exist in such a system: a homogeneous liquid solution and a solid solution. Therefore, from the reduced phase rule, $F' = 1$, so an invariant system cannot result. One of three possible types of equilibrium diagram can be exhibited by systems of this kind. In the first type, illustrated in Figure 3.8a, all mixtures of the two components have freezing or melting points intermediate between the melting points of the pure components. In the second type shown in Figure 3.8b, a minimum is produced in the freezing and melting point curves. In the third type of diagram, a maximum is exhibited in the curves, which is rare.

Figure 3.8a shows the temperature-concentration phase diagram for the system naphthalene-β-naphthol, which forms a continuous series of solid solutions. The melting points of pure naphthalene and β-naphthol are 80 and 120°C, respectively. The upper curve is the liquidus or freezing point curve, the lower the solidus or melting point curve. Any system represented by a point above the liquidus is completely molten, and any point below the solidus represents a completely solidified mass. A point within the area enclosed by the liquidus and solidus curves indicates an equilibrium mixture of liquid and solid solution. Point *X,* for instance, denotes a liquid of composition *L* in equilibrium with a solid solution of composition S, and point *Y* a liquid *L'* in equilibrium with a solid *S'*.

The phase reactions occurring on the cooling of a given mixture can be traced as follows. If a homogeneous liquid represented by point *A* (60 per cent β-naphthol) is cooled slowly, it starts to crystallize when point *L* (105°C) is reached. The composition of the first crystals is given by point *S* (82 per cent *β*-naphthol). As the temperature is lowered further, more crystals are deposited but their composition changes successively along curve *SS'*, and the liquid composition changes along curve *LL'*. When the temperature is reduced to 94 °C (points *L'* and *S'),* the system solidifies completely. The over-all composition of the solid system at some temperature represented by, say, point *A'* is the same as that of the original homogeneous melt, if no crystals have been removed during the cooling process, the system is no longer homogeneous because of the successive depositions of crystals of varying composition. The changes occurring when a solid mixture *A'* is heated can be traced in a manner similar to that of the cooling operation.

Figure 3.8b shows the relatively uncommon, but not rare, type of binary system in which a common minimum temperature is reached by both the upper liquidus and lower solidus curves. These two curves approach and touch at point *M.* The example shown in Figure 3.8b is the system naphthalene-β-naphthylamine. Freezing and melting points of mixtures of this system do not necessarily lie between the melting points of the pure components. Three sharp melting points are observed: 80°C (pure naphthalene), 110°C (pure p-naphthylamine) and 72.5°C (mixture *M,* 0.3 mass fraction β-naphthylamine). Although the solid solution deposited at point *M* has a definite composition, it is not a chemical compound. The components of such a minimum melting point mixture are rarely, if ever, present in stoichiometric proportions. Point *M,* therefore, is not a eutectic point: the liquidus curve is completely continuous; it only approaches and touches the solidus at *M.* The phase reactions occurring when mixtures of this system are cooled can be traced in the same manner as that described for the continuous series solid solutions (Mullin, 2001).

MELT AND SOLUTION EQUILIBRIUM

At this point, the difference between the terms "melt" and "solution equilibrium" must be clarified. Particularly driven from the application, that is, whether one refers to melt or to solution crystallization (the term "solution" used here to characterize a liquid homogeneous mixture with a classical solvent as one component), it is frequently distinguished between both, what is not mandatory from thermodynamic viewpoint. Generally, a solution is a homogeneous mixture of two or more substances that can be gaseous, liquid, or solid. In this respect, a *melt* is a liquid close to its freezing point and, hence, can be treated as a solution. The formal analogy of melt and solution equilibrium with respect to the phase diagram is illustrated in Figure 3.9.

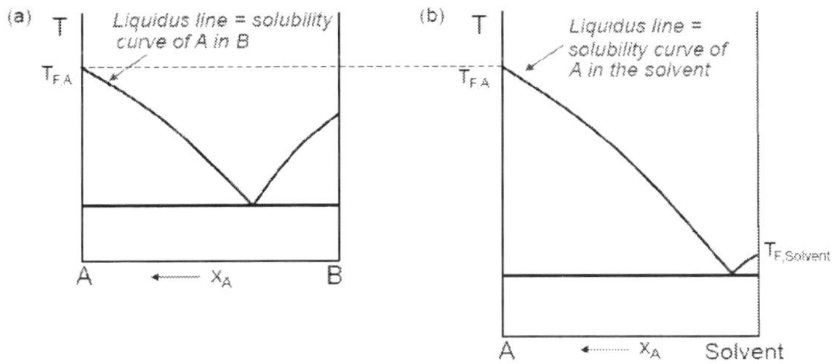

Figure 3.9: Binary phase diagrams of (a) two compounds A and B and (b) compound A and a solvent demonstrating the formal analogy between melt and solution equilibrium (Beckmann, 2013).

Figure 3.9a shows the melt (or melting point) phase diagram of two compounds A and B. The liquidus lines characterize the solubility of one compound in the other. Thus, the marked liquidus line is the solubility curve of A in B. Figure 3.9b presents the phase diagram of compound A and a solvent, and thus, the solution equilibrium of A in the solvent. The liquidus line of A also starts at the melting temperature of A, $T_{F,A}$, and in ideal case is identical to the liquidus line of A in Figure 3.9a since the freezing point depression depends only on composition, not on identity of the second component in the binary system. Thus, both phase diagrams, one representing melt equilibrium and the other solution equilibrium of a component in a solvent, can be treated in the same way. Moreover, the liquidus line of A in the solvent in Figure 3.9b is nothing else than the "classical" solubility curve of A in this solvent. When rotating the phase diagram in Figure 3.9b by 90° to the right (see Figure 3.10), the typical representation of the solubility curve (concentration/composition as function of temperature) results.

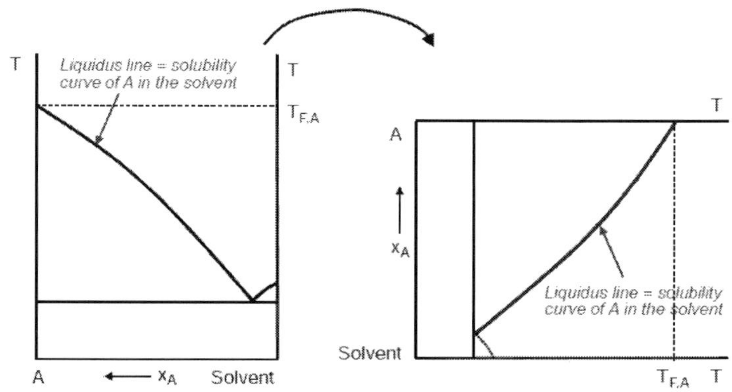

Figure 3.10: Graphical demonstration of the relationship between the liquidus lines of A in the binary phase diagram A – solvent and the solubility curve of A in the solvent (Beckmann, 2013).

PHASE CHANGE DETECTION

Thermal analysis

A phase reaction is always accompanied by an enthalpy change, and this heat effect can readily be observed if a cooling curve is plotted for the system. In many cases a very simple apparatus can be used. A large glass test tube, fitted with a stirrer and a thermometer graduated in increments of 0.1°C and held in a temperature-controlled environment, will often suffice. The temperature of the system is recorded at regular intervals of say 1 min.

A smooth cooling curve is followed until a phase reaction takes place, when the accompanying heat effect causes an arrest or change in slope. Figure 3.11a shows a typical example for a pure substance. AB is the cooling curve for the homogeneous liquid phase. At point B, the substance starts to freeze and the system remains at constant temperature, the freezing point, until solidification is complete at point C. The solid then cools at a rate indicated by curve CD. It is possible, of course, for the liquid phase to cool below the freezing point, and some systems may withstand appreciable degrees of super cooling. The dotted curve in Figure 3.11a denotes the sort of path followed if super cooling occurs. Seeding of the system will minimize these effects.

Figure 3.11b shows the type of cooling curve obtained for a binary system in which eutectic or compound formation occurs. The temperature of the homogeneous liquid phase falls steadily along curve *EF* until, at point *F*, deposition of the solid phase commences.

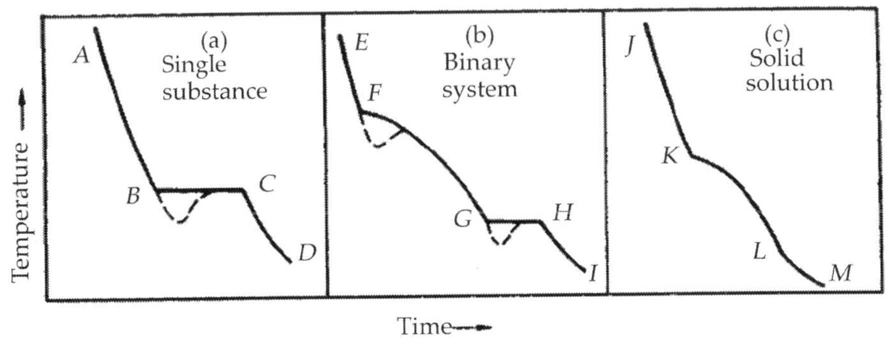

Figure 3.11: Some typical cooling curves (Mullin, 2001).

The rate of cooling changes along curve *FG* as more solid is deposited. The composition of the remaining solution changes until the composition of the eutectic is reached, then crystallization or freezing continues at constant temperature (line *GH),* i.e. the eutectic behaves as a single pure substance. The completely solidified system cools along curve *HL.* Super cooling, denoted by the dotted lines, may be encountered at both arrest points if the system is not seeded.

Figure 3.11c shows a typical cooling curve for a binary mixture that forms a series of solid solutions. The first arrest, *K,* in the curve corresponds to the onset of freezing, and this represents a point on the liquidus. The second arrest, *L,* occurs on the completion of freezing and represents a point on the solidus. It will be noted that no constant-temperature freezing point occurs in such a system (Mullin, 2001).

Differential thermal analysis (DTA)

Differential thermal analysis is a method used for observing phase changes and measuring the associated changes in enthalpy. A small test sample, often only a few milligrams, is heated near a sample of reference material in an identical container. The reference material, chosen for its similarity to the test sample, must not exhibit any phase change over the temperature range under consideration.

When the test sample undergoes a phase change, there will be heat release or absorption. For example, if it melts it will absorb heat and its temperature will lag that of the reference material (Figure 3.12a). The difference in temperature between the two samples is detected by a pair of matched thermocouples and recorded as a function of time. The area between the differential curve (Figure 3.12b) and the base line is a function of the enthalpy associated with the phase change.

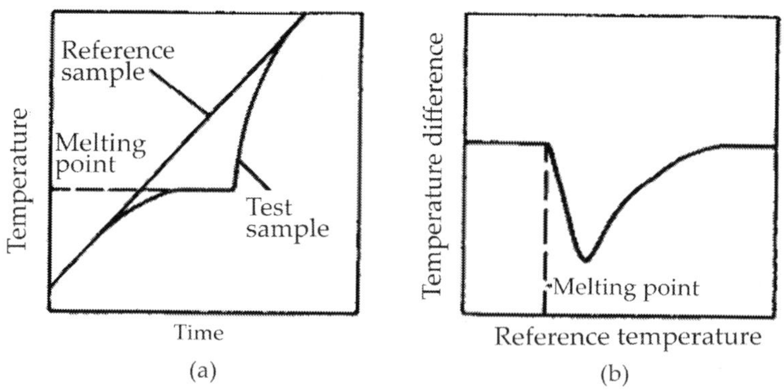

Figure 3.12: Differential thermal analysis: (a) comparative heating curves, (b) differential temperature curve for a single substance (Mullin, 2001).

THREE-COMPONENT SYSTEMS

Construction of ternary diagrams

The phase equilibrium in ternary systems can be affected by four variables, viz. temperature, pressure and the concentration of any two of the three components. This fact can be deduced from the phase rule (from equation???):

$$P + F = 3 + 2$$

which indicates that a one-phase ternary system will have four degrees of freedom. It is impossible to represent the effects of the four possible variables in a ternary system on a two-dimensional graph. For solid-liquid systems, however, the pressure variable may be neglected, and the effect of temperature will be considered later.

The composition of a ternary system can be represented graphically on a triangular diagram. Two methods are in common use. The first utilizes the equilateral triangle, and the method of construction is shown in Figure 3.13a. The apexes of the triangle represent the pure components *A, B* and *C.* A point on a side of the triangle stands for a binary system, *AB, BC* or *AC;* a point within the triangle represents a ternary system *ABC.* The scales may be constructed in any convenient units, e.g. weight or mole percent, weight or mole fraction, etc., and any point on the diagram must satisfy the equation $A + B + C = 1$ or 100. The quantities of the components *A, B* and C in each mixture *M* are represented by the perpendicular distance from the sides of the triangle.

Special triangular graph paper is required if the equilateral diagram is to be used, and for this reason many workers prefer to employ the right-angled triangular diagram which can be drawn on ordinary squared graph paper. The construction of the right-angled isosceles triangle is shown in Figure 3.13b. Again, as in the case of the equilateral triangle, each apex represents a pure component *A, B* or *C,* a point on a side a binary system, and a point within the triangle a ternary system; in all cases $A + B + C = 1$ or 100. The quantities of *A, B* and C in each mixture *M* are represented by the perpendicular distances to the sides of the triangle. If two compositions, *A* and *B, B* and *C,* or *A* and *C* are known, the composition of the third component is fixed on both triangular diagrams.

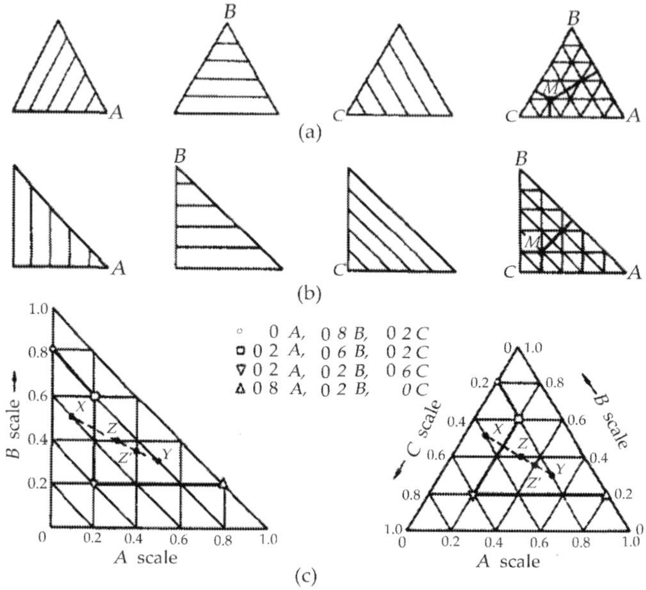

Figure 3.13: Construction of equilateral triangular diagrams (Mullin, 2001).

Two actual plots are shown in *Figure 4.19*~to illustrate the interpretation of these diagrams. For clarity, the C scale has been omitted from the right-angled diagram; the C values can be obtained from the expression C = *1 – (A + B).* The 'mixture rule' is also illustrated in Figure 3.13c. When any two mixtures *X* and Yare mixed together, the composition of the final mixture *Z* is represented by a point on the diagram located on a straight line drawn between the points representing the initial mixtures. The position of *Z* is located by the expression:

$$\frac{\text{mass of mixture } X}{\text{mass of mixture } Y} = \frac{\text{distance } YZ}{\text{distance } XZ}$$

For example, if one part of a mixture *X (0.1A, 0.5B, 0.4C)* is mixed with one part of a mixture *Y (0.5A, 0.3B, 0.2C),* the composition of the final mixture *Z (0.3A, 0.4B, 0.3C)* is found on the line *XY* where *XZ* = *YZ.* Again, if *3* parts of Yare mixed with 1 part of *X,* the mixture composition *2' (0.4A, 0.35B, 0.25C)* is found on the line XY where *XZ'* = 3(YZ'). The mixture rule also applies to the removal of one or more constituents from a system. Thus, one part of a mixture *X* removed from 2 parts of a mixture *2* would yield one part of a mixture ***Y*** given by:

$$\frac{\text{mass of original } Z}{\text{mass of } X \text{ removed}} = \frac{YX}{YZ} = \frac{2}{1}$$ **use equation editor

Similarly, one part of *X* removed from 4 parts of *2'* would yield 3 parts of a mixture *Y* given by

$$\frac{\text{mass of original } Z'}{\text{mass of } X \text{ removed}} = \frac{YX}{YZ'} = \frac{4}{1}$$ **use equation editor

The principle of the mixture rule is the same as that employed in the operation of lever-arm problems, i.e. $m_1 l_1 = m_2 l_2$, where m is a mass and *l* is the distance between the line of action of the mass and the fulcrum. For this reason, the mixture rule is often referred to as the lever-arm or center of gravity principle (Mullin, 2001).

CHAPTER 4

NUCLEATION

INTRODUCTION

The formation of crystals takes place when nuclei are formed and then grow. Usually crystals can grow from supersaturated solutions, super cooled melts and vapors. Crystal formation in metals occurred in two steps:

- Nucleation: may be defined as arranging of a few atoms to form a three-dimensional periodic array called nucleus which already shows faces, although it is only a few unit cells in size as shown in Figure 4.1.
- Crystal Growth: after nucleation happened, nucleus grows to form a crystal. Nucleus attracts more atoms, they take up positions on its faces in accordance with its three-dimensional periodicity (Borchardt-Ott, 2011). Crystal growth phenomena will be discussed in detail in chapter 5.

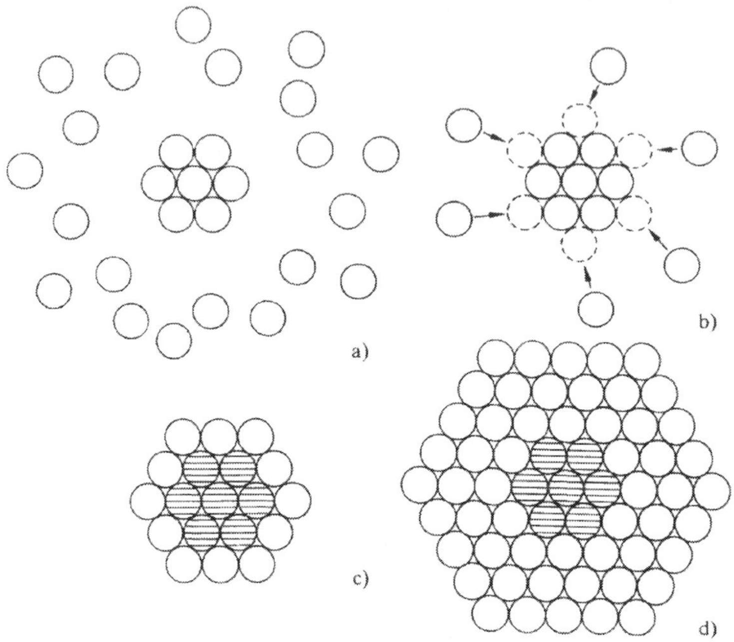

Figure 4.1: The phenomena of nucleation and growth of the nucleus to a macro crystal illustrated in two steps. (a) Nucleus formed from melt. (b) Atoms stick to the nucleus. (c) Growth of a new layer on the faces of a nucleus. (d) The formation of a macro crystal by the addition of further layers of atoms (Borchardt-Ott , 2011).

NUCLEATION DIFFERENT TECHNIQUES

The kinetic processes of nucleation and crystal growth require super saturation, which can generally be obtained by a change in temperature, by removing the solvent (usually by evaporation), or by adding a drowning-out agent or reaction partners. The system then attempts to achieve thermodynamic equilibrium through nucleation and the growth of nuclei. Homogeneous nucleation happened if a solution is totally pure i.e. contains neither solid foreign particles nor crystals of its own type. If foreign particles are present, nucleation is facilitated and the process is known as heterogeneous nucleation. It should be noted that both types mentioned of nucleation; homogeneous and heterogeneous occur in the absence of solution-own crystals and are collectively known as primary nucleation. This occurs when a specific super saturation, known as the metastable super saturation ΔC_{met}, is obtained in the system. On the other hand, in secondary nucleation which happened usually in semi-commercial and industrial crystallizers, it has often been observed that nuclei occur even at a very low super saturation $\Delta C < \Delta C_{met}$ when solution-own crystals are present (e.g., in the form of attrition fragments or added seed crystals). However, it should be noted that a distinction is made between nucleation resulting from contact; shearing action, breakage, abrasion, and needle fraction as shown in Figure 4.2 (Mersmann, 2001).

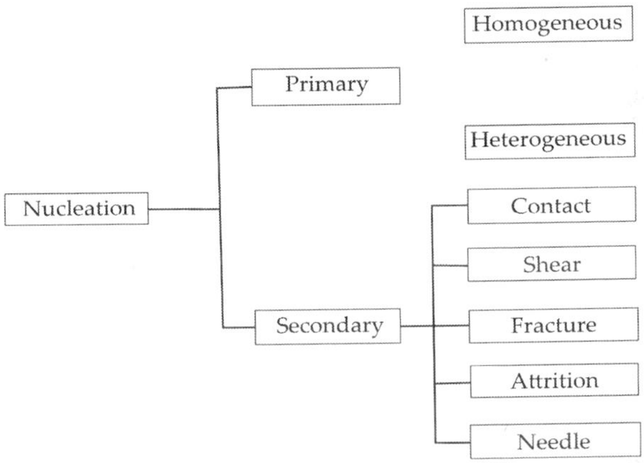

Figure 4.2: Mechanisms of primary and secondary nucleation (Jones, 2002).

Nucleation can be divided into two separate types as shown in Figure 4.2:

Primary nucleation (nucleation without crystalline matter)

- homogeneous nucleation (natural nucleation from clear solution)
- heterogeneous nucleation (made by foreign particles)

Secondary nucleation (made by presence of existing crystals)

- contact (with other crystals or the crystallizer parts)
- shear (due to fluid flow)

- attrition (due to particle impact or fluid flow)
- fracture (due to particle impact)
- needle (due to particle disruption) (Jones, 2002).

Figure 4.3 illustrates the dependence of super saturation on several types of nucleation process plotted against solubility (Mersmann, 2001).

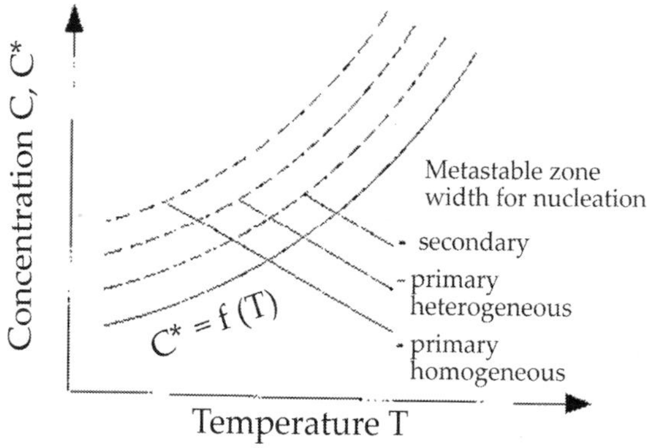

Figure 4.3: Metastable super saturation as a function of temperature for many types of nucleation process (Mersmann, 2001).

In the following paragraphs, we will discuss each technique of nucleation in details.

PRIMARY NUCLEATION

Primary nucleation occurs at high levels of super saturation and is considered the classical form of nucleation which is most prevalent during unseeded crystallization or precipitation. This type of nucleation is subdivided into either homogeneous, i.e. spontaneously from clear solution, or heterogeneous which happened in the presence of 'dust' particles in suspension, or solid surfaces (Jones, 2002).

Homogeneous nucleation

Exactly how a stable crystal nucleus is formed within a homogeneous fluid is not known with any degree of certainty. To take a simple example, the condensation of a supersaturated vapor to the liquid phase is only possible after the appearance of microscopic droplets, called condensation nuclei, on the condensing surface. However, as the vapor pressure at the surface of these minute droplets is exceedingly high, they evaporate rapidly even though the surrounding vapor is supersaturated. New nuclei form while old ones evaporate, until eventually stable droplets are formed either by coagulation or under conditions of very high vapor super saturation.

The formation of crystal nuclei is an even more difficult process to envisage. Not only have the constituent molecules to coagulate, resisting the tendency to re-dissolve, but they also must become orientated into a fixed lattice. The number of molecules in a stable crystal nucleus can vary from about ten to several thousand: water (ice) nuclei, for instance, may contain about 100 molecules. However, a stable nucleus could hardly result from the simultaneous collision of the required number of molecules since this would constitute an extremely rare event. More likely, it could arise from a sequence of bimolecular additions according to the scheme:

$$A + A \rightleftharpoons A_2$$

$$A_2 + A \rightleftharpoons A_3$$

$$A_{n-1} + A \rightleftharpoons A_n \quad \text{(Critical cluster)}$$

(give all these 3 equation numbers)

Further molecular additions to the critical cluster would result in nucleation and subsequent growth of the nucleus. Similarly, ions or molecules in a solution can interact to form short-lived clusters. Short chains may be formed initially or flat monolayers, and eventually a crystalline lattice structure is built up. The construction process, which occurs very rapidly, can only continue in local regions of very high super saturation and many of the embryos or 'sub-nuclei' fail to achieve maturity; they simply re-dissolve because they are extremely unstable. If, however, the nucleus grows beyond a certain critical size, as explained below, it becomes stable under the average conditions of super saturation is achieved in the bulk fluid.

The structure of the assembly of molecules or ions which we call a critical nucleus is not known, and it is too small to observe directly. It could be a miniature crystal, nearly perfect in form. On the other hand, it could be a rather diffuse body with molecules or solvated ions in a state not too different from that in the bulk fluid, with no clearly defined surface. The classical theory of nucleation is based on the condensation of a vapor to a liquid, and this treatment may be extended to crystallization from melts and solutions. The free energy changes associated with the process of homogeneous nucleation may be considered as follows.

The overall excess free energy, ΔG, between a small solid particle of solute (assumed here, for simplicity, to be a sphere of radius *r)* and the solute in solution is equal to the sum of the surface excess free energy, ΔGs, i.e. the excess free energy between the surface of the particle and the bulk of the particle, and the volume excess free energy, ΔGv, i.e. the excess free energy between a very large particle ($r = \infty$) and the solute in solution. The ΔGs is a positive quantity, the magnitude of which is proportional to r^2. In a supersaturated solution G_v is a negative quantity proportional to r^3. Thus

$$\Delta G = \Delta Gs + \Delta Gv$$

$$\Delta G = 4\pi r^2 \gamma + \frac{4}{3}\pi r^3 \Delta Gv \qquad (4.1)$$

Where ΔG, is the Gibbs free energy change of the transformation per unit volume and γ is the interfacial tension, i.e., between the developing crystalline surface and the supersaturated solution in which it is located. The term 'surface energy' is often used as an alternative to interfacial tension, but the latter term will be used throughout here for consistency. The two terms on the right-hand side of equation 4.1 are of opposite sign and depend differently on *r*, so the Gibbs free energy of formation, ΔG, passes through a maximum (as shown in figure 4.4**)**. This maximum value, ΔG_{crit}, corresponds to the critical nucleus radius, r_c, and for a spherical cluster is obtained by maximizing equation 4.1, setting $\frac{d\Delta G}{dr} = 0$

$$\frac{d\Delta G}{dr} = 8\pi r\gamma + 4\pi r^2 \Delta Gv = 0 \qquad (4.2)$$

Therefore,

$$r_c = \frac{-2\tilde{a}}{\Delta Gv} \qquad (4.3)$$

Where ΔGv is a negative quantity. From equations 4.1 and 4.3 we get

$$\Delta G_{crit} = \frac{16\pi\gamma^3}{3(\Delta Gv)^2} = \frac{4\pi\gamma r_c^2}{3} \quad (4.4)$$

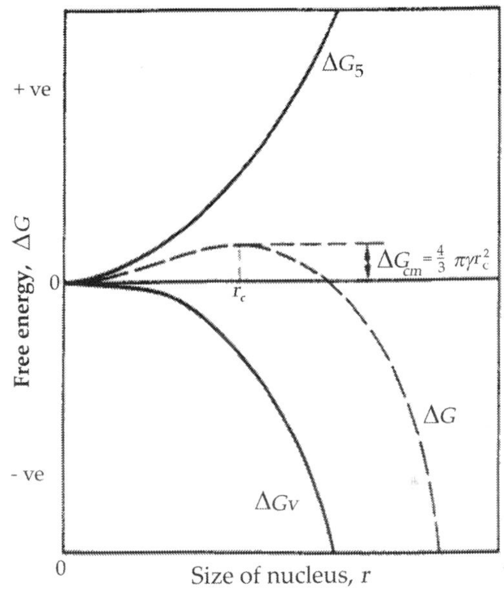

Figure 4.4: Free energy diagram **for** nucleation explaining the existence of a 'critical nucleus' (Mullin, 2001).

The behavior of a newly created crystalline lattice structure in a supersaturated solution depends on its size; it can either grow or re-dissolve, but the process which it undergoes should result in the decrease in the free energy of the particle. The critical size *r*, therefore, represents the minimum size of a stable nucleus. Particles smaller than r_c will dissolve, or evaporate if the particle is a liquid in a supersaturated vapor, because only in this way can the particle achieve a reduction in its free energy. Similarly, particles larger than r_c will continue to grow.

Although it can be seen from the free energy diagram why a particle of size greater than the critical size is stable, it does not explain the amount of energy, ΔG_{crit} that is produced to necessarily form a stable nucleus. This may be explained as follows. The energy of a fluid system at constant temperature and pressure is constant, but this does not mean that the energy level is the same in all parts of the fluid. There will be fluctuations in the energy about the constant mean value, i.e. there will be a statistical distribution of energy, or molecular velocity, in the molecules constituting the system, and in those supersaturated regions where the energy level rises temporarily to a high value nucleation will be favored. The rate of nucleation, *J,* e.g. the number of nuclei formed per unit time per unit volume, can be expressed in the form of the Arrhenius reaction velocity equation commonly used for the rate of a thermally activated process:

$$J = A exp\left(-\Delta G / kT\right) \qquad (4.5)$$

Where k is the Boltzmann constant, the gas constant per molecule (1.3805 $\times 10^{-23}$ JK^{-1} = R/N, where R is the gas constant = 8.314 JK^{-1} mol^{-1} and **N** = the Avogadro number = 6.023 x 10^{23} mol^{-1}) (Mullin, 2001).

INDUSTRIAL APPLICATION OF HOMOGENEOUS NUCLEATION

So far, the theoretical background of homogeneous nucleation has been discussed. In industrial crystallizers, however, homogeneous nucleation is usually not desired at all, and to produce large crystals , it has to be avoided. Only for the crystallization of very fine or even nano-sized materials may this mechanism be useful. In each case, it is essential to know the super saturation $\Delta C_{met;hom}$ that must be obtained to produce a certain number of homogeneous nuclei. This super saturation $\Delta C_{met;hom}$ is known as the metastable zone width in homogeneous nucleation. Although homogeneous nuclei are not formed in the Ostwald–Miers range of $0 < \Delta C < \Delta C_{met;hom}$, crystals can grow at $\Delta C > 0$. The curve $C^{*}\rceil + \Delta C_{met;hom}$ as a function of temperature ϐ is known as the super solubility curve and depends on substance values and generally on the concentration C. Figure 4.5 illustrates the relation between solubility and super saturation for homogeneous nucleation (Mersmann, 2001).

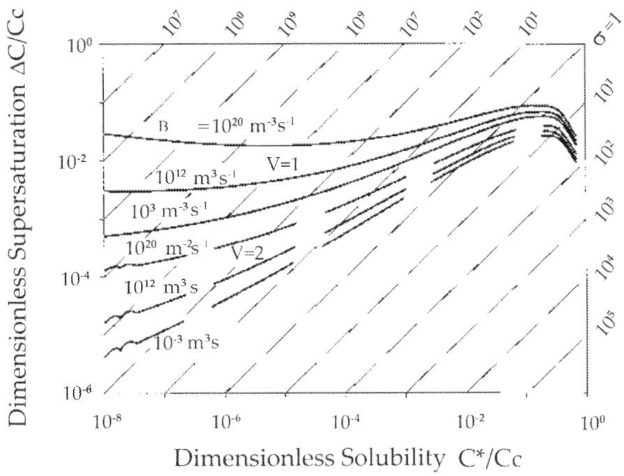

Figure 4.5: Dimensionless metastable super saturation against dimensionless solubility for homogeneous nucleation for υ = 1 (molecules) and υ = 2 (Mersmann, 2001).

HETEROGENEOUS NUCLEATION

The rate of nucleation of a solution or melt can be affected considerably by the presence of mere traces of impurities in the system. However, an impurity that acts as a nucleation inhibitor in one case may not necessarily be effective in another; indeed, it may even act as an accelerator. No general rule applies and each case must be considered separately.

Many reported cases of spontaneous (homogeneous) nucleation are found on careful examination to have been induced in some way. Indeed, it is generally accepted that true homogeneous nucleation is not a common event. For example, a super cooled system can be seeded unknowingly by the presence of atmospheric dust which may contain 'active' particles (hetero nuclei). Aqueous solutions as normally prepared in the laboratory may contain $>10^6$ solid particles per cm^3 of sizes < 1 μm. It is virtually impossible to achieve a solution completely free of foreign bodies, although careful filtration can reduce the numbers to $<10^3$ cm^{-3} and may render the solution immune to spontaneous nucleation (Mullin, 2001).

Cases are often reported of large volumes of a given system nucleating spontaneously at smaller degrees of super cooling than small volumes. A plausible explanation is that the larger samples stand a greater chance of being contaminated with active hetero nuclei. The size of the solid foreign bodies is important and there is evidence to suggest that the most active hetero nuclei in liquid solutions lie in the range 0.1 to 1 μm.

Hetero nuclei play an important role in atmospheric water condensation or ice formation. Atmospheric nuclei have been classified as 'giant' (10 to 1 μm) which remain airborne for limited periods only, 'large' (1 to 0.2 μm) and 'Aitken' (0.2 to 0.005 μm). Particles smaller than about 10^{-3} μm are not normally found in air because they readily aggregate. Aitken nuclei are so called because they are active at the super saturations produced in an Aitken counter, an apparatus in which a known volume of air is rapidly expanded; water droplets, formed on the particles, settle and are counted microscopically. Aitken nuclei, which occur ~ 10^4 to 10^5 cm^{-3} in the atmosphere, result from industrial smokes and vapors, ocean salts arising from spindrift, land dusts, particles from volcanic eruptions and even from outer space (Mullin, 2001).

As the presence of a suitable foreign body or 'sympathetic' surface can induce nucleation at degrees of super cooling lower than those required for spontaneous nucleation, the overall free energy change associated with the formation of a critical nucleus under heterogeneous conditions $\Delta G'_{crit}$, must be less than the corresponding free energy change, ΔG_{crit}, associated with homogeneous nucleation:

$$\Delta G'crit = \varnothing \Delta Gcrit \quad (4.6)$$

where the factor φ is less than unity.

It has been indicated that the interfacial tension, γ, is one of the important factors controlling the nucleation process. Figure 4.6 shows an interfacial energy diagram for three phases in contact; in this case, however, the three phases are not the more familiar solid, liquid and gas, but two solids and a liquid. The three interfacial tensions are denoted by γ_{Cl} (between the solid crystalline phase, c, and the liquid l), γ_{Sl} (between another foreign solid surface, s, and the liquid) and γ_{Cs} (between the solid crystalline phase and the foreign solid surface). Resolving these forces in a horizontal direction is less than unity.

$$\gamma_{sl} = \gamma_{cs} + \gamma_{cl} \cos\theta \quad (4.7)$$

$$\cos\theta = \frac{\gamma_{sl} - \gamma_{cs}}{\gamma_{cl}} \qquad (4.8)$$

The angle θ, the angle of contact between the crystalline deposit and the foreign solid surface, corresponds to the angle of wetting in liquid-solid systems.

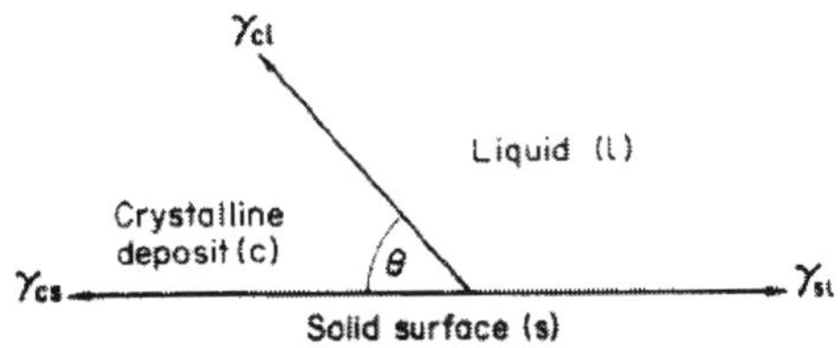

Figure 4.6: Interfacial tensions at the boundaries between three phases (two solids, one liquid) (Mullin, 2001).

The factor φ in equation 4.6 can be expressed as:

$$\varnothing = \frac{(2+\cos\theta)(1-\cos\theta)^2}{4} \qquad (4.9)$$

Thus, when **θ** = 180°, cos θ = -1 and φ = 1, equation 4.6 becomes

$$\Delta G'crit = \Delta Gcrit \qquad (4.10)$$

When θ lies between 0 and 180º, φ < 1; therefore

$$\Delta G'crit < \Delta Gcrit \qquad (4.11)$$

When θ= 0, φ = 0, and

$$\Delta G'crit = 0 \qquad (4.12)$$

The three cases represented by equations 4.10-4.12 can be interpreted as follows. For the case of complete non-affinity between the crystalline solid and the foreign solid surface (corresponding to that of complete non-wetting in liquid-solid systems), θ = 180°, and equation 4.10 applies, i.e. the overall free energy of nucleation is the same as that required for homogeneous or spontaneous nucleation. For the case of partial affinity (cf. the partial wetting of a solid with a liquid), $0 < \theta < 180°$, and equation 4.11 applies, which indicates that nucleation is easier to achieve because the overall excess free energy required is less than that for homogeneous nucleation. For the case of complete affinity (cf. complete wetting) $\theta = 0$, and the free energy of nucleation is zero. This case corresponds to the seeding of a supersaturated solution with crystals of the required crystalline product, i.e. no nuclei must be formed in the solution. Figure 4.7 indicates the relationship between φ and θ (Mullin, 2001).

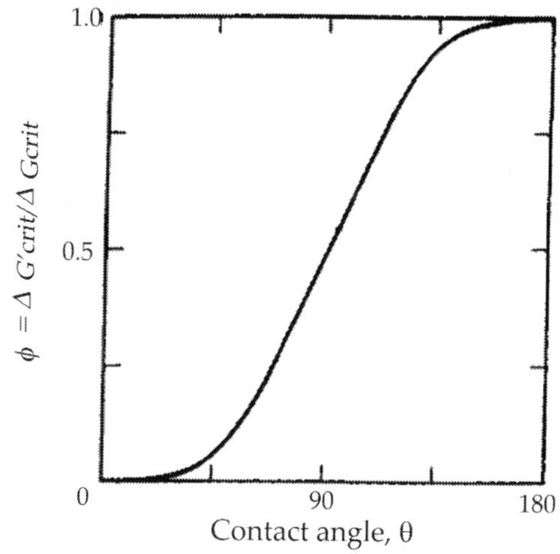

Figure 4.7: Ratio of free energies of homogeneous and heterogeneous nucleation as a function of the contact angle (Mullin, 2001).

As mentioned above, the heterogeneous nucleation of a solution can occur by seeding from embryos retained in cavities, e.g. in foreign bodies or the walls of the retaining vessel, under conditions in which the embryos would normally be unstable on a flat surface. This problem has been analyzed by Turnbull (1950) for different types of cavity. The maximum diameter of a cylindrical cavity which will retain a stable embryo is given by:

$$d_{max} = \frac{4\gamma_{cl}\cos\theta}{\Delta G_v} \tag{4.13}$$

Where ΔG_v is the volume free energy for the phase transformation. If the system is heated, this reduces the super saturation or super cooling and eliminating all embryos in cavities larger than d_{max}, and when subsequently cooled, the embryos retained in the cavities smaller than d_{max} will grow to the mouth of the cavity. They will then act as nuclei only if the cavity size $d_{max} \geq 2r_c$, where r_c is the size of a critical nucleus (Mullin, 2001).

Secondary nucleation

A supersaturated solution nucleates much more readily, i.e. at a lower super saturation, when crystals of the solute are already present or deliberately added. The term secondary nucleation will be used here for this pattern of behavior to distinguish it from so-called primary nucleation (no crystals initially present) discussed in section 4.1 (Mullin, 2001).

As described before in section 4.2 (Figure 4.2) there are many types and techniques of secondary nucleation, in the following sections we will describe each technique in details.

Contact nucleation

Clontz and McCabe (1971) showed that at moderate levels of super saturation, crystal contacts readily caused secondary nucleation of $MgSO_4 \cdot 7H_2O$, but crystal-crystal contacts gave up to five times as many nuclei as did crystal metal rod contacts. Furthermore, the faster growing faces produced fewer nuclei than did the slower growing faces indicating a connection between secondary nucleation and the crystal growth process. Clontz and McCabe (1971) showed that at moderate levels of super saturation, crystal contacts readily caused secondary nucleation of $MgSO_4 \cdot 7H_2O$, but crystal-crystal contacts gave up to five times as many nuclei as did crystal-metal rod contacts. Furthermore, the faster growing faces produced fewer nuclei than did the slower growing faces indicating a connection between secondary nucleation and the crystal growth process (Mullin, 2001).

Collisions in a liquid medium can initiate complex behaviors. Fracture may occur at the point of contact, but substantial hydrodynamic forces can operate over the surfaces near the point of contact, giving rise to plastic and elastic deformation in the parent crystal. Due to energy absorption, a small fragment broken off a crystal by collision could be in a considerably disordered state, with many dislocations and mismatch surfaces: in fact, it may be nearer to an amorphous glassy condition than to a crystal. It is not surprising, therefore, that these small crystalline fragments often grow much more slowly than macro crystals. Indeed, cases have been recorded where they do not grow at all. Ristic, Sherwood and Shripathi (1991) suggest that the formation of varying numbers of dislocations and the development of elastic strain in the new interface are the two main reasons for growth rate dispersion in attrition fragments smaller than about 150 μm (Mullin, 2001).

Crystal-agitator contacts are prime suspects for causing secondary nucleation in crystallizers, although only those crystals that manage to penetrate the fluid boundary layer around the blade will be hit. The probability of such an impact is directly proportional to the rotational speed of the agitator. The relative hardness of the contacting bodies is also a factor to consider: a metal impeller gives a much higher nucleation rate than one coated with a soft material such as polyethylene.

Energy-impact models have been developed from the results of attrition and breakage studies in agitated vessels using crystals suspended in inert liquids. A generalized model to quantify nucleation by mechanical attrition, based on Rittinger's law for the energy required for producing new surface and the additivity of two attrition processes due to crystal-crystal and crystal-impeller collisions, has been proposed by Kuboi, Nienow and Conti (1984).

Several hydrodynamic models of secondary nucleation in agitated crystallizers were applied to experimental data obtained from a 6-L agitated batch crystallizer using potassium sulfate by Shamlou, Jones and Djamarani (1 990). They concluded that the secondary nuclei were produced by an attrition process with a turbulent fluid-induced mechanism with critical eddies in the viscous dissipation subrange of the turbulent energy spectrum. An empirical attrition model, which relates crystal attrition to crystal size and hold-up, was developed by Jager *et al.* (1991) from data obtained with a 20-L continuous evaporative crystallizer using ammonium sulfate.

Metastable Zone and Induction Time for Nucleation

Due to the steep increase of the nucleation rate with super saturation, primary nucleation in technical system can be differentiated into two categories: has-not-yet occurred and has-occurred. Two types of descriptions are used to quantify the effects of nucleation: for a system with a constantly increased super saturation, the metastable zone width is used; for a system with constant super saturation, the induction time is used.

The metastable zone describes the zone in the solubility diagram where a mother phase is supersaturated, but has not yet responded with spontaneous nucleation, within a certain finite time. The two boundaries of the metastable zone are the solubility line and a somewhat hypothetical line at which spontaneous nucleation has occurred. This second line is kinetic value and depends on many conditions to be discussed below (Beckmann, 2013).

In a cooling crystallization, the system first follows along the path of constant concentration (arrowed line in Figure 4.8). The system crosses the saturation line, dotted line, becomes supersaturated, and the nucleation rate will be greater than zero; the dependence is reproduced as insert. At small super saturations, the nucleation rate will be practically zero; however, when the super saturation reaches a value for which the nucleation rate deviates measurably from zero, any further small increase in super saturation will dramatically increase the nucleation rate, so nucleation will be macroscopically observed.

The width of this metastable zone depends on several experimental conditions and is characteristic of the system under considerations. The most influential parameter is the rate of supersaturating the system, i.e. the rate of cooling, evaporation of solvent or addition of anti-solvent addition. An equally influential parameter is the concentration of additives or impurities in the mother phase. Another influence is the thermal history of the phase, namely, how long and to which degree has the phase been heated above saturation. Case is known, where the metastable zone width does not vary between laboratory and plant, while other accounts are also given.

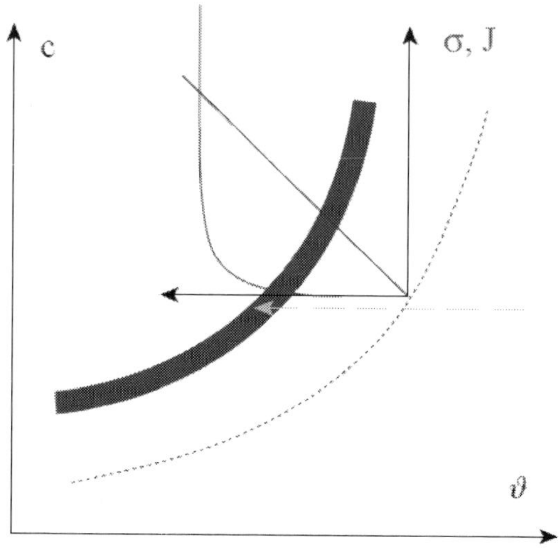

Figure 4.8. Schematic representation of the width of the metastable zone for the case of a cooling crystallization along the path represented by the arrowed line. The system crosses the saturation line, dotted line, and becomes supersaturated. The nucleation rate will be greater than zero (Beckmann, 2013).

Figure 4.9 shows the width of the metastable zone as a function of cooling rate for a sugar in water. The width is on the order of 10–50 K and strongly depends on the rate of cooling. The sub-cooling necessary for spontaneous nucleation is plotted in Figure 4.10 as a function of the cooling rate. For very low cooling rates, the width tends to 0 K. Conversely, the time the system is metastable, that is, the time between crossing the solubility line and spontaneous nucleation, decreases with increasing cooling rate. In Figure 4.10 as a function of the cooling rate. For very low cooling rates, the width tends to 0 K.

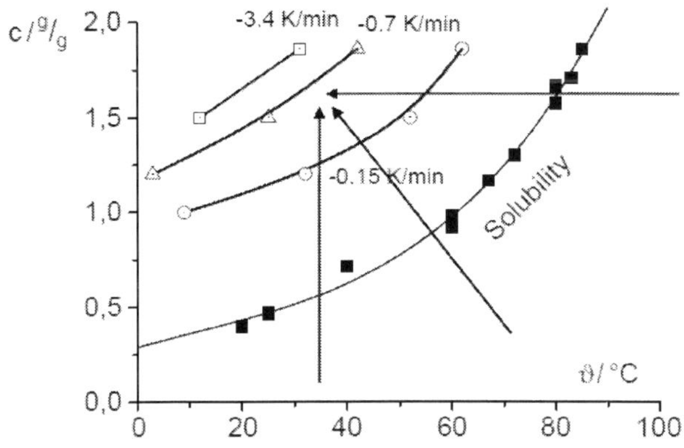

Figure 4.9. Width of the metastable zone for a sugar in water. The width is relatively large, on the order of 10–50 K, and that the width strongly depends on the rate of cooling (Beckmann, 2013).

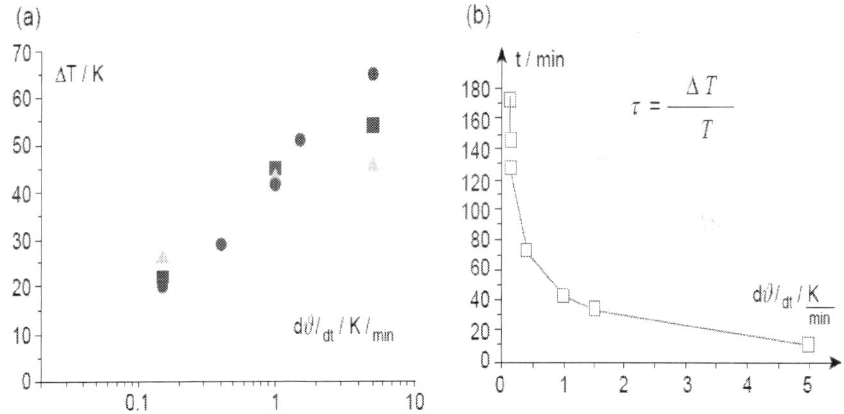

Figure 4.10 (a) Width of the metastable zone expressed as sub-cooling DT for a cooling crystallization of a sugar as function of the cooling rate. (b) For very low cooling rates, the width tends to 0 K. The time the system is metastable and decreases with the cooling rate (Beckmann, 2013).

The metastable zone width is in case independent of the technique with which the super saturation is generated. Figure 4.11 shows a system in which the width is nearly the same for a cooling and evaporative crystallization. Note the dispersion between repeated experiments, which is typical for the width of the metastable zone. It should be kept in mind that the nucleation process has a statistical nature and that the rate of the nucleation does depend exponentially on super saturation. Usually, aqueous solutions of salts will exhibit a small width of the metastable zone, typically on the order of 1–2 K, but values of 10 K are not an exception. Organic substances dissolved in organic solvents usually have a width of the metastable zone on the order of 10 K. Finally, sugars in aqueous solutions exhibit metastability typically on the order of 50 K.

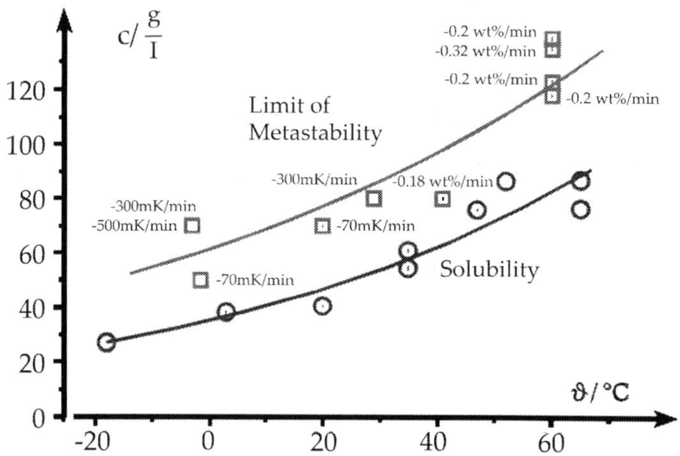

Figure 4.11. Width of the metastable zone for an organic moiety nucleating in an organic solvent. The width is nearly the same for a cooling and evaporative crystallization (Beckmann, 2013).

Thus, the super saturation necessary for spontaneous nucleation can be considerable, which in turn will result in a substantial amount of material to crystallize upon spontaneous nucleation. This crystallization mainly occurs at high super saturation, which is often undesirable.

For systems with retarded nucleation, the time necessary to build up a certain super saturation can be short compared to the time necessary for spontaneous nucleation to occur. Thus, nucleation can be observed for constant super saturation. The time elapsed before nucleation occurs is called induction time τ_{ind} and is given by the time necessary for the formation of nuclei and the time required for the nuclei to reach observable sizes Equation 4.14:

$$\tau \alpha \frac{1}{j} \qquad (4.14)$$

By assuming that the number of nuclei necessary for observing the nucleation, N, is independent of the experimental conditions, the induction time becomes inversely proportional to the nucleation rate (Equation 4.14), that is, the induction time should be the inverse of super saturation.

The number of visual and sub visual particles in a solution is known to influence the nucleation process. The induction time of hexose in water has been measured as a function of super saturation at room temperature. Solutions prepared from pure sugar and deionized water tend to nucleate within minutes at high super saturations and within less than an hour at low super saturations (Figure 4.12). However, filtering the solutions through a 0.2 μm filter into flasks that have been rinsed with particle-free water increases the induction time by one to two orders of magnitude. In addition, the dispersion in the induction times increases considerably.

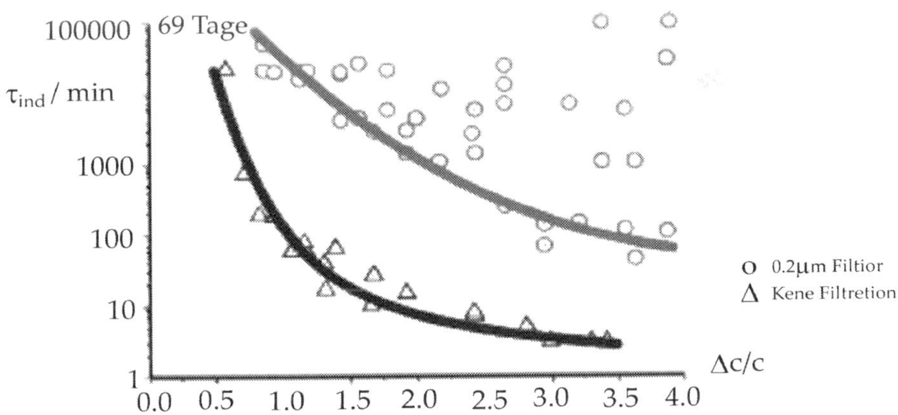

Figure 4.12: Induction times τ_{ind} for a sugar in water at room temperature. The solutions were used as is (Δ) and filtered using a 200-nm filter (O) (Beckmann, 2013).

Crystal nucleation, as crystal growth, involves the transport of building blocks from the bulk of the mother phase to the nucleus. Thus, the nucleation rate is influenced by the viscosity of the mother phase. For the nucleation of citric acid from aqueous solutions, the number of nuclei was counted as a function of super saturation (Figure 4.13). Two solutions with different concentrations and thus saturation temperatures are compared. The nucleation rate first increase with decreasing temperature and thus increasing super saturation. However, with decreasing temperature, the viscosity of the solution increases drastically, entailing a drastic decrease in nuclei counted, as the diffusion of building blocks becomes more difficult. The viscosity is also given; note that due to the high solubility of citric acid in water, the viscosity has values on the order of 10 P.

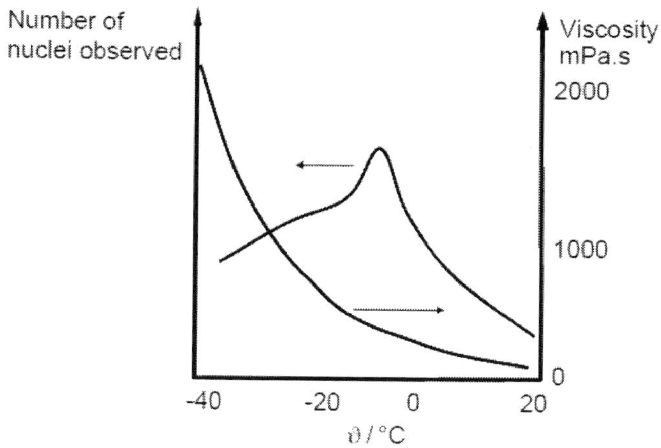

Figure 4.13: Number of nuclei counted as a function of temperature for two solutions with different saturation temperatures. The nucleation rate first increases with decreasing temperature and thus increasing super saturation (Beckmann, 2013).

CHAPTER 5

CRYSTAL GROWTH

INTRODUCTION

It is easier to understand the morphology of a crystal if the formation and growth of crystals is considered. Crystals grow from, among other things, supersaturated solutions, super cooled melts and vapors. The formation of a crystal may be considered in two steps:

- Nucleation: as described in chapter 4, nucleus is formed from coming together atoms to form a three-dimensional periodic array.
- Crystal growth: after nucleus have been formed, nucleus attracts further atoms, they take up positions on its faces in accordance with its three-dimensional periodicity. In this way, new lattice planes are formed (Figure 4.1b-d). Note that the illustration is two-dimensional only. The growth of the nucleus, and then of the crystal, is characterized by a parallel displacement of its faces. The rate of this displacement is called the rate of crystal growth, and is a characteristic, anisotropic property of a crystal (Borchardt-Ott , 2011).

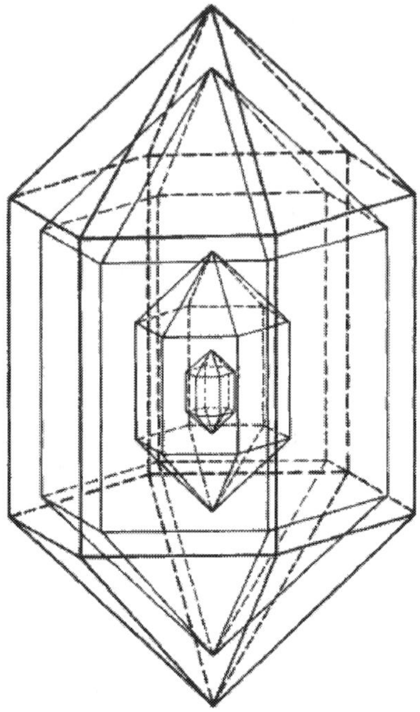

Figure 5.1: Quartz crystal showing its stepwise growth (Borchardt-Ott , 2011).

Crystal growth phenomena

Up till now, the phenomena of crystal growth in a supersaturated solution has not been well understood, as it is considered a very complex process. The explanation for this complexity is that several successive steps must occur before incorporation of a growth unit from the bulk solution into the crystal lattice. Since only a few of these steps are satisfactorily understood, a predictive calculation of the growth rates is allowed. Furthermore, a variability of growth units (atoms, molecules, ions, hydrated solute molecules, dimers, trimers, polymers, clusters, etc.) does exist depending on the crystallizing system and the solvent, which complicates the situation even more. On dealing with crystal growth of an ionizing solute, the next steps can be distinguished (Mullin, 2001):

- Bulk diffusion of solvated ions through the diffusion boundary layer
- Bulk diffusion of solvated ions through the adsorption layer

- Surface diffusion of solvated or unsolvated ions
- Partial or total desolvation of ions
- Integration of ions into the lattice
- Counter diffusion through the adsorption layer of the water released
- Counter diffusion of water through the boundary layer

According to this list, a certain role is played by diffusion and reaction steps, such as dissociation and integration in a kink or at a step. The slowest of these subsequent steps are rate determining. A crystal surface grows in such a way that units (atoms, molecules, ions) in a supersaturated solution (or generally in a supersaturated fluid) are first transported by diffusion and convection and then built into the surface of the crystal by integration or an integration reaction, with the super saturation, ΔC or Δc, being the driving force. Based on all system, state of flow, and super saturation, the first or second step can regulate the whole process, or both steps can determine growth to variable degrees. This is revealed by Figure 5.2, in which a crystal surface and a continuous solution are represented with concentration profiles of the solute (Mersmann, 2001).

Accordingly, the entire concentration gradient, $\Delta c = c\text{-}c^*$, is divided into two parts. The first part, C-C_I, within a diffusive–convective layer causes diffusive–convective transport, whereas the second, $C_I - C^*$ (I, is the interface), is decisive for the integration reaction within a reaction boundary layer. For growth that is completely determined by diffusion and convection, C_I $C^* << C\text{-}C_I$ or $((C_I\text{-}C^*))$ $((C\text{-}C_I) \ll 1)$. Alternatively, $((C\text{-}C_I))$ $((C_I\text{-}C^*) \ll 1)$, when crystal growth is controlled by the integration reaction. The mass flux densitym˙, directed toward the crystal surface is

$$\dot{m} = K_d\left(C - C_I\right) = K_r\left(C_I - C^*\right)^r \qquad (5.1)$$

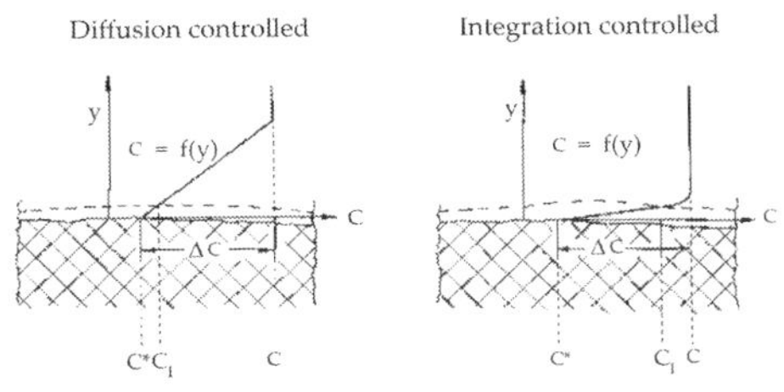

Figure 5.2: Concentration profiles for growth that is limited by diffusion or integration (Mersmann, 2001).

Here, is the mass transfer coefficient is the reaction rate constant, and r is the order of the integration reaction. The temperature dependency of the reaction rate constant is generally described according to the Arrhenius formulation:

$$K_r = K_{r0} \exp\left(-\frac{\Delta E_r}{RT}\right) \quad (5.2)$$

where is the reaction constant and ΔE_r is the activation energy.

Instead of describing crystal growth by mass flux density, $\dot{m}$ (with dimensions of kg/m^2s), it can be described by the displacement rate, v, of a crystal surface [referred to as v_{hkl} for surfaces indexed with (hkl), etc.] or the overall crystal growth rate, G = dL/dt (both with dimensions of m/s). The overall linear crystal growth rate, G, denotes, in principle, to any characteristic length, L, where a diameter of the crystal is generally used. When r is the radius and L = 2r, the diameter of a sphere corresponding to geometrically similar crystals with the volume shape factor $\alpha = V_p/L^3$ and the surface shape factor $\beta = A_p/L^2$, the following conversion is obtained between the mass flux density, $\dot{m}$, the mean displacement rate, $\bar{v}$, of the crystal surfaces, and the crystal growth rate $G = 2\bar{v}$:

$$\dot{m} = \frac{1}{A_p}\frac{dM}{dt} = \frac{6\alpha}{\beta}\rho_C\frac{dr}{dt} = \frac{6\alpha}{\beta}\rho_C\bar{v} = \frac{3\alpha}{\beta}\rho_C G$$

use equation editor for this equation, at present state, it is a figure, (5.3)

Each of the given definitions is useful for a different approach to crystal growth, such as from a mass balance, an observation of the crystal size distribution with time, or a focus on the crystal shape. According to Figure 5.2, to calculate the growth rate a priori, the concentration profile and the corresponding coefficients would have to be identified, which is not the case. Prior to reflecting on the general case of a superposition of mass transport and integration mechanisms, two special regimes should be considered first: the control of crystal growth by diffusion/convection or by the integration reaction (Mersmann, 2001).

DIFFUSION-CONTROLLED CRYSTAL GROWTH

If the integration reaction is indefinitely fast or $k_r \rightarrow \infty$, then crystal growth could be determined by the diffusive–convective transport of units. In this case,

$$\left(C - C_I\right) \approx \left(C - C^*\right) = \Delta c$$ (equation number)

And equation 5.3 is obtained when the mass flux density is low:

$$\dot{m} = K_d \Delta c \qquad (5.3)$$

Or

$$G = \frac{\beta}{3\alpha} K_d \frac{\Delta c}{\rho c} \qquad (5.4)$$

Regarding the mass transfer coefficients k_d given in the literature and the equations provided for them, it is necessary to determine whether they concern equimolar diffusion or mass transfer to a semipermeable interface. Additionally, it is essential to realize whether transport is only diffusive or both diffusive and convective. Increase of the difference occurs in association with mass flux density and can be significant if the substances possess a high solubility. If k_d is used to denote purely diffusive or true mass transfer coefficients and $k_{d,s}$ the mass transfer coefficients at a semipermeable interface, which is generally the case in crystal growth, the following applies:

$$K_{d,s} = \frac{K_d}{1 - W_i} \qquad (5.5)$$

If the mass fraction, W_i, tends toward zero, $W_i \rightarrow 0$, then $k_{d,s} = k_d$.

Determination of the slope of the linear relationship between the growth rate and super saturation is conducted through the mass transfer coefficient, which may be linked to the Sherwood number, $Sh = k_d L / D_{AB}$, and consequently to fluid dynamics, particle size, diffusivity, and viscosity. Different factors that diffusivity in liquids depends on include the volume of the diffusion species or its diameter, the viscosity, η_L of the solution and the kinetic energy kT. Figure 5.3 provides a sketch of some principle dependencies of diffusion-controlled growth on operating conditions. The dashed lines for minor levels of super saturation indicate the initial limitation by integration reactions. It should be noted that the growth rate calculated from equation (5.4) always gives the maximum growth rate, as no growth unit may be integrated into the crystal lattice that was not first transported from the bulk solution.

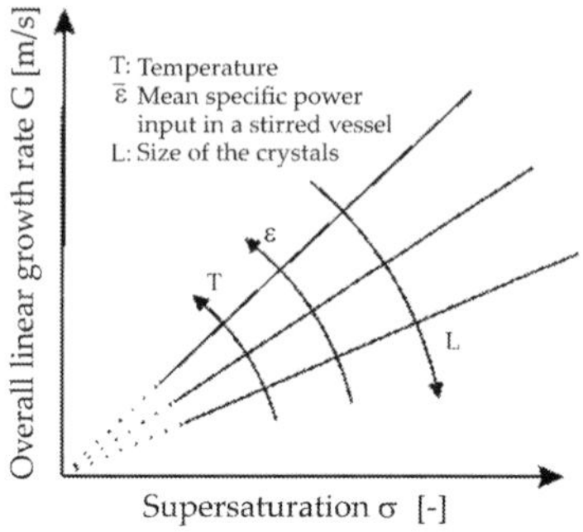

Figure 5.3: Influence of some principal operating conditions on diffusion controlled growth (Mersmann, 2001).

INTEGRATION-CONTROLLED CRYSTAL GROWTH

Once the diffusive/convective transport of units occurs quickly [i.e., mass transfer coefficient $k_d \rightarrow \infty$ or the integration reaction takes place very slowly $\left(K_r \rightarrow 0\right)$], determination of the crystal growth is performed by the integration reaction taking place on the crystal surface. The individual processes involved can be different and complex and as a result are difficult to understand. A possible integration inhibition may consist, for instance, of foreign particles or impurities being adsorbed on the crystal surface. Then, the particles need to be desorbed or overgrown, which are processes that occasionally require a higher super saturation of the crystallizing substance. Large integration units of, for instance, organic substances are transported very slowly by surface diffusion to the crystal surface and must revolve around the integration site until they are in a suitable position for integration (Mersmann, 2001).

ENERGETICALLY FAVORABLE INTEGRATION SITES

Despite of the difficulty of the integration of a growth unit into the crystal lattice, there has been many attempts trying to theoretically define the crystal growth. The topology, the structure of the crystal–solution interface, needs to be taken into consideration here. Once a unit collides with the crystal and does not instantaneously rebound but is adsorbed, it diffuses on the surface. The chemical bonds of the unit to a plane face of the crystal have enough strength for the sorption, but do not have enough strength to hold it tightly at one place of the lattice. Therefore, the growth unit must find an energetically favorable site, into which it becomes permanently integrated; otherwise, the growth unit will return to the solution after a certain diffusion distance. Such favorable integration sites arise from the topology of the crystal surface, in other words from the molecular crystal surface structure, and can be classified into kinks and steps (Figure 5.4) (Mersmann, 2001).

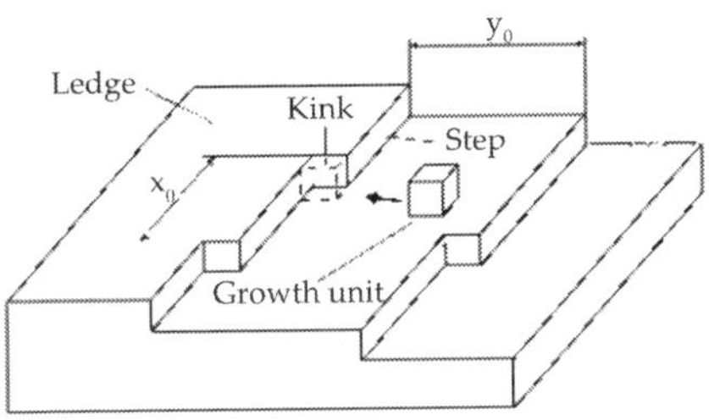

Figure 5.4: Energetically favorable sites for integration (Mersmann, 2001).

In molecular terms, if the surface is characterized by roughness and possesses several kink sites, "continuous growth" may take place, on condition that it is not hindered by the thermodynamic barriers. Following a short distance of surface diffusion, every growth unit, could reach the surface in a supersaturated solution, is incorporated into the crystal lattice. On a molecular scale, if the crystal surface turns out to be smoother, growth comes to be noticeably more difficult, since energetically favorable sites are limited. For kink and step sites, variable sources are available such as surface nucleation and screw dislocations. They should be taken into consideration to have a quantitative description of growth rates in this case.

Overall, the mechanism of crystal growth on the crystal surface can be described by the dimensionless driving force $\Delta\mu^* = \Delta\mu / kT$ and the surface energy $\varnothing^*$. The parameter $\varnothing^*$ is a thermodynamic surface energy parameter defined for a (001) surface of a Kossel crystal by the following equation:

$$\Delta\varnothing^* = \frac{4\varnothing_{CL} - 2(\varnothing_{CC} - \varnothing_{LL})}{kT} \qquad (5.1)$$

The energies $\varnothing_{CL}$ $\varnothing_{CC}$, and $\varnothing_{LL}$ represent the binding energy of a first neighboring solid–liquid bond (index CL), a first neighboring solid–solid bond (index CC), or a first-neighboring liquid–liquid bond (index LL), respectively.

As the energy gain Δϕ in the formation of a solid–liquid bond is

$$\Delta\varnothing = \varnothing_{CL} = \frac{1}{2}(\varnothing_{CC} + \varnothing_{LL}) \tag{5.2}$$

The following is obtained:

$$\Delta\varnothing^* = \frac{4\Delta\varnothing}{kT} = \frac{4}{kT}\left[\varnothing_{CL} - \frac{1}{2}(\varnothing_{CC} + \varnothing_{LL})\right] \tag{5.3}$$

Usually, low values of $\Delta\varnothing^*$ permits continuous growth to occur as it produces a rough surface with integration sites covering the whole surface. On the other hand, increasing $\Delta\varnothing^*$ generates a smoother surface, and above a critical value of $\Delta\varnothing^*$, the surface can be considered even (i.e., it is smooth on a molecular scale).

Determining $\Delta\varnothing^*$ accurately to substitute in equation 5.3 is very difficult actually. Examination of the growth of hexamethylene tetramine from aqueous and ethanol solutions as well as from the gaseous phase is done by Bourne and Davey. With the estimated $\Delta\varnothing^*$ values that result from interactions of varying degrees between the solvent and the dissolved substance, growth from the aqueous solution should be continuous, from the alcohol solution by mechanisms of nucleation and step growth and from the gaseous phase by flaw mechanisms and experimental results look to confirm this. In brief, the case of growth from solutions, a nucleation mechanism or a continuous-step growth mechanism usually best matches the experimental result (Mersmann, 2001).

SCREW DISLOCATION MECHANISMS

As we discussed in the above paragraphs, high surface energies often permit very smooth and even surface. So, it does not provide any integration site for any more arriving growth unit. Crystals have lattice imperfections or defects which preventing these ideally smooth surfaces mentioned above. First it was assumed that the existences of spiral dislocations, which usually end someplace on the crystal surface will create steps which considered a favorable integration sites for continuous growth. After that a step model was designed by Burton, Cabrera, and Frank (BCF) for crystal growth in which the crystal surfaces grow by the addition of growth units to kink sites on an endless series of steps an equal distance apart.

These spiral steps are characterized by the average distance, y_0, between neighboring turns and by the average distance, x_0, between neighboring kinks in the steps. These distances can be described by the equations:

$$\frac{X_0}{d_m} = \exp\left(\frac{\gamma CLd_m^2}{kT}\right)\frac{1}{\sqrt{s}} \approx \frac{\left(C_C / C^*\right)^K}{\sqrt{S}} \tag{5.4}$$

$$\frac{y_0}{d_m} = 19\frac{r_s^*}{d_m} = 19\frac{\gamma CLd_m^2}{KL\upsilon \ln S} \approx 19\frac{K\ln\left(C_C / C^*\right)}{\upsilon \ln S} \tag{5.5}$$

Where υ is the number of ions dissociating from a molecule, and with the expression:

$$\frac{\gamma CLd_m^2}{KT} \approx K\ln\left(\frac{C_C}{C^*}\right) = 0.414\ln\left(\frac{C_C}{C^*}\right) \tag{5.6}$$

As shown in Figure 5.4, the average distance between steps, y_0, and kinks along a step, x_0 together indicate the average density of surface sites on a crystal face. As shown in Figure 5.5, the dimensionless super saturation $\Delta C/C_C$ is plotted versus the dimensionless solubility C^*/C_C, and lines of constant x_0/d_m and y_0/d_m which represent lines of constant kink spacing and lines of constant step spacing are illustrated, respectively. Also in Figure 5.5, the distances x_0 and y_0 decrease with increasing super saturation. At lower values of solubility C*, super saturation S become higher to obtain the same x_0 and y_0 values (Mersmann, 2001).

Spiral dislocations which are considered the source of these steps, and the steps, which are remote from the centers of these spiral dislocations, are parallel to each other and the same distance apart from each other. Regarding the surface-diffusion process, which can be expressed by Fick's law, BCF derived equations that describe this surface diffusion. They show that the linear growth rate of a crystal surface from the vapor phase is given by the following type of equation:

$$V_{BCF} = \beta' X_s f \frac{KT}{2\pi\gamma} \exp\left(-\frac{W}{KT}\right)(\upsilon \ln S)\sigma \tanh\left(\frac{2\pi\gamma a}{X_s KT}\frac{1}{\upsilon \ln S}\right) \tag{5.7}$$

With $\ln S \approx \sigma$ valid for $\sigma < 0.5$, we obtain

$$V_{BCF} = \beta' X_s f \frac{KT}{2\pi\gamma} \exp\left(-\frac{W}{KT}\right)\upsilon\sigma^2 \tanh\left(\frac{2\pi\gamma a}{X_s KT}\frac{1}{\upsilon\sigma}\right) \tag{5.8}$$

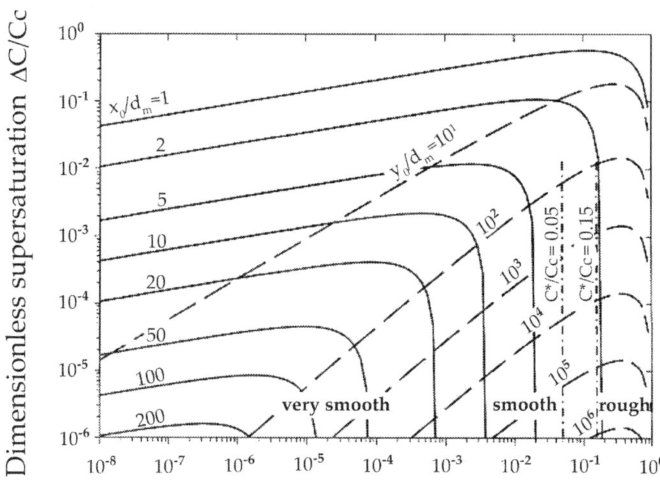

Figure 5.5: Dimensionless super saturation $\Delta C/C_C$ against the dimensionless solubility C^*/C_C; shaded areas relate to rough, smooth, and very smooth surfaces of crystals (if there are no collisions between crystals) (Mersmann, 2001).

In these equations, β' is a correction factor $\beta' \approx 1 or \beta' < 1$, f is a frequency of the order of the atomic frequency of vibration, W is the total energy of evaporation, x_s is the mean displacement of adsorbed units, γ is the edge energy and a is the distance between two neighboring equilibrium points on the crystal surface or the intermolecular distance.

For low super saturations $\sigma \ll \sigma_C \equiv 2\pi\gamma a / X_s KT\upsilon$, a parabolic law

$$V_{BCF} \sim \sigma^2 \tag{5.9a}$$

is obtained, but for high super saturations $\sigma \gg \sigma_c$, the linear law

$$V_{BCF} \sim \sigma \qquad (5.9b)$$

is valid as confirmed experimentally by numerous authors (Mersmann, 2001).

SURFACE NUCLEATION MECHANISMS

As discussed before, the screw-dislocation mechanism already origins a certain roughening of the crystal surface but the density of surface sites with favorable energies is still small and as a result the increase in the growth rate with increasing super saturation in a parabolic law is still not high. With rising super saturation, there is a possible change in growth mechanisms and surface nucleation becomes a source of much higher densities of kink sites, which contrasts with an exponential dependency on super saturation. As a result, adsorbed growth units join to form clusters and, finally, the nuclei. To have stable two-dimensional cluster which is larger than the critical nucleus diameter, sufficient growth units must join to form a nucleus of critical size on the crystal surface. After that, other growth units can join onto the corners of the nucleus so that crystal growth takes place over the whole surface area. There are numerous assumptions about the spreading rate of this type of nucleus form the basis of theoretical considerations. In brief, growth depends on the ratio of the spreading rate to the time required for another nucleus to be formed on the smooth surface. The number of growth units required to form a critical nucleus can also vary considerably (Mersmann, 2001).

There are three models, all denoting the relationship between the growth rate υ and the super saturation σ as follows:

$$V = A'\sigma^{P}\exp\left(-\frac{B'}{\sigma}\right) \qquad (5.10)$$

A borderline case is the so-called mononuclear model, whereas the spreading rate of this layer is very rapid compared to the surface nucleation rate and $P = \frac{1}{2}$. The other extreme configuration is the poly nuclear mechanism, where the spreading rate of this layer is slow compared to always new nuclei formed; in this case, $P = -3/2$. In between these two borderline cases is the birth and spread (B+S) model with $P \approx \frac{5}{6}$, considered for the first time by Hilling (Mersmann, 2001) and known as the nuclei-above-nuclei (NAN) model. With the assumptions of nuclei possibly borne on incomplete layers and growing at a constant step advancement, v_{step}, independently from each other (i.e., nuclei may slip over the edge of other nuclei), the growth rate of this B+S mechanism rate v_{B+S} is given by:

$$v_{B+S} = h v_{step}^{2/3} B_{surf}^{1/3} \quad (5.11)$$

Where, B_{surf} and h refer to the two-dimensional nucleation rate and to the height of the nuclei, respectively. With assumed rate-controlling surface diffusion, the step advancement, reads

$$v_{step} = \beta' \frac{2 V_m \Gamma^* D_{surf}}{x_s h} \sigma \quad (5.12)$$

In this equation, β' is a correction factor, as was that introduced with equation (5.7) ($\beta' \approx 1 or \beta' < 1$), xs is the mean displacement of adsorbed units, D_{surf} is the surface-diffusion coefficient, V_m is the volume of a unit, and $\tilde{A}^*$ is the equilibrium concentration of surface adsorbed units or molecules (molecules per unit area) at the temperature in question if the bulk super saturation were unity. The two-dimensional nucleation rate derives analogous to the considerations in Chapter 4 from a number concentration of nuclei with an impinging rate of units from the surface and the Zeldovich factor:

$$B_{surf} = \frac{2}{\pi} (\Gamma N_A)^2 \bar{v} \left(\frac{V_m}{h} \nu \ln S \right)^{1/2} \exp\left(\frac{-\Delta G_{max}^S}{kT} \right) \quad (5.13)$$

This poly nuclear (PN) or activated growth mechanism exhibits a very strong dependency of growth on super saturation, because with increasing super saturation, the time elapsing between two nucleation events decreases rapidly. For a distinct range of small super saturation, almost no growth appears, and when super saturation rises above a critical value, the growth increases rapidly to the domain of limiting bulk diffusion. The rate V_{PN} can be derived from the rate of surface nucleation and can be described by equation 5.14.

$$v_{PN} = \frac{D_{AB} d_m}{3} (N_A C^*)^{2/3} \sigma^{2/3} \exp\left(-\frac{\Delta G^S_{max}}{kT}\right) \tag{5.14}$$

In the case of cylindrical surface nuclei, the ratio of the surface nucleation energy ΔG^s_{max} and kT can be written as:

$$\frac{\Delta G^S_{max}}{kT} = \pi \left(\frac{\gamma_{CL} d_m^2}{kT}\right)^2 \frac{1}{\nu \ln S} \tag{5.15}$$

For squares instead of cylinders, the factor π must be replaced by 4, combining the last two equations 5.14 and 5.15 leads to:

$$v_{PN} = \frac{D_{AB}}{3d_m} \left(\frac{\Delta C}{C_C}\right)^{2/3} \exp\left(-\pi \frac{[K \ln(C_C/C^*)]^2}{\nu \ln S}\right) \tag{5.16}$$

With the rapid increase of the growth rate with super saturation assumed from equation (5.16), it is possible to make a general prediction regarding the intersection of the two different regimes of integration-controlled and diffusion-controlled growth. Equating V_{PN} with the growth rate calculated from diffusion-controlled growth allows a critical super saturation to be calculated at which the change from a smooth surface to a rough surface appears. Below this super saturation, the growth rate follows the BCF and B+S mechanisms. Note that due to the strain of the crystals and/or impurities, the growth rate in the domain of integration limitation may be reduced, essentially even down to zero. In this case, growth is attainable only with a super saturation sufficiently high for a surface roughening through equation (5.16) (Mersmann, 2001).

Poly nuclear growth was found for several highly soluble materials as well as for sparingly soluble substances. For example, for $BaSO_4$ with the maximum super saturation $S < 10$, the growth rates were described by equation (5.16) with $K = 0.296$ for cylindrical nuclei, with π in the exponential term (Mersmann, 2001).

Figure 5.6 summarize the principle of the growth rate. From a molecular scale, crystal growth preferably takes place at energetically favorable surface sites. For very low super saturations, only the omnipresent imperfections at the crystal surface, such as screw dislocations, act as possible sources for integration sites. With rising super saturation, the formation of surface nuclei from adsorbed growth units becomes more probable and new sources for integration sites are possibly created by a birth and spread mechanism. Generally, the surface nucleation mechanisms exhibit a very strong dependency of the growth rate on super saturation. By further increasing the super saturation, the spreading of the nuclei loses ground compared to new nuclei continuously roughening the surface (poly nuclear mechanism) and thus forming numerous energetically favorable integration sites. With a rough surface providing enough favorable sites for approaching growth units, the course of the growth rate with super saturation continues to be controlled solely by bulk diffusion (Mersmann, 2001).

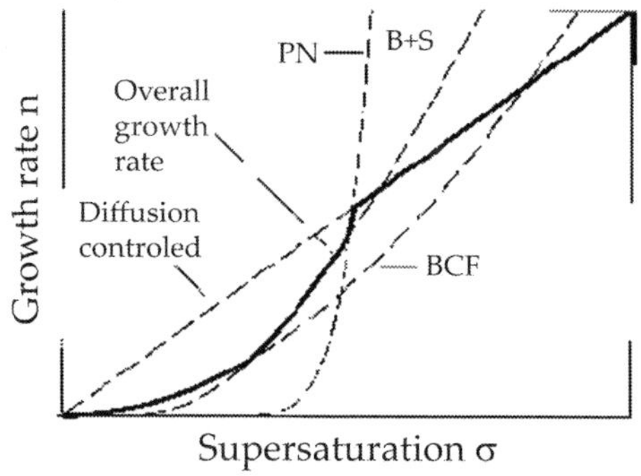

Figure 5.6: Scheme of competing growth regimes with rising super saturation (Mersmann, 2001).

ESTIMATION OF CRYSTAL GROWTH RATES

Although the models presented earlier give a good insight into the physics of crystal growth, they imply many parameters such as surface-diffusion coefficients and kink densities that are difficult or even impossible to determine or predict, and even though this might be possible one day, it would lead to a single-face growth rate of an ideal crystal rather than to an overall growth rate of a crystal collective in an industrial crystallizer. Thus, usually the application of these models is restricted to accompanying experimental studies of growth rates to determine the missing parameters. For a predictive estimation of growth rates, further-reaching assumptions are necessary, where the crucial question is to predict the different regimes of the integration-controlled and diffusion-controlled growth mechanisms (Mersmann, 2001).

Analogy to Diffusion-Reaction Theory

As suggested in the preceding sections, there is a variety of mechanisms of crystal growth, which take place in competition and lead to different growth regimes. Table 5.1 provides a summary of these growth models. In addition to the B+S (birth and spread) model, the BCF (Burton–Cabrera–Frank) model, and the diffusion-integration model with the special cases of a reaction order r of r = 1 and r = 2, the following very simple equation for the overall molar flux density, $\dot{n}$, is often used:

$$\dot{n} = k_g(\Delta C)^g \tag{5.17}$$

The results of these relationships coincide with the two borderline cases of the BCF theory, where g = 2 at low super saturations and g = 1 at high super saturations. If both the diffusive–convection step and the integration step determine growth, the concentration C_I at the interface must be used in the models mentioned. For example, the decisive factor in the BCF equation is the super saturation $\sigma_I = (C - C_I)/C^*$ at the interface between the surface diffusion or adsorption layer and the volume diffusion layer. This also applies to equations that describe only the integration step (Mersmann, 2001).

In reaction kinetics, the relationship between the diffusion and reaction rates is commonly described by characteristic numbers (e.g., the Damkohler number, Da, the Hatta number for chemical adsorption, or the catalytic efficiency or Thiele module for chemical catalysis) (Mersmann, 2001). Accordingly, an effectiveness factor η was also introduced by Garside for crystal growth according to:

$$\eta = \frac{G_{\text{exp}}}{G_{\text{int}}} \tag{5.18}$$

where G_{exp} is the experimentally determined growth rate and G_{int} the growth rate at an indefinitely fast and, thus, negligible volume diffusion.

Table 5.1 Growth-Rate Models (Mersmann, 2001).

Total growth rate	Mass transport	$\dot{n} = k_g(C - C^*)^g$ with $C - C^* = \Delta C$
Physicochemical models	Only convection + diffusion	$\dot{n} = k_d(C - C_I)$
	Only surface integration	$\dot{n} = k_r(C_I - C^*)^r$ with $k_r = k_{r0}\exp\left(-\frac{\Delta E_r}{RT}\right)$
	Elimination of C_I	$\dot{n} = k_r\left(\Delta C - \frac{\dot{n}}{k_d}\right)^r$
	Special case $r = 1$	$\dot{n} = \frac{\Delta C}{1/k_d + 1/k_r}$
	Special case $r = 2$	$\dot{n} = k_d\Delta C + \frac{k_d^2}{2k_r} - \sqrt{\frac{k_d^4}{4k_r^2} + \frac{k_d^3\Delta C}{k_r}}$
Surface-integration models	B + S model	$v_{B+S} = K_{B+S}\left(\frac{\Delta C}{C^*}\right)^{5/6}\exp\left(-\frac{K'_{B+S}}{T}\frac{C^*}{\Delta C}\right)$
	BCF model	$v_{BCF} = K_{BCF}T\left(\frac{\Delta C}{C^*}\right)^2\tanh\left(\frac{K''_{BCF}}{T}\frac{C^*}{\Delta C}\right)$
	PN model	$v_{PN} = \frac{D_{AB}}{3d_m}\left(\frac{\Delta C}{C_C}\right)^{2/3}\exp\left(-\pi\frac{K\ln(C_C/C^*)]^2}{v\ln S}\right)$ with $v = \frac{\dot{n}M}{\rho_C}$

When the physical mass transfer coefficients become very large ($K_d \to \infty$) or the integration reaction is very slow compared to the diffusive–convective transport of the crystal components, we obtain ($\eta \to \infty$). In this case, crystal growth is controlled by the reaction and can be described by $\Delta C = C_I - C^* \approx C - C^*$, and in the BCF equation, $\sigma_I = C_I - C^* / C^* \approx C - C^* / C^*$ must be used. On the other hand, when the integration reaction is indefinitely fast ($K_d \to \infty$) and thus negligible, or diffusive–convective transport is very slow (e.g., when the solution is highly viscous) compared to integration, we obtain $\eta \to 0$. In Figure 5.7, the effectiveness factor η is plotted against the ratio h:

$$h = \frac{Mass\ flow\, density\, at\ compl.integration\, limitation}{Mass\ flow\, density\, at\ compl.diffusion\, limitation}$$

The parameter h can be formulated for the integration reaction as follows:

$$h = \frac{K_r \left(C - C^*\right)^r}{K_d \left(C - C^*\right)} \qquad (5.19)$$

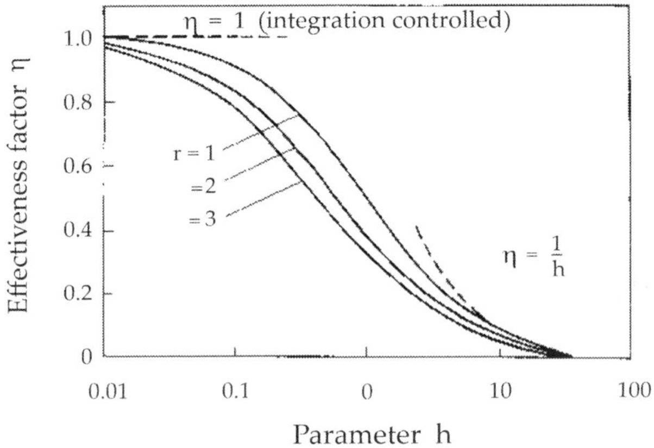

Figure 5.7: Curve for the effectiveness factor against parameter h (Mersmann, 2001).

If the parameter h is very small ($0 < h \lll 1$), crystal growth is completely controlled by the reaction (η=1). If parameter h is large ($1 \ll h < \infty$), crystal growth is increasingly limited ($\eta \rightarrow 0$) by diffusion.

As the reaction rate constant k_r generally increases more quickly with temperature than the mass transfer coefficient k_d , the abscissa value increases with increasing temperature. The effectiveness factor tends toward zero and growth finally becomes completely limited by diffusion. This has been proven, for example, for potash alum, magnesium sulfate heptahydrate, and ammonium sulfate in the range 5°C < υ < 60°C. However, it must be noted that impurities and additives can have a considerable influence on crystal growth. As the temperature increases, the crystal surface generally becomes less covered with foreign molecules (i.e., its adsorptive coverage decreases). Unfortunately, adsorption isotherms of impurities, admixtures, and additives are usually not known in solution–crystal systems. If the crystal growth rate increases with temperature, it is essential to check whether this is due to an increase in the reaction-rate constants (e.g., according to Arrhenius's formulation) in the "pure" system or due desorption of foreign particles in the real system. It must be emphasized that it is difficult to divide systems into those that are limited by diffusion and those that are limited by integration with respect to crystal growth. First, the growth limitation depends on super saturation. Furthermore, small crystals under 100 μm often grow under integration limitation, whereas the growth of large crystals of the same system can be described by the assumption of a purely diffusive–convective resistance and an indefinitely fast integration reaction. This behavior can be explained, for example, by the BCF equation, according to which the growth rate depends on the number of dislocations s on the crystal. Therefore, it can be assumed that large crystals exhibit sufficient dislocations and, therefore, sufficient integration sites so that growth will not be limited by the integration reaction. On the other hand, it is precisely the integration reaction that can limit the growth of small crystals if these have only a few dislocations and are largely "smooth." It is also possible for layers of units on the surface of the crystal to be so deformed that absolutely no growth takes place at all, at least at low super saturation. Such a crystal may be stimulated to grow (i.e., activated) only at relatively high super saturation by the growth of surface nuclei. This explains why crystals of the same size in completely the same environment (super saturation, fluid dynamics, temperature, concentration of impurities) grow at different rates. Such effects are observed and described by the term growth dispersion and will be considered in the following sections (Mersmann, 2001).

GENERAL PREDICTION OF MEAN GROWTH RATES

In the literature, experimentally determined growth rates can be found for more than 50 systems. In most cases, the system studied was an aqueous solution of inorganic solutes, but measurements carried out with organic components, such as solvents or solutes, do not show major differences. Moreover, all theoretical derivations are based on the physical properties of the materials under discussion and do not distinguish between organic and inorganic substances. The growth-rate measurements of the 50 systems mentioned earlier refer to a wide range of super saturations and temperatures. If super saturation is too high, the quality of products is inferior because of the excessive nucleation rate. In addition, admixtures and impurities are incorporated into the crystal lattices with rising super saturation. If super saturation is too low, however, the crystallization process is not economical because of the low crystal growth rate. A comparison of the super saturation and growth rates published in a great number of articles shows that the most suitable, and probably the most economical growth rates are usually limited by bulk diffusion and integration. Therefore, when dealing with crystallization technology, there is tremendous interest in deriving, with tolerable accuracy, equations that describe mean crystal growth rates for any system in the transition range of diffusion–integration. Due to growth-rate dispersion, these equations should be based on several crystals (as in the case of MSMPR crystallizers) and provide statistically mean values rather than be derived from single-crystal experiments. The multitude of experimental results for a variety of systems has enabled us to develop such a relationship.

The overall molar flux density $\dot{n} = \bar{\upsilon} C_C$ is given by:

$$\dot{n} = K_g \left(\ddot{A}C\right)^g \qquad (5.20)$$

with the exponent $1 < g < 2$ in most cases. Growth that is limited only by convection of the solution and bulk diffusion can be described by:

$$\dot{n}_{\text{dif}} = k_d(C - C_I) \qquad (5.21)$$

If growth is limited only by integration, the molar flux density $\dot{n}_{int}$ is expressed as

$$\dot{n}_{\text{int}} = k_r(C_I - C^*)^r \qquad (5.22)$$

As mentioned at the very beginning of this section, the entire concentration gradient $\Delta C = C - C^*$ is thus—with the concentration at the interface CI—divided into two parts. It is difficult to predict C_I, which alters with the super saturation ΔC. Therefore, as a first approach for a material with unknown growth kinetics, it is interesting to know at what level of super-saturation the superposition of transport limitation and limitation by integration has to be taken into account; in other words, the range of super saturation in which the growth rate might be described in a good approximation solely by transport limitation (i.e., $C - C_I \approx C - C^* = \Delta C$) or solely by integration limitation (i.e., $C_I - C^* \approx C - C^* = \Delta C$). To demarcate these two regimes, the parameter h, as introduced in equation (5.19) can be set to unity, assuming that growth is completely controlled by integration up to the specific super saturation $\Delta C_{h=1}$ or $\sigma_{h=1}$ and completely controlled by mass transport above this level:

$$1 = h = \frac{k_r \Delta C^r_{h=1}}{k_d \Delta C_{h=1}} = \frac{k_r \sigma^r_{h=1}}{k_d \sigma_{h=1}} C^{*(r-1)} \qquad (5.23)$$

The poly nuclear growth mechanism can be considered as a borderline case between integration-controlled and diffusion-controlled growth, because with the strong dependency on super saturation, the number of energetically favorable integration sites increases dramatically. A relation denoting the maximum super saturation $\sigma_{h=1}$ for the regime of integration controlled growth:

$$1 = \frac{D_{AB}}{3d_m k_d} \left(\frac{C^*}{C_C} \sigma_{h=1} \right)^{-1/3} \exp\left(-\frac{\pi K^2}{\nu} \frac{[\ln(C_C/C^*)]^2}{\ln(1+\sigma_{h=1})} \right) \qquad (5.24)$$

Unfortunately, equation (5.24) cannot be solved for $\sigma_{h=1}$. The predictive character of this equation for the demarcation of the integration-controlled and diffusion-controlled growth regimes has been compared with experimental data for growth rates of different materials with dimensionless solubility C^*/C_C varying over five decades and stoichiometric coefficients between $\upsilon = 1$ and $\upsilon = 3$. The comparison shows that the course of the predicted specific super saturations with the dimensionless solubility is in good agreement with experimental findings, where the predicted values are always about three times higher than the experimentally determined values. This can be explained by the fact that equation (5.24) denotes the maximum value of $\sigma_{h=1}$, where the diffusion-limited range may be found for smaller values, when the resistance of the other integration mechanisms (e.g., BCF, B+S) is low. Another explanation for this deviation arises from the assumption of complete dissociation of the ions assumed for the experimental data. Values of $\sigma_{h=1}$ calculated for different solubility and stoichiometric coefficients are plotted in Figure 5.8 (Mersmann, 2001).

Note that the diagrams given in concentrations. It is also valid for the more general case with respect to non-ideal behavior of the thermodynamic driving force, when replacing the concentrations with activities, if ideal behavior of the solid crystal and the interfacial tension can be assumed (Mersmann, 2001).

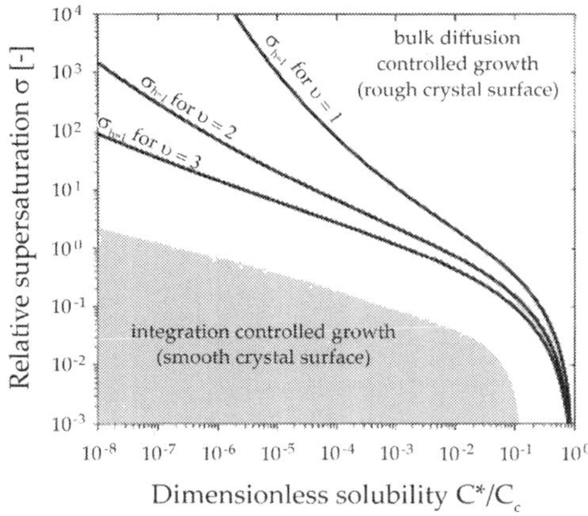

Figure 5.8: Calculated borderline cases for the transition of smooth growth faces to rough growth faces with the dimensionless solubility for different degrees completely dissociating $\upsilon = 2$, $\upsilon = 3$ materials (Mersmann, 2001).

CHAPTER 6

RECRYSTALLIZATION

INTRODUCTION

Usually, removal of the impurities from a crystalline mass could be possible by dissolving the crystals in a minor amount of fresh hot solvent and then cooling the solution to create a fresh crop of purer crystals, on condition that the impurities are characterized by being more soluble in the solvent than is the main product. It may be essential to repeat this step several times before obtaining a yield of crystals of the required purity, depending on the nature of the phase equilibrium displayed by the multicomponent system (Mullin, 2001).

Figure 6.1 shows a simple crystallization scheme. An impure crystalline mass AB (A is the less soluble pure component, B the more soluble impurity) is dissolved in the least amount of hot solvent S and then cooled. The crop of crystals X_1 will have a lesser amount of impurity B than the original combination: however, if, the wanted degree of purity has not been reached, it is possible to repeat the procedure: crystals X_1 are dissolved in fresher solvent S and then recrystallized to provide a crop X_2, and so on.

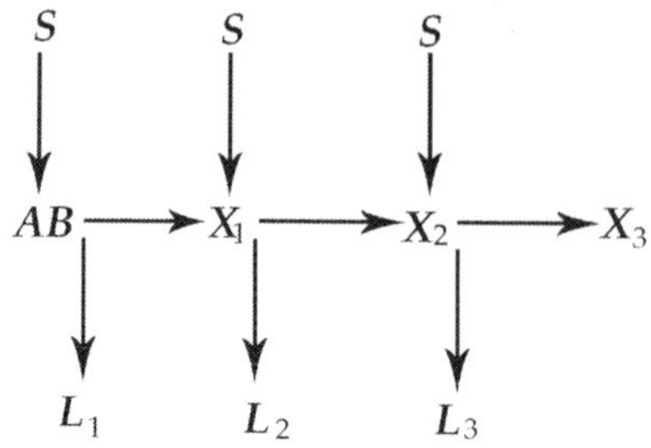

Figure 6.1: Simple recrystallization (Mullin, 2001)

After repeating this operations for many times (heating and cooling or recrystallization), the losses of the desired component A can be considerable, and the final amount of 'pure' crystals may easily be a small portion of the starting mixture AB. In industrial application, recrystallization process and it's number of consequence operation needed for yielding high purity and suitable separation is very important. The choice of solvent depends on the nature of the required substance A and the impurity B. Ideally, B should be very soluble in the solvent at the lowest temperature employed, and A should have a high temperature coefficient of solubility so that high yields of A can be obtained from operation within a small temperature range.

Figure 6.2 shows a modification of a simple recrystallization process. In this case some additions are added to the original impure mixture AB at certain intervals. The first two stages are the same as described above in recrystallization process in Figure 6.1, but added to that, the crystals X_2 are set aside and fresh feedstock AB is dissolved by heating in basic liquid L_2. During cooling process, a portion of crystals X_3 is gained and a basic liquid L_3, which is discarded. Then crystals X_3 are recrystallized from fresh solvent to give a crop X_4, which is set aside; the basic liquid L_4 can, if required, be used as a solvent from which more feedstock can be crystallized. If the crops X_2, X_4, X_6, etc. are still not pure enough, they can be bulked together and recrystallized from fresh solvent.

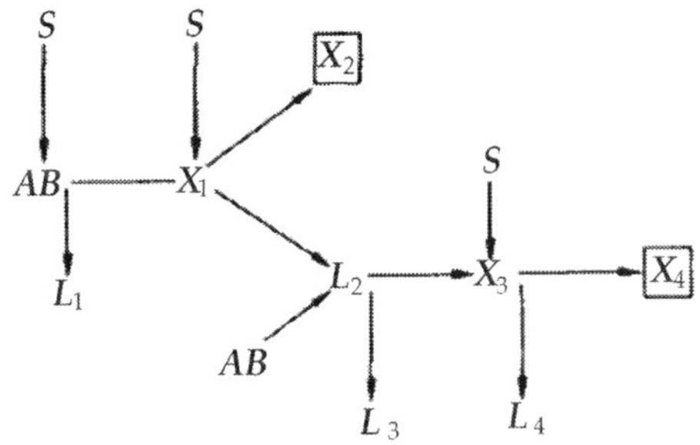

Figure 6.2: Modification of a simple recrystallization with further additions of feedstock (Mullin, 2001)

Another scheme of recrystallization is shown in Figure 6.3, which makes better use of the successive mother liquid fractions than recrystallization process described in Figure 6.2. Also, A and B are defined as A is the less soluble pure component, B the more soluble impurity. The mixture AB is dissolved in the minimum amount of hot solvent S, and then cooled. The crop of crystals X_1 which is deposited is separated from the basic liquid L_1 and then dissolved in fresh solvent. Then, repeating of the cooling and separating operations is done, providing an additional crop of crystals X_2 and basic liquid L_2. The first basic liquid L_1 is concentrated to permit a crop of crystals X_3 and basic liquid L_3. Dissolving of crystals X_3 is done by warming in basic liquid L2 and then cooled to give crystals X_5 and liquid L_5. Regarding crystals X_2, they are dissolved in fresh hot solvent and cooled to produce crystals X_4 and liquid L_4. Crystals X_6 and liquid L_6 result from concentration of liquid L_3. Continuation of the scheme can be done till the necessary degree of separation is affected. The less soluble substance A is concentrated in the fractions on the left-hand side of the diagram, the more soluble constituent B on the right-hand side. If any other substances with intermediate solubility were existing, they would be concentrated in the fractions in the center of the diagram (Mullin, 2001).

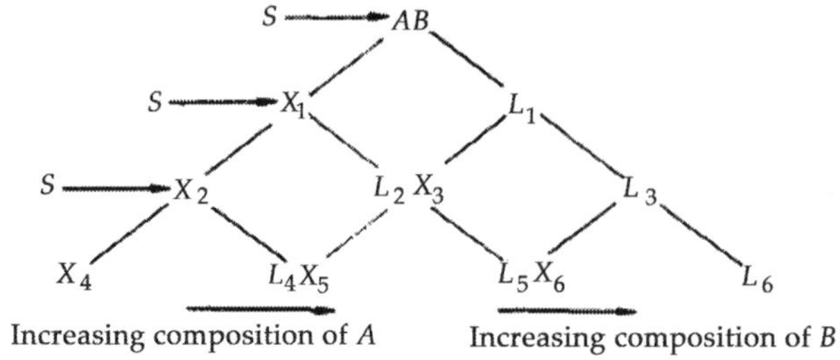

Figure 6.3: Fractional recrystallization of **a** solution (triangular scheme) (Mullin, 2001)

If a unit quantity of component A was contained in the starting material, and every crystallization step gave rise to the deposition of a proportion P of this component, the proportions of A which would appear at any point in the triangular scheme (see Figure 6.4) would be given by a term in the binomial expansion:

$$[P+(1-P)]^n = 1 \quad (6.1)$$

$$p_{n,r} = \frac{n!}{r!(n-r)!} \cdot P^{n-r}(1-P)^r \quad (6.2)$$

Where = proportion of A at a point represented by row r and stage n. Thus, for example,

$$\begin{aligned} p_{3,2} &= \frac{3!}{2!} \cdot P^{(3-2)}(1-P)^2 \\ &= 3P(1-P)^2 \end{aligned} \quad (6.3)$$

Figure 6.5 shows the modification of the triangular scheme. In such case more quantities of the feedstock AB are added to the system by dissolving it in sequential basic liquids on the right-hand side of the diagram. This scheme is principally useful if component A possess a high temperature coefficient of solubility (Mullin, 2001).

Many other much more difficult schemes can be used for fractional recrystallization. Their objective is to increase the production of the preferred component by additional re-use of the basic liquids. Two of them are illustrated in *Figure* 6.6. For the 'diamond' scheme (Figure 6.6a), when the outermost fractions have reached a predetermined degree of purity, they are set aside, and continuation of fractionation is conducted till the whole material is gained either in a crystalline form or in solution in final basic liquid. If necessary, various crystal fractions can be bulked together and recrystallized, and in a similar manner the mother liquids can receive further treatment.

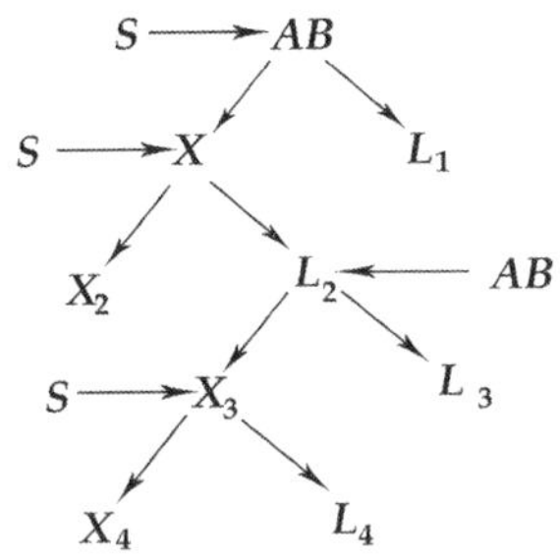

Figure 6.5: Simple recrystallization with further additions of feedstock (Mullin, 2001)

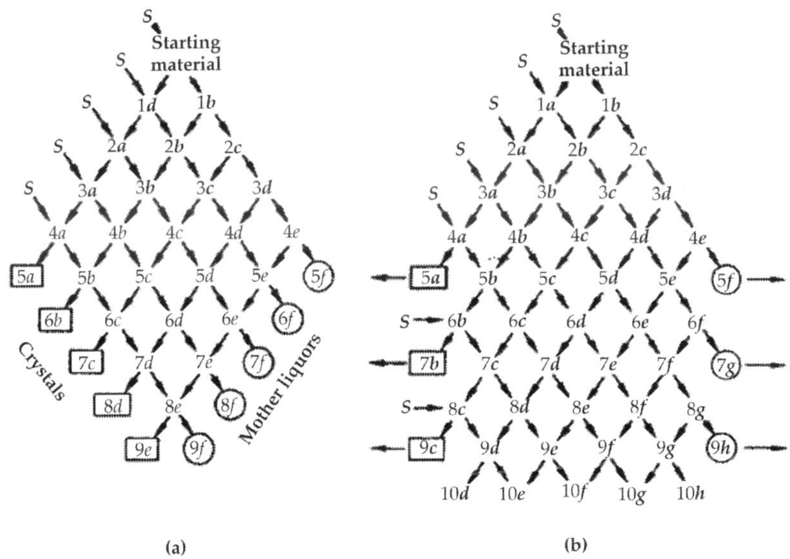

Figure 6.6: Fractional recrystallization schemes: (a) diamond; (b) double withdrawal (Mullin, 2001)

Unluckily, the several fractions gained via the diamond scheme will be different in composition, and a mixture of relatively pure crystals and relatively impure ones will result. Overcome of this difficulty can be performed using the ‘double withdrawal’ scheme illustrated in Figure

6.6b. The procedure is the same as that used in the diamond scheme up to the point where no further fresh solvent is used (line 5 in Figure 6.6a). At this instant, it is supposed that crystals 5a and basic liquid 5f have reached the preferred level of purity, and both are set aside from the system. Then fresh solvent is added to the crystal crop 6b to give a purer crop 7b, which is organized to have a degree of purity like that of crop 5a. Crop 7b is set aside, crop 8c is crystallized from pure solvent, and so on.

RECRYSTALLIZATION OF DRUGS

Manufacturing of most active pharmaceutical ingredients takes place in the crystalline form for their chemical stability during transportation, packaging and storage. The process of recrystallization is considered a simple and an inexpensive technique for purifying solid organic materials, whereby obtaining a crystalline form of a compound may be done from other solid-state forms of the same substance. For the pharmaceutics, usually recrystallization is the process that is conducted for the purification of solid active pharmaceutical ingredients (APIs). The two main aspects in the recrystallization process of drugs are production of another crystalline form and purification. In general, preparation of a hot saturated solution of the drug is done using a solvent with suitable boiling point and then the solution is cooled and as a result a new crystalline form of the drug is separated. The explanation of this process is due to the lower solubility of new crystalline form of the drug in the appropriate solvent at lower temperatures. Schematic illustration of the recrystallization process of a drug has been presented in Figure 6.7 (Glebovsky, 2015).

Because of recrystallization of the drug in a relevant solvent, alternation of physical and chemical properties of attained particles takes place. These properties are consisting of the wettability, melting point, solubility, dissolution rate, flowability and compactibility. Very effective and efficient results were obtained by using this technique, which was applied for both small and huge quantities of materials.

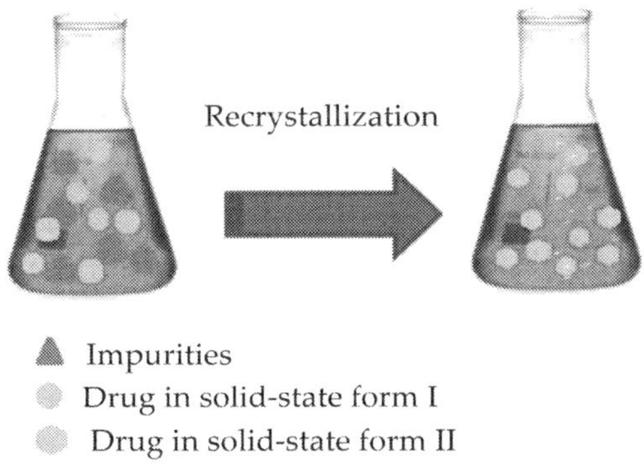

Figure 6.7: Schematic illustration of the drug recrystallization process (Glebovsky, 2015).

Several factors including thermodynamic (*e.g.*, solubility, solid-liquid interfacial tension, solvent activity, temperature, *etc.*), kinetic (*e.g.*, super saturation, molecular mobility, metastable zone width), and molecular recognition (hydrogen bonds, non-covalent bonds, molecular networks) may have an influence on the rate and mechanisms by which crystals are formed from liquid solutions. Super saturation is considered the driving force for crystallization. Super saturation can be generated by increasing the solute concentration (solvent evaporation) or decreasing the solute solubility (*e.g.*, temperature change, anti-solvent addition, pH change, salting out). The following step is known as crystal growth, which means the formation of macroscopic crystals from stable nuclei. This step is regulated by internal (crystal structure) and external factors (temperature, impurities, super saturation, solvent type) and controls the particle morphology (Glebovsky, 2015).

It has been proved that altering the crystal habit produced by recrystallization has important effect on dissolution rate. Nogami and Kato (year) revealed that the recrystallized thin crystal of the aspirin displays better dissolution and excretion behavior in comparison with the commercial aspirin (Yadhav, 2007).

Valsartan (VAL) is an antihypertensive drug, which is characterized by being potent, highly selective and orally active with sparingly solubility in aqueous media (particularly in gastric fluids). For that reason, its absorption is limited by its dissolution rate (Nalluri, Krishna, Rao, & Crooks, 2012). Nalluri et al. (year) prepared and investigated the recrystallized form of this drug. On the word of their results (proved by the DSC and powder X-RD studies), via application of the recrystallization technique, transposition of VAL can be done into either amorphous phases or amorphous dispersions. Compared to the untreated drug, the recrystallized VAL revealed an improvement in terms of solubility and dissolution properties and potentially therapeutic efficacy (Glebovsky, 2015).

Carbamazepine is prescribed as first line monotherapy for seizures in elderly patients. It is characterized by poor water solubility and its absorption may be unpredictable. Two main advantages of carbamazepine are proven efficacy and less cost. Mahalaxmi et al. (year) conducted the process of recrystallization for carbamazepine by using two solvents and at different cooling conditions. Their results exhibited enhanced dissolution behavior of the treated drug (Glebovsky, 2015).

Adhiyaman et al. (Year) described different properties of dipyridamole crystals that were prepared through the recrystallization process. Dipyridamole is classified as a critical antiplatelet and peripheral vasodilator drug that possesses poor water solubility and thus bioavailability; this causes limitation of its effectiveness in clinical tries. Their outcomes revealed that the prepared crystals of dipyridamole under adjusted conditions, considerably improved the dissolution rate of the drug compared to untreated one (Glebovsky, 2015).

On the other hand, applying of solvents or high temperatures did not have a modifying effect on the crystalline form of ketoconazole. It was stated that adjustment of growth of crystal faces can influence the physicochemical properties of the ketoconazole crystals. Also, it was revealed that habit modification and different degrees of crystallinity are considered important parameters in the improvement of the dissolution rate (Glebovsky, 2015).

RECRYSTALLIZATION STEPS

The first step of the recrystallization process is to choose a suitable solvent. Once a proper solvent has been selected, dissolving of an impure compound is done in the least amount of a hot solvent. Then, decolorizing the solution is made using activated charcoal (if required) and any insoluble material is filtered off. If the crystal formation process occurs too rapidly, impurities may become trapped in the crystals. The next step is slow cooling of the resulting solution to induce crystallization. Lastly, the crystals are collected, washed and dried. Figure 6.8 shows steps of the recrystallization process.

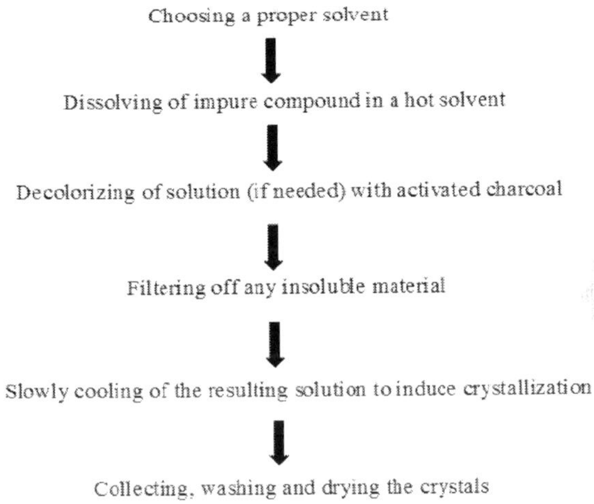

Figure 6.8: Steps of the recrystallization process (Glebovsky, 2015).

Choosing a proper solvent

The most important feature of the recrystallization technique is the selection of the solvent. There are two main facts in purification of solid organic materials by the recrystallization method: different solids have different solubility in each solvent and most solids are more soluble in hot than in cold solvents. Numerous solvents are employed for recrystallization of drug compounds. Selection of the recrystallization solvent should be performed on a case-to-case basis. This is due to the variant solubility of different organic compounds in these solvents. At first, the material should be introduced into solution and then must be able to come back out of solution. Choosing the appropriate solvent is very important in recrystallization, because it leads to increasing our yield of product. This selected solvent should have a lower boiling point than the melting point of the solid that being recrystallized. Furthermore, the drug should be soluble in the boiling solvent but have poor solubility in the same solvent after cooling (Glebovsky, 2015).

Melting point of drug and solvent

A melting point is employed to identify a substance and to get an indication of its purity. The melting point of a solid is the temperature at which the vapor pressure of the solid phase becomes equal to the vapor pressure of the liquid phase. Purity of a recrystallized drug is evaluated by observing its color and by measuring its melting point range. According to literature, the boiling point of the recrystallization solvent should be lower than the melting point of the compound to be recrystallized. If it is higher than the melting point of drug, the drug will "oil out", the process in which a drug is insoluble in a solution at a temperature above its melting point (Glebovsky, 2015).

COMMON PROCEDURES FOR RECRYSTALLIZATION

Different polymorphs of active pharmaceutical ingredients can be obtained through recrystallization processes based on solvents, solid-state transitions and vapor deposition.

In the controlled kinetic or thermodynamic conditions, crystallization of drug polymorphs can be performed from a solvent system. Supercritical fluid techniques are used for micronization and recrystallization of different drugs. According to the reports, modification of solid state characteristics, such as crystal habit, crystallinity and polymorphism, has been successfully achieved through recrystallization of drug particles using various supercritical anti-solvent (SAS) processes. This technology by using carbon dioxide gas is able to modify the solid-state properties of drugs, such as characteristics of particles (size, shape, surface, crystal structure and morphology), crystallinity and polymorphism leading to affecting their dissolution rate and bioavailability. The supercritical anti solvent technique is based on injection of an organic solution (containing of drug crystals) into supercritical carbon dioxide. During mixing, supercritical CO_2 is quickly dissolved in the organic solution and leads to the precipitation of solutes (drug) by an anti-solvent effect. Then supercritical CO_2 efficiently extracts the organic solvent, leads to achieving completely solvent-free products. Figure 6.9 shows recrystallization of a drug by using the supercritical anti-solvent (SAS) process schematically (Glebovsky, 2015).

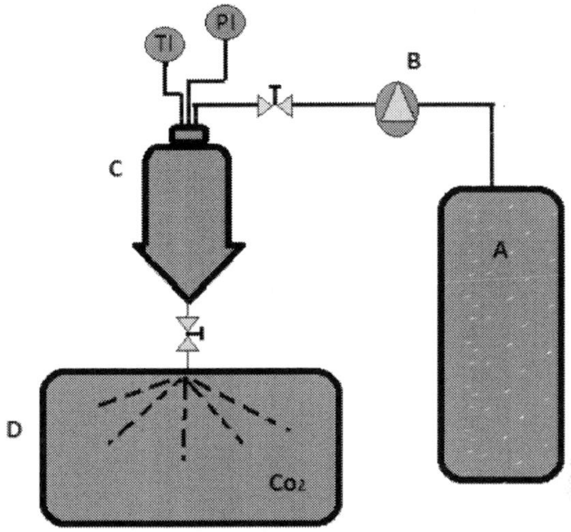

Figure 6.9: Recrystallization by using the supercritical anti-solvent (SAS) process. A) Drug solution in good solvent; B) Peristaltic pump; C) Spraying gun for drug solution; D) Chamber with saturated CO_2 gas (Glebovsky, 2015).

Some drugs have also been recrystallized by the spherical agglomeration technique to modify their dissolution properties. For poorly water-soluble drugs, a water immiscible organic solvent can act as an external phase and a 20 % calcium chloride as a bridging liquid. The spherical crystallization is a particle size enlargement method that includes crystallization and agglomeration using bridging liquid. Indeed, this method is a particle design technique that crystallization and agglomeration can be performed simultaneously in one step to transform crystals directly into the compacted spherical form. By using of this method, direct tableting of drug instead of further processing like mixing, granulation, sieving and drying is possible. In 1986, Kawashima applied the spherical crystallization method for size enlargement of drugs in the pharmaceutical field. Martino et al. (year) prepared spherical propyphenazone crystals by the agglomeration recrystallization technique using a three solvent system. Figure 6.10 shows the mechanism of the recrystallized agglomerates formation by the spherical agglomeration method.

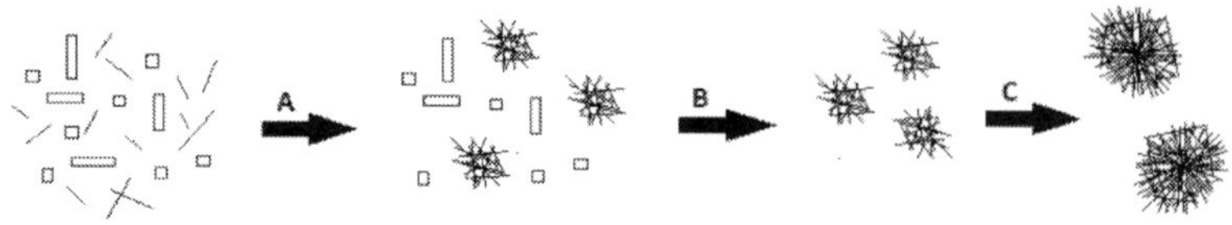

Figure 6.10: Mechanism of the recrystallized agglomerates formation by the spherical agglomeration technique; A) Recrystallization by addition of bridging liquid, B) Random coalescence, C) Enlarged spherical agglomerates (Glebovsky, 2015).

The most commonly used spherical crystallization techniques are spherical agglomeration and quasi-emulsion solvent diffusion. In both processes, a good solvent that dissolves the compound to be crystallized is used. A poor solvent (as an anti-solvent) is used to generate the required super saturation. Polymorphs may also be generated through solid-state transition by recrystallization of metastable polymorphs and amorphous forms, or through desolvation of solvates (including hydrates). Physical vapor deposition (PVD) method is an atomistic deposition procedure in which material is vaporized from a solid or liquid source in the form of atoms or molecules and transported in the form of vapor through a low-pressure environment to the substrate where it condenses (Glebovsky, 2015).

MECHANISM OF RECRYSTALLIZATION THAT EFFECTS ON DISSOLUTION RATE

In the nature, crystals can be categorized into seven crystal systems with various properties. Each of these crystal systems has differences in the relative sizes of their faces as well. This variation is called crystal habit modifications. In fact, the morphology of a crystal depends on the growth rates of the different faces of it. Fast growing faces have little or no effect on the growth form while slow growing faces have more influence. Environmental conditions and structure and defects of crystal governed the growth of a given face. The interaction of the solvent at the different crystal solution interfaces may cause to modified roundness of growing crystal faces (or edges), changes in crystal growth kinetics and enhancement or inhibition of crystal growth at certain faces. Figure 6.11 shows the crystal habits arising from growth inhibition at crystal faces.

In pharmaceutical manufacturing, a crystal habit of a drug molecule is an important characteristic. Different crystals have different planes, specific surface and free surface energy. Therefore, they show different physicochemical properties such as dissolution rate, flowability and compressibility. The existence of the drug structure in various crystalline shape influences on the bioavailability and biological efficacy of a drug. The crystalline form that has the improved dissolution rate and solubility is useful for developing the efficient pharmaceutical product of drugs (Glebovsky, 2015).

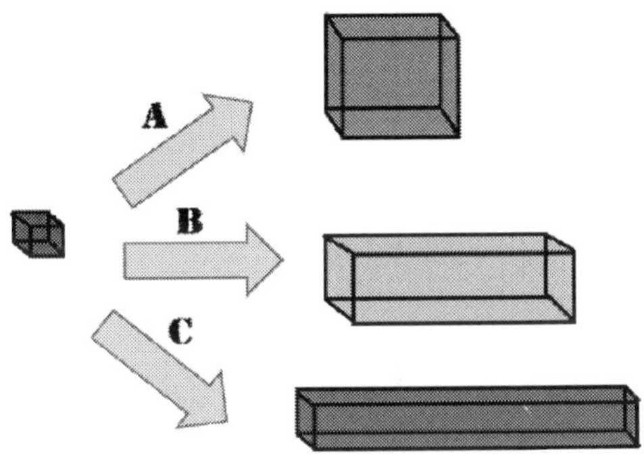

Figure 6.11: Crystal habits arising from growth inhibition at crystal faces; A) Growth equal in all dimensions, B) Growth inhibited in one dimension, C) Growth inhibited in two dimensions (Glebovsky, 2015).

The primary mechanism of recrystallization is the long-range motion of grain boundaries that removes dislocations from the material. According to numerous reports wettability and consequent dissolution rate of different crystalline forms of drugs could be determined by exposure of diverse crystal faces. The constituent molecules of a crystalline material are part of an orderly three-dimensional structure. Depending on the crystallization conditions, these molecules arrange in different forms. This crystalline form of material is known as polymorphic and each of its crystal ordering is called a polymorph. Indeed, the polymorphs are substances with the same chemical composition but presenting different crystalline forms. Different crystalline forms include various crystallographic systems, lattice parameters and position of the atoms in the elementary cell. Under different conditions, a crystalline drug is crystallized into different polymorphic forms. Polymorphs of a substance show different physical properties, such as solubility and dissolution properties. In fact, crystals with higher free energy show higher solubility. Since the molecules in crystals with high energy have weaker solid-solid interactions compared to crystals with lower energy, they exhibit higher affinity for the solvent in the surroundings. Figure 6.12 shows the probable consequences of a different crystal form in solubility (Glebovsky, 2015).

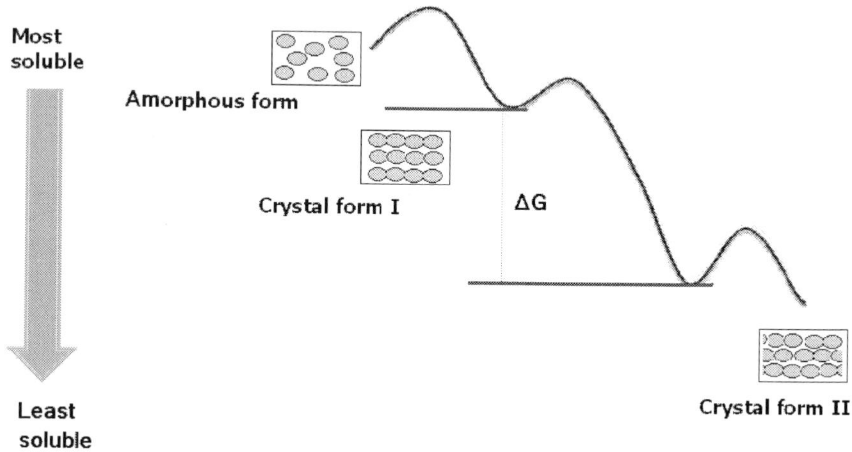

Figure 6.12: Probable consequences of a different crystal form in solubility (Glebovsky, 2015).

Figure 6.13 show formation of different crystals following by recrystallization of the initial crystalline form schematically. First, the reaction between solid and solvent break the solid crystalline substances (opens the amorphous lattice for amorphous substances). This creates cavities in the solvent (known as phase change) and, hence, molecules of solid become molecules of solute. Then solute molecules recrystallized into another crystalline form(s).

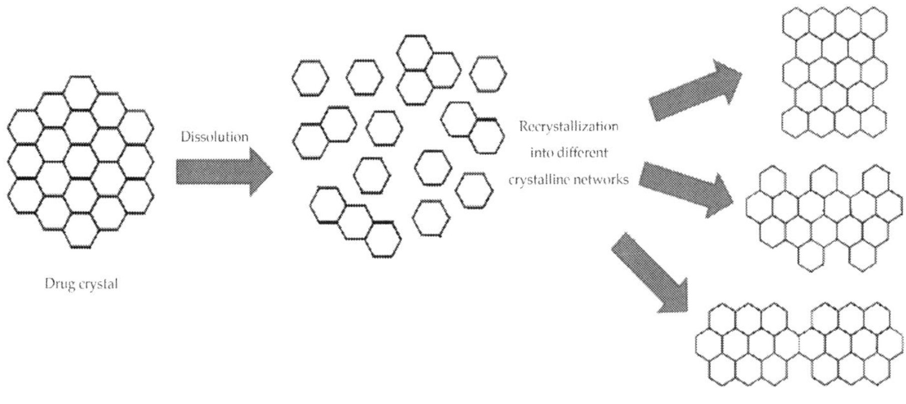

Figure 6.13: First the reaction between solid and solvent breaks the solid crystal for crystalline substances (opens the amorphous lattice for amorphous substances). This creates cavities in the solvent (known as phase change) and, hence, molecules of solid become molecules of solute. Then solute molecules recrystallized into another crystalline form(s) (Glebovsky, 2015).

Additional information is necessary for control of the nucleation and crystal growth. Dunitz and Bernstein (year) have evaluated the disappearing polymorphs by considering various examples. They state that "once a polymorph has been obtained, it is always possible to obtain it again; it is only a matter of finding the right experimental conditions".

EFFECT OF EXCIPIENTS AND ADDITIVES ON THE RECRYSTALLIZATION PROCESS

Crystal morphology engineering is a useful tool for improving tableting efficiency of pharmaceutical solids. Pharmaceutically accepted excipients can be applied as crystal habit modifiers to overcome the toxicity and/or environmental concerns. Excipients and polymers have a major role to prepare the recrystallized form of a drug. The difference in the physicochemical properties of the drug molecules and the excipient is the main factor in selection of a solvent system for the recrystallization process. The presence of additives such as the polymeric material and surface-active agents can influence molecular aggregation during crystallization. Viscosity of the medium and surface tension are reduced by the surfactants that affect the nucleation process. The presence of these additives in the spherical agglomeration process may also reduce the processing time and improves the bioavailability and the micrometric properties of drug.

Various polymers like hydroxy propyl methylcellulose (HPMC), poly ethylene glycol (PEG), ethyl cellulose (EC) and poly vinyl pyruvate (PVP) can be used to improve poor compressibility and handling qualities of pure drugs by crystallization techniques. This improves the micromeritic and drug release properties of the crystals as well. Crystallization is inhibited by some of the polymers such as methylcellulose, hydroxypropyl methyl cellulose (HPMC), polyvinylpyrrolidone (PVP). Among these, PVP has been found to be the most effective in inhibiting crystallization of drugs. These anti-nucleant polymers are incompatible in both size and size of the host molecules of the growing crystals surface. Then, their incorporation into the lattice alters growth characteristics of the host molecules.

Recrystallization of the amorphous drug may be avoided by entry of stabilizing hydrophilic excipients in the composition. Indeed, stabilizing excipients are prevented recrystallization of amorphous drug particles by steric hindrance and/or the formation of hydrogen bond with drug molecules. As a noticeable point, the stabilizing excipients should be hydrophilic for improving the wetting properties of drug particles and these hydrophilic excipients should not be hygroscopic as well (Glebovsky, 2015).

RECRYSTALLIZATION OF SINGLE-PHASE ALLOYS

Recrystallization of the deformed microstructure is often called primary recrystallization to distinguish it from the process of abnormal grain growth which may occur in fully recrystallized material and which is sometimes called secondary recrystallization. The unqualified term recrystallization is always used to mean primary recrystallization. It is convenient to divide primary recrystallization into two regimes, nucleation which corresponds to the first appearance of new grains in the microstructure and growth during which the new grains replace deformed material. Although these two events occur consecutively for any grain, both nucleation and growth may be occurring at any time throughout the specimen. The kinetics of recrystallization is therefore superficially like that of a phase transformation which occurs by nucleation and growth. In discussing the origin of recrystallization is not appropriate because nucleation in the classic thermodynamic sense does not occur. A more accurate description of the process would be 'initiation', although, for simplicity, the accepted terminology will be used (Rollett, Humphreys, Rohrer, & Hatherly, 2004).

The progress of recrystallization with time during isothermal annealing is commonly represented by a plot of the volume fraction of material recrystallized (X_V) as a function of log(time). This plot usually has the characteristic sigmoidal form of Figure 6.14 and typically shows an apparent incubation time before recrystallization is detected. This is followed by an increasing rate of recrystallization, a linear region, and finally a decreasing rate of recrystallization.

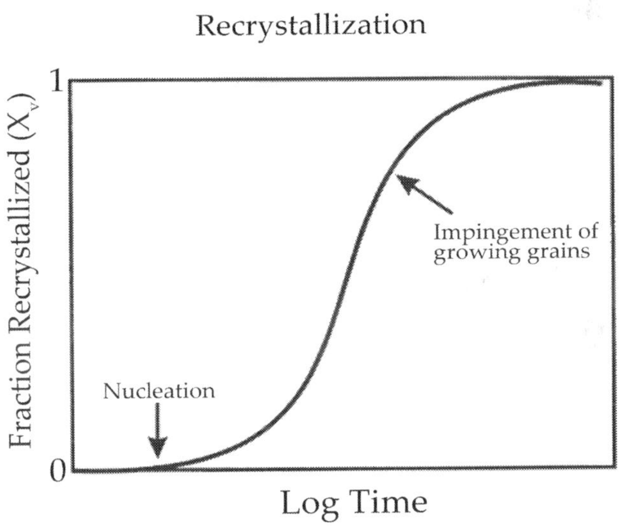

Figure 6.14: Typical recrystallization kinetics during isothermal annealing (Rollett, Humphreys, Rohrer, & Hatherly, 2004).

QUANTIFYING RECRYSTALLIZATION

Recrystallization is a microstructural transformation, which is most directly measured by quantitative metallography. However, it is also possible to follow the progress of recrystallization by the measurement of various physical or mechanical properties.

- The overall transformation

The extent of recrystallization is often described by Xv, and for isothermal experiments it is convenient to use the time at which recrystallization is 50% complete ($t_{0.5}$), as a measure of the rate of recrystallization. For isochronal experiments, where annealing is carried out at various temperatures for a constant time, e.g., 1 hr., then the recrystallization temperature, is usually defined as the temperature at which the material is 50% recrystallized.

Although such measurements of the recrystallization process have practical value, more fundamental measurements of recrystallization are obtained by separate reference to the constituent nucleation and growth processes.

- Nucleation

Our first difficulty lies in defining what we mean by a recrystallization nucleus. A working definition might be a crystallite of low internal energy growing into deformed or recovered material from which it is separated by a high angle grain boundary. If the number of nuclei per unit volume (N) remains constant during recrystallization, then this is the key parameter. However, if this is not the case, the nucleation rate also needs to be considered. As nucleation of recrystallization is a very complex process, the nucleation rate is not expected to be a simple parameter. Early work indicated that the rate was not always constant during recrystallization as shown in Figure 6.15.

However, such measurements need to be regarded with caution, particularly at the early stages of recrystallization, because the number of nuclei observed depends strongly on the technique used to detect them. For example, in Figure 6.15a are there really no nuclei present at times shorter than 1000 sec; several reasons can be cited and one is that they just too small to be detected by optical microscopy. Thus, the initial rising nucleation rate shown in Figure 6.15b can be a real effect or a reflection of the difficulty of detecting small nuclei; in which case, this figure represents a monotonically decreasing nucleation rate.

We must recognize that although formal models of the recrystallization kinetics usually refer to nucleation rates, these are often very difficult to measure experimentally. When we come to consider the mechanisms and sites of nucleation, it will be clear that in most cases, it is not yet possible to obtain a realistic quantitative model for the number of nuclei and its variation with time and other parameters (Rollett, Humphreys, Rohrer, & Hatherly, 2004).

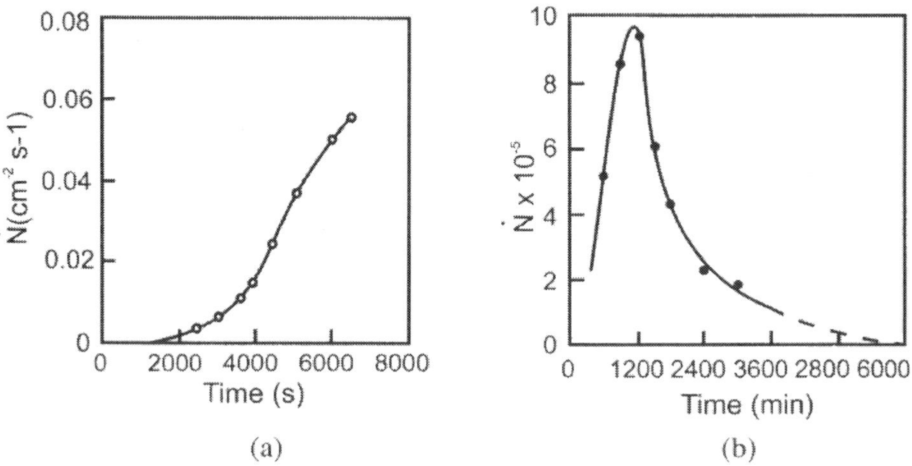

Figure 6.15: The variation of nucleation rate () with time during the annealing of 5% deformed aluminum, annealed at 350°C. (a) Initial grain size 45 μm, (b) initial grain size 130 μm (Rollett, Humphreys, Rohrer, & Hatherly, 2004).

Growth

The growth of the new grains during recrystallization is more readily analyzed than is their nucleation. It is generally accepted that the velocity (v) of a high angle grain boundary, which is also the growth rate () is given by:

$$v = \dot{G} = MP$$

Where, M is the boundary mobility and P is the net pressure on the boundary

*Driving pressure.*The driving pressure (P_d) for recrystallization is provided by the dislocation density (ρ), which results in a stored energy (E_D). The resultant driving force for recrystallization is then given approximately as:

$$Pd = ED = \alpha\rho Gb^2$$

Where, α is a constant of the order of 0.5.

For a dislocation density of 10^{15} m^{-2}, typical of a metal, and taking Gb^2 as 10^{-9} N, the driving pressure for primary recrystallization is of the order of 1MPa.

Boundary curvature. If a small, spherical, new grain of radius R, growing into the deformed structure is considered, there is an opposing force which comes from the curvature of a high angle grain boundary of specific energy . The grain boundary area will be reduced and the energy lowered if the grain were to shrink, and there is thus a retarding pressure on the boundary given by the Gibbs-Thomson relationship:

$$P_C = \frac{2\gamma_b}{R}$$

This pressure is however only significant in the early stages of recrystallization when the new grains are small. For grain boundary energy of 0.5 Jm^{-2} and the driving force discussed above, it is found that P_c is equal to P_d when R is ~1 μm. Below this grain size there would therefore be no net driving force for recrystallization. During the recrystallization of alloys, other retarding forces arising from solutes or second-phase particles may be important, and these will be discussed elsewhere.

The pressure on a grain boundary and therefore the boundary velocity may not remain constant during recrystallization. As what will be discussed later, the driving force may be lowered by concurrent recovery and both the driving force and the boundary mobility may vary through the specimen. Therefore, the growth rate , which is an important parameter in any model of recrystallization is also a complex function of the material and the deformation and annealing conditions.

THE LAWS OF RECRYSTALLIZATION

One of the earliest attempts to explain the recrystallization behavior of materials is the formulation of the so-called laws of recrystallization (Burke and Turnbull, 1952; Mehl 1948). This series of qualitative statements, based on the results of a large body of experimental work, predicts the effects of the initial microstructure (grain size), and processing parameters (deformation strain and annealing temperature), on the time for recrystallization and on the grain size after recrystallization. These rules are obeyed in most cases and are easily rationalized if recrystallization is a nucleation and growth phenomenon, controlled by thermally activated processes, whose driving force is provided by the stored energy of deformation (Rollett, Humphreys, Rohrer, & Hatherly, 2004).

- A minimum deformation is needed to initiate recrystallization. The deformation must be sufficient to provide a nucleus for the recrystallization and to provide the necessary driving force to sustain its growth.
- The temperature at which recrystallization occurs decreases as the time of anneal increases. This follows from the fact that the microscopic mechanisms controlling recrystallization are thermally activated and the relationship between the recrystallization rate and the temperature is given by the Arrhenius equation.
- The temperature at which recrystallization occurs decreases as strain increases. The stored energy, which provides the driving force for recrystallization, increases with strain. Both nucleation and growth are therefore more rapid or occur at a lower temperature in a more highly deformed material.
- The recrystallized grain size depends primarily on the amount of deformation, being smaller for large amounts of deformation. The number of nuclei or the nucleation rate is more affected by strain than the growth rate. Therefore, a higher strain will provide more nuclei per unit volume and hence a smaller final grain size.
- For a given amount of deformation the recrystallization temperature will be increased by:
- **A larger starting grain size**. Grain boundaries are favored sites for nucleation, therefore a large initial grain size provides fewer nucleation sites, the nucleation rate is lowered, and recrystallization is slower or occurs at higher temperatures.
- **A higher deformation temperature.** At higher temperatures of deformation, more recovery occurs during the deformation (dynamic recovery), and the stored energy is thus lower than

for a similar strain at a lower deformation temperature (Rollett, Humphreys, Rohrer, & Hatherly, 2004).

STUDIES ON FACTORS AFFECTING THE RATE OF RECRYSTALLIZATION

In this section, some of the important factors which are known to affect the rate at which recrystallization occurs in materials will be considered.

The deformed structure

The amount of strain

The amount, and to some extent the type of deformation, affect the rate of recrystallization, because the deformation alters the amount of stored energy and the number of effective nuclei. The type of nucleation site may also be a function of strain. There is a minimum amount of strain, typically 1–3%, below which recrystallization will not occur. Above this strain the rate of recrystallization increases, leveling out to a maximum value at true strains of ~2–4. The effect of tensile strain on the recrystallization kinetics of aluminum is shown in Figure 6.16.

The mode of deformation

The mode of deformation also affects the recrystallization rate. Single crystals deformed in single glide, recover on annealing, but may not recrystallize, because the dislocation structure does not contain the heterogeneities and orientation gradients needed to provide a nucleation site. The best examples of this effect are found in the hexagonal metals which deform by glide on the basal plane.

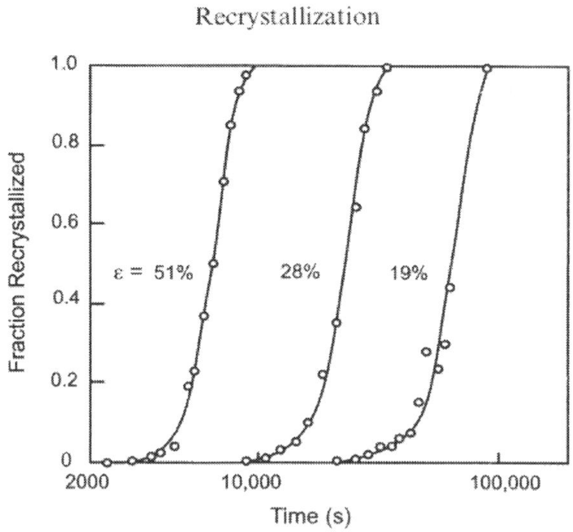

Figure 6.16. The effect of tensile strain on the recrystallization kinetics of aluminum annealed at 350°C (Rollett, Humphreys, Rohrer, & Hatherly, 2004)

Strain path changes

In many metal working operations, the strain path is not constant, and may even be reversed. An example of this is in rolling, where shear deformation near the surface changes sign. Although the prime interest in strain path effects is their effect on formability, there is also interest in determining how such strain path effects influence the subsequent recrystallization behavior.

Experiments to determine the role of strain path on the recrystallization of high purity copper have been reported by Lindh et al. (1993). They carried out tension and combined tension and compression experiments and measured the recrystallization temperature. For specimens deformed solely by tension, the recrystallization temperature decreased with increasing tensile strain (as shown in Figure 6.17) as would be expected. However, specimens deformed to the same total permanent strain () by tension and compression were found to recrystallize at higher temperatures. By reading across from the data points for the tension/compression samples to the curve for the tensile specimens of Figure

6.17, an equivalent recrystallization strain could be determined for the tensile/compression samples, and this was given by:

$$\varepsilon_{equ}^{R} = \varepsilon_{perm} + \eta \varepsilon_{red}$$

Where is the redundant strain and η a constant found to be 0.65. This shows that the redundant strain is only 0.65 times as effective in promoting recrystallization as is the permanent strain.

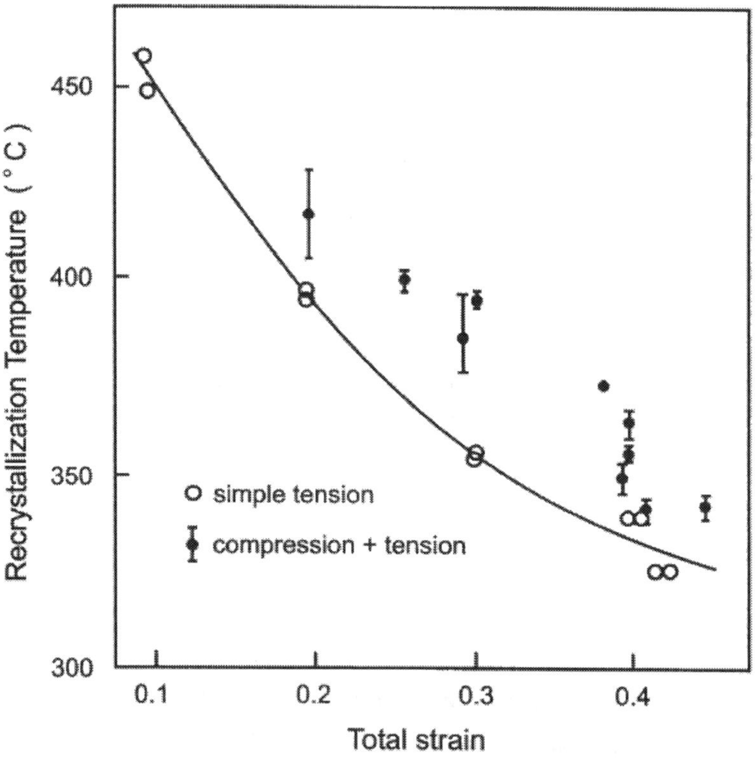

Figure 6.17: The recrystallization temperature of copper as a function of the total applied strain in specimens deformed by tension or by a combination of tension and compression, (Lindh et al. 1993).

Embury et al. (1992) compared the recrystallization behavior of cubes of pure aluminum deformed in uniaxial compression with similar specimens which were deformed incrementally in different directions to produce no overall shape change. Their results (Fig. 6.18a) show that for the same equivalent strain the specimens deformed by multi-axis compression recrystallized more slowly than the specimens deformed by monotonic compression, in agreement with the results of Figure 6.17. However, if the specimens were deformed to the same level of stress, as shown in Figure 6.18b, the recrystallization kinetics of the specimens were similar (Rollett, Humphreys, Rohrer, & Hatherly, 2004).

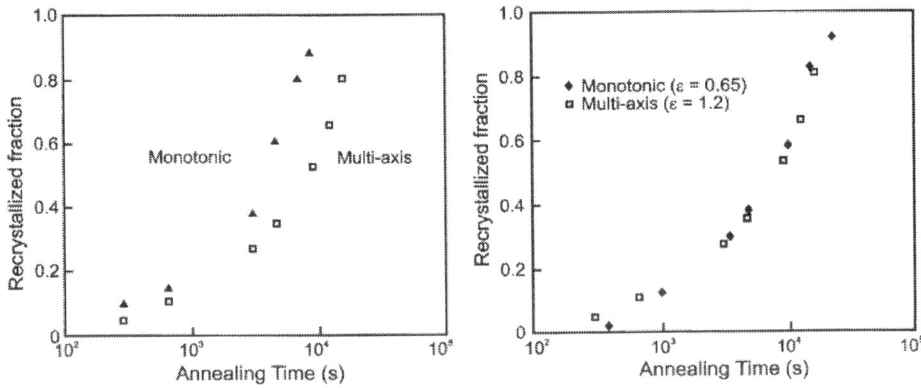

Figure 6.18: The effect of redundant deformation on the recrystallization of pure aluminum. The recrystallization kinetics for specimens deformed monotonically is compared with those of specimens deformed by multi-axis compression. (a) Specimens deformed to the same equivalent strain (0.7), (b) specimens deformed to the same equivalent stress.

The grain orientations

Single crystals

The microstructure and stored energy of a deformed grain depends strongly on the active slip systems and hence on its orientation. Therefore, the initial orientation as well as the 'orientation path' during deformation, will affect both the nucleation sites and the driving force for recrystallization.

Hibbard and Tully (1961) measured the recrystallization kinetics of both copper and silicon-iron single crystals of various orientations, which had been cold rolled 80%, and found significant differences in rates of recrystallization as shown for silicon-iron in Table 6.1.

Brown and Hatherly (1970) investigated the effect of orientation on the recrystallization kinetics of copper single crystals which had been rolled to a reduction of 98.6%. The recrystallization times on annealing at 300°C varied between 5 and 1000 min. These compared with a recrystallization time of 1 minute for a polycrystalline specimen deformed to the same strain. It was found, in agreement with the first two samples in Table 6.1, that the recrystallization kinetics could not be clearly associated with the texture resulting from deformation. For example, both {110}<112> and {110}<001> crystals developed similar {110}<112> rolling textures, but due to the differences in the nature of the deformed structure, the former recrystallized 50 times more slowly. Differences between the recrystallization kinetics and textures of the single crystals and the reference poly crystal in the above investigation, serve to highlight the importance of grain boundaries during recrystallization and show that care must be taken in using data from single crystal experiments to predict the behavior of poly crystals.

Table 6.1: Recrystallization of silicon–iron single crystals at 600°C (Rollett, Humphreys, Rohrer, & Hatherly, 2004)

Initial orientation	Final orientation	Time for 50% recrystallization (s)	Orientation after recrystallization
{111} < 112 >	{111} < 112 >	200	{110} < 001 >
{110} < 001 >	{111} < 112 >	1000	{110} < 001 >
{100} < 001 >	{001} < 210 >	7000	{001} < 210 >
{100} < 011 >	{100} < 011 >	No recrystallization	{100} < 011 >

**contruct a true table for Table 6.1, it is a figure at its present form.

Poly crystals

In polycrystalline materials, there is evidence that the overall recrystallization rate may be dependent on both the starting texture and the final deformation texture. In iron, the orientation dependence of the stored energy after deformation has been determined, and it has been predicted on this basis that the various texture components will recrystallize in the sequence shown in Figure 6.19. Recent work (Hutchinson and Ryde, 1995) has confirmed the trends shown in this figure. In aluminum alloys, Blade and Morris (1975) have shown that differences in the texture formed after hot rolling and annealing lead to differences in recrystallization kinetics when the material is subsequently cold rolled and annealed. The observations of inhomogeneous recrystallization discussed in provide further evidence of the variation of recrystallization rate for different components of the deformation texture.

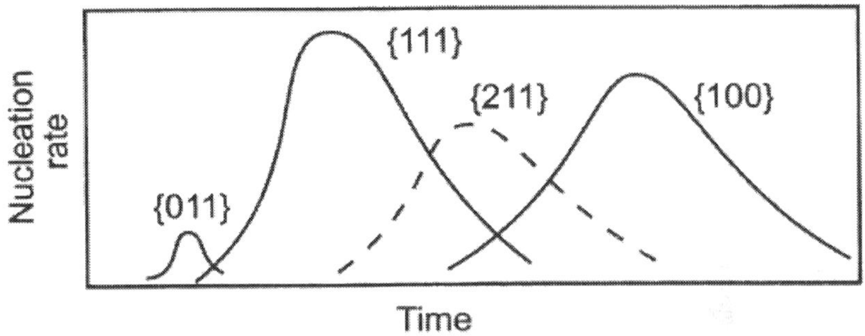

Figure 6.19: Schematic representation of the proposed orientation dependence of the nucleation rate in deformed iron (Rollett, Humphreys, Rohrer, & Hatherly, 2004)

The effect of boundary character on growth rate

The orientation relationships between deformed and recrystallizing grains may also have a strong effect on the growth rate during recrystallization. This is a consequence of the dependence of grain boundary mobility (M) on boundary character that this directly influences the growth rate. Grain boundaries of high mobility will not only cause more rapid recrystallization, but may strongly influence the recrystallization texture. It has been pointed out by Juul Jensen (1995) that oriented growth effects may also result from low angle boundaries of low mobility, if recrystallizing grains encounter deformed regions of a similar orientation.

The important general conclusions to be drawn from this section are:

- Recrystallization of different texture components will occur at different rates, and will inevitably lead to inhomogeneous recrystallization.
- The strain path history affects both the stored energy and the heterogeneities of the microstructure, and therefore apparently identical texture components may recrystallize quite differently. Without knowledge of the starting texture, final texture and the orientation path linking them, it may not be possible to predict

the recrystallization behavior of a particular component of the deformation texture.

- The spatial distribution of orientations, which affects the grain boundary character distribution, will influence both the initiation and growth of recrystallized grains (Rollett, Humphreys, Rohrer, & Hatherly, 2004).

The original grain sizes

It is generally found that a fine-grained material will recrystallize more rapidly than a coarse-grained material, as seen in Figure 6.20. There are several ways in which the initial grain size may be expected to affect the rate of recrystallization.

- The stored energy of a metal deformed to low strains ($\varepsilon<0.5$) tends to increase with a decrease in grain size.
- Inhomogeneities such as deformation and shear bands are more readily formed in coarse-grained material and thus the number of these features, which are also sites for nucleation, increases with increasing grain size.
- Grain boundaries are favored nucleation sites, and therefore the number of available nucleation sites is greater for a fine-grained material.
- The deformation texture and hence the recrystallization texture may be influenced by the initial grain size, and as discussed above, the recrystallization kinetics will then be affected by the grain orientations.

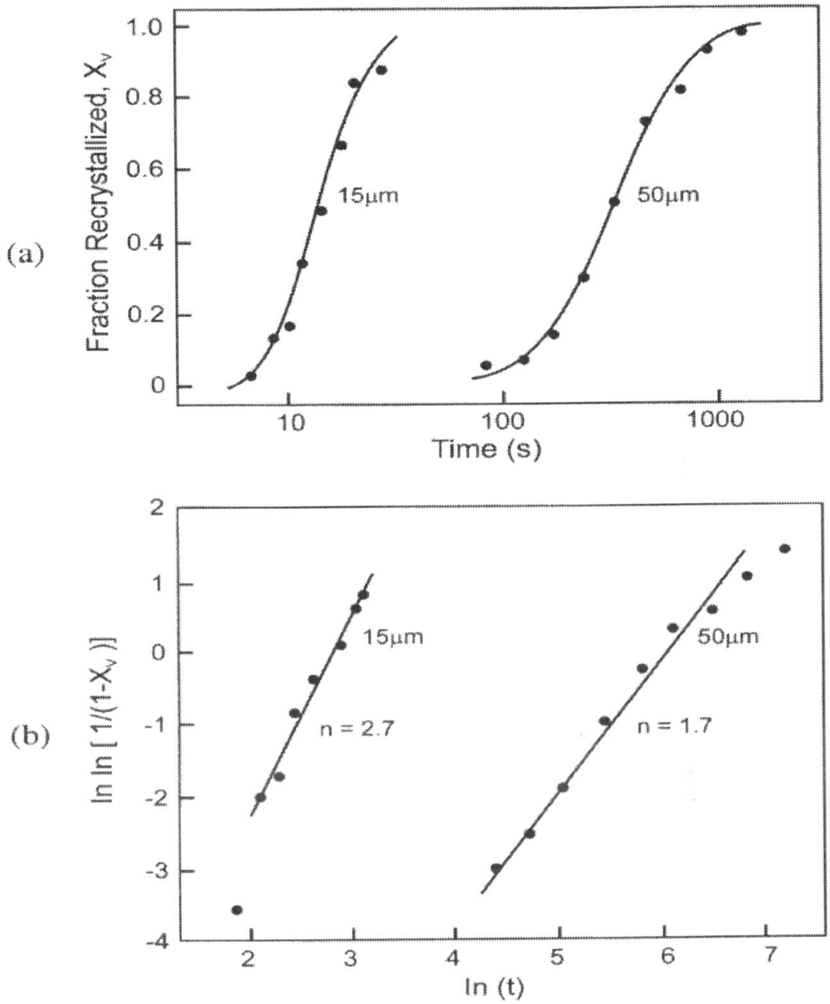

Figure 6.20: The recrystallization kinetics at 225°C of copper of different initial grain sizes cold rolled 93%: (a) fraction recrystallized, (b) JMAK plot (Rollett, Humphreys, Rohrer, & Hatherly, 2004).

Evidence for these effects may be found in the literature, although it should be noted that in the investigation of Hutchinson et al. (1989) shown in Figure 6.20, where the growth rate of the grains in the fine-grained material was some 20 times faster than that of the coarse-grained material at the same temperature, the difference in kinetics was ascribed primarily to orientation effects.

Solutes

The usual effect of solutes is to hinder recrystallization. Table 6.2 shows the effect of zone refining on the recrystallization temperatures of various metals after large rolling reductions. Although the amount and type of impurities in the commercial materials is not specified, and in some cases, they may contain second-phase particles, the marked effect of purifying the metals is clearly seen. The quantitative effect of a solute on the recrystallization behavior is dependent on the specific solvent/solute pair. A good example of a very potent solute which has an important influence on commercial alloys is iron in aluminum. Figure 6.21 shows how very small concentrations of iron in solid solution in high purity aluminum may increase the recrystallization temperature by ~100°C.

Table 6.2: The effect of purity on the temperature of recrystallization (Rollett, Humphreys, Rohrer, & Hatherly, 2004)

Metal	Recrystallization temperature (°C)	
	Commercial purity	Zone refined
Aluminium	200	−50
Copper	180	80
Iron	480	300
Nickel	600	300
Zirconium	450	170

** rewrite this Table as Table 6.1, it is a figure at its present form.

Impurities may affect both the nucleation and growth of recrystallizing grains. It is difficult to measure or even define a nucleation rate for re-crystallization, and the effect of impurities has not been quantified. Sol-utes may in some cases have a strong influence on the recovery processes which are an integral part of nucleation and which may affect the driving force for recrystallization. However, most experimental work suggests that the main influence of solutes is on the grain boundary mobility and hence on the growth rate of recrystallizing grains (Rollett, Humphreys, Rohrer, & Hatherly, 2004).

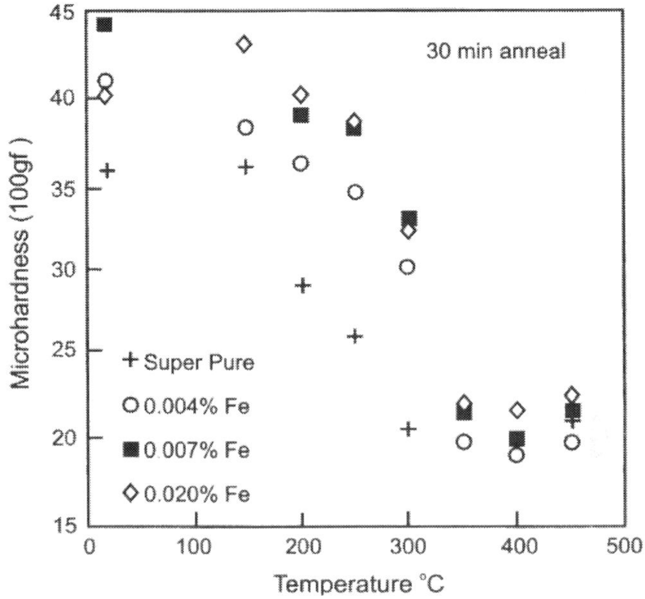

Figure 6.21: The effect of trace amounts of iron in solid solution on the annealing response of aluminum deformed 80% by cold rolling (Rollett, Humphreys, Rohrer, & Hatherly, 2004)

The deformation temperature and strain rate

At temperatures where thermally activated restoration processes such as dislocation climb occur during deformation, the microstructure will be dependent on the deformation temperature and strain rate in addition to the strain. After deformation at high temperatures and low strain rates, the stored energy will be reduced and recrystallization will occur less readily than after deformation to a similar strain at low temperature (Rollett, Humphreys, Rohrer, & Hatherly, 2004).

The annealing conditions

The annealing temperature

As may be seen from Figure 6.22a, the annealing temperature has a profound effect on the recrystallization kinetics. If we consider the total transformation and take the time for 50% recrystallization ($t_{0.5}$) to be a measure of the rate of recrystallization, the relationship is of the type:

$$Rate = \frac{1}{t_{0.5}} = C\exp\left(-\frac{Q}{KT}\right) \quad \text{(a)}$$

(a)

(b)

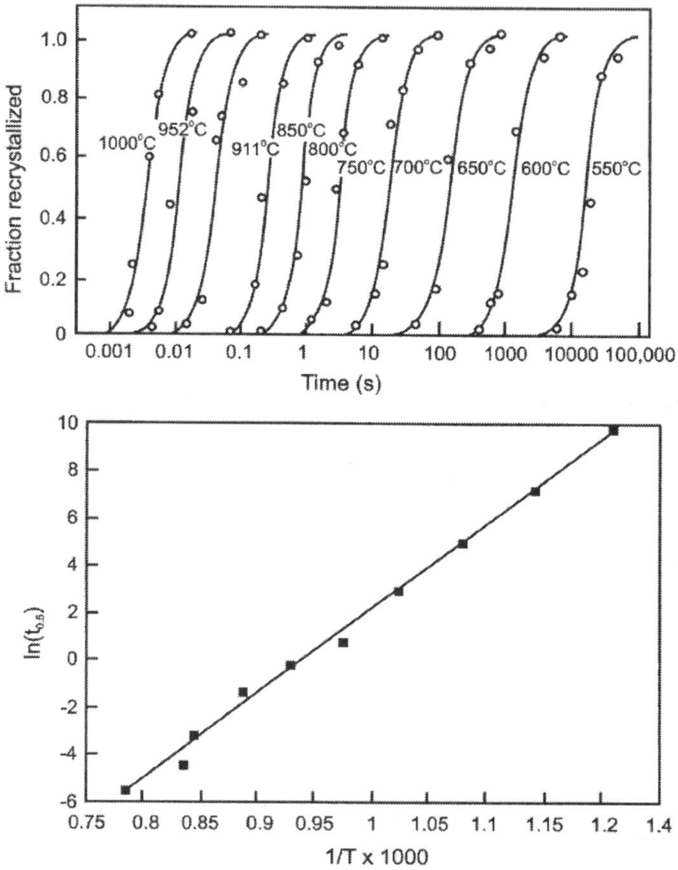

Figure 6.22: (a) The effect of annealing temperature on the annealing of Fe-3.5%Si deformed 60%, (b) Arrhenius plot of the time for 50% recrystallization as a function of temperature (Rollett, Humphreys, Rohrer, & Hatherly, 2004).

In Figure 7.11b, we see that a plot of $\ln(t_{0.5})$ vs. 1/T gives a good straight line in accordance with equation (a), with a slope corresponding to an activation energy of 290 kJ/mol. Although such an analysis may be of practical use, an activation energy obtained from it is not easy to interpret, as it refers to the wholistic transformation, and is unlikely to be constant. For example, such activation energies have been found to depend on strain and to vary significantly with small changes in material purity. Vandermeer and Gordon (1963) found that an addition of only

0.0068 at% of copper to pure aluminum raised the activation energy for recrystallization from 60 to 120 kJ/mol.

Activation energy is only interpretable if it can be related to thermally activated processes at the atomic level. Consider the separate nucleation and growth processes which constitute recrystallization.

$$\dot{N} = C_1 \exp(-\frac{Q_N}{kT}) \qquad \text{(b)}$$

$$\dot{G} = C_2 \exp(-\frac{Q_G}{kT}) \qquad \text{(c)}$$

Although equation (c) has some validity because grain boundary migration rates are known to follow this type of relationship, equation (b) is of little use because we have already noted that a constant nucleation rate is rarely found in practice. For example, the rate controlling process in recrystallization nucleation may be sub grain growth, in which case the activation energy for that process is the most relevant to nucleation (Rollett, Humphreys, Rohrer, & Hatherly, 2004).

The heating rates

The rate of heating of the specimen to the annealing temperature can also affect the rate of recrystallization. As proposed, rapid or slow heating rates might produce conditions for a notorious change of the nature of recovery, enhancing faster or slower recrystallization. The rapid heating rate for instance might produce an undesirable change recovery, by enhancing faster recrystallization, as observed in Al-based alloys and Tantalum, as reported by Atkinson in 1999. A more recent study by Stockemer qt al., (2003) and Chowdhury et al., (2008) emphasizes the tendency to raise the onset of recrystallization as the heating rate is increased, particularly in carbon steels. For this to occur, it is necessary for some other process to interact with recrystallization, and for this to have different temperature dependence. Consider the following two cases:

The heating rate affects the relative rates of recovery and recrystallization.

If slow heating increases the amount of recovery occurring before recrystallization, the driving pressure will be reduced and recrystallization retarded. Although this effect is frequently cited in the literature, there is no experimental evidence to support it for single-phase alloys. Recovery would be enhanced at low heating rates only if it had lower activation energy than recrystallization. In most alloys, the rates of both processes are controlled by lattice diffusion and have similar temperature dependencies; therefore, heating rate is not predicted to affect one more than the other. In a very pure metal where High Angle Grain Boundary (HAGB) migration is controlled by boundary diffusion and recovery by lattice diffusion, it would be expected that a low heating rate would in fact promote recrystallization, although this has not been reported.

Phase transformations can occur.

Precipitation reactions have different temperature dependence to recrystallization. If a supersaturated solid solution is deformed and annealed at a low heating rate, precipitation may occur before recrystallization and hinder it, whereas on rapid heating, recrystallization may be completed before any precipitation occurs. This effect may influence the rate of recrystallization in industrial alloys, where it is often found that a rapid heating rate accelerates recrystallization and results in a smaller grain size, e.g. Al-Li alloys and Al–Zn–Mg alloys. Conversely, it is found that in steels, rapid heating may slow down recrystallization. In this case the heating rate determines the amount of carbon in solid solution, and this in turn affects the recrystallization (Rollett, Humphreys, Rohrer, & Hatherly, 2004).

Single crystals

Nowadays, there is an increasing demand for single, pure and defect-free crystals of uncountable substances, e.g. for the scientific appraisal of the chemical and physical properties of pure solids, for use in solid-state electronics, where crystals of specific substances are needed for their dielectric, piezoelectric, paramagnetic and semiconductor properties, and for optical and laser materials and synthetic gem-stones (see *Table* 6.3).

The production of large single crystals demands exacting techniques, but broadly speaking there are three general methods available, that is, growth from solutions, from melts, and from vapors. There are many variations; however, in the processing techniques available within these three groups (Mullin, 2001).

Table 6.3: some uses of single crystals (Mullin, 2001).

Uses	*Substances*
Piezoelectric	quartz, Rochelle salt, ammonium and potassium dihydrogen phosphates, ethylenediamine tartrate, triglycine phosphate
Optical materials	CaF_2, α-SiO2
Lasers	ruby, sapphire
Paramagnetic materials	α-Al_2O_3, TiO_2, CaF_2
Semiconductors	Ge, Si, GaAs
Luminescence and fluorescence	ZnS and CsS activated with Ag, Tl, Cr, Mn and Cu salts; various organic crystals
Gem-stones	alumina (sapphire and ruby), diamond

**rewrite as a true table

Low temperature solution growth

For long periods, the standard method for growing large pure crystals of inorganic and organic substances was slow crystallization from solution in water or organic solvents. Principally, a small crystal seed is immersed in a supersaturated solution of the known substance and its growth is controlled by a careful regulation of the temperature, concentration and degree of agitation of the system. For example, a tiny selected crystal may be mounted on a suitable support and then suspended in a vessel containing a solution of the substance, maintained at a fixed temperature. An adequate movement of the solution around the crystal will be obtained from slow rotation of the vessel, and slow, controlled evaporation of the solvent will produce the degree of super saturation essential for crystal growth. On the other hand, the vessel may persist stationary and the growing crystal, or several suitably mounted crystals, may be smoothly rotated in the solution (Mullin, 2001).

Variability of the definite operating conditions is possible in relation to the nature of the crystallizing substance and solvent; the optimum super saturation and solution movement past the crystal must be found by trial and error. In general, the degree of super saturation must not be high, and in any case the solution must never be permitted to reach the labile condition. The degree of super saturation should be preserved as constant as possible to ensure a constant rate of deposition of solute on the crystal seed. Holden (1949) proved that, for salts with solubility in the approximate range 20-50 per cent by mass, the maximum linear growth rate that can be tolerated by the fastest-growing faces of a crystal is about 1-3 mm per day; faster rates tend to give imperfections. Single crystals are better grown by a cooling than by evaporation process, as super saturation can then be much more closely controlled (Mullin, 2001).

A typical example of a cooling-type growth unit is shown in Figure 6.23a. The carefully selected seed crystals are supported on a rotating arm, which is submerged in the gently agitated system, and grow under carefully controlled conditions. A very slow cooling rate is used, often less than 0.1°Ch^{-1}.

High temperature solution growth

Many substances normally considered insoluble in water have an appreciable solubility at elevated temperatures and pressures. This property is utilized in the technique called 'hydrothermal crystallization', which is basically crystallization from aqueous solution at high temperature (350-550 °C) and pressure (1-3 k bar). The operation is carried out in a steel autoclave (Figure 6.23b), which can be provided with a silver or platinum liner for protection. The technique has proved satisfactory for the growth of silica and aluminosilicates from aqueous alkaline solutions. Quartz crystal, the ideal piezoelectric material, is now grown in this manner on a commercial scale.

Molten salt solvents may also be used for high-temperature solution growth in the technique normally called 'flux growth'. Oxide crystals, like yttrium aluminum garnet, $Y_3Al_5O_{12}$, for example, may be grown from ionic solvents such as PbO-PbF_2 -B_20_3, while metallic solvents are suitable for covalently bonded semiconductors such as Si, GaAs and InP, while diamond can be crystallized from molten Ni-Fe alloys (Mullin, 2001).

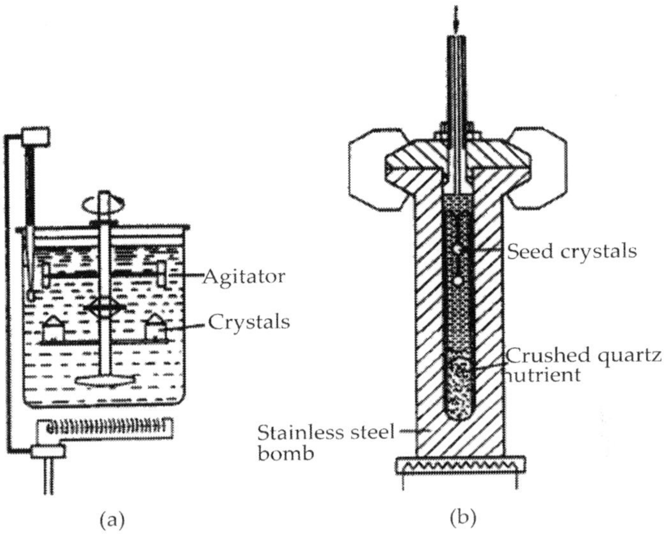

Figure 6.23 Crystallization from solution: (a) temperature-lowering methods, (b) hydrothermal growth (Mullin, 2001)

Growth in gels

A room-temperature method, particularly useful to produce single crystals of substances that are thermally unstable or have a very low solubility in water, is the technique of growth in a gel. The gel provides a kind of protective barrier for the growing crystals; it eliminates turbulence, yet it imposes no strain on the crystals, and permits a steady diffusion of the growth components. The diffusion and reaction processes are retarded and very much larger crystals are grown than could be produced by normal reaction and precipitation techniques (Mullin, 2001).

Growth from the melt

A single crystal can be grown in a pure melt in a manner similar to that described earlier for growth in a solution. For example, a small crystal seed could be rotated in a super cooled melt, or the melt could flow past a fixed seed in a growth cell. However, these methods do not find any significant application. The most widely used techniques for the production of single crystals from the melt can be grouped into four categories: the withdrawal or pulling techniques, the crucible methods, flame fusion and zone refining. The latter method has been described in some detail above.

The pulling techniques are typified by the Kyropoulos and Czochralski methods. In both cases a small, carefully selected seed crystal is partially immersed in a melt, kept just above its melting point, and the growing crystal is maintained just below its melting point by holding it on a water-cooled rod or tube (see Figure 6.24). The apparatus is usually kept under reduced pressure or supplied with an inert atmosphere. Silicon, germanium, bismuth, tin, aluminum, cadmium, zinc, intermetallic compounds, and many organic substances have been grown as large strain free single crystals in this manner.

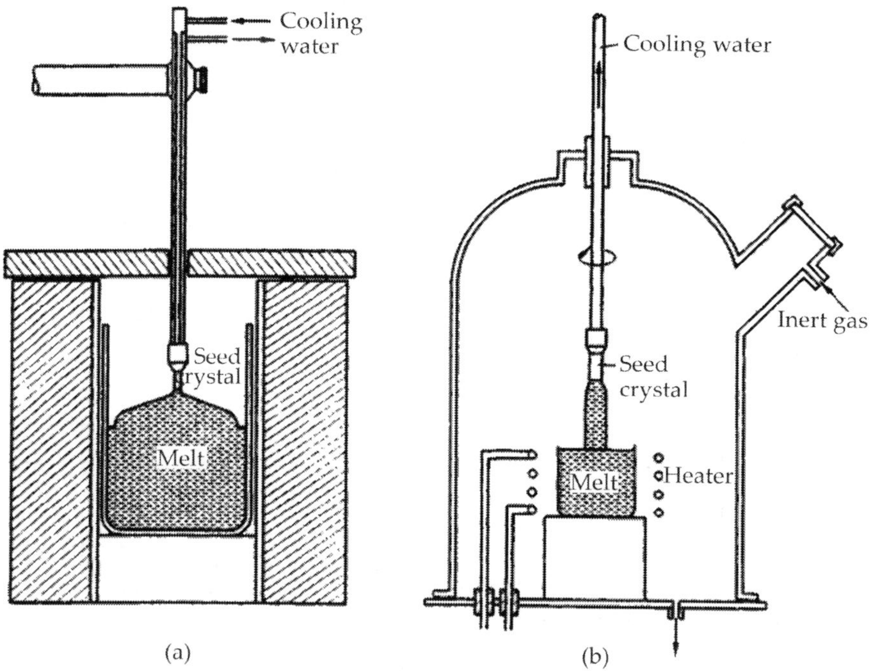

Figure 6.24: Crystal pulling and withdrawal techniques: (a) Kyropoulos, (b) Czochralski (Mullin, 2001)

The main difference between the methods of Kyropoulos (see Figure 6.24a) and Czochralski (see Figure 6.24b) is that in the former the seed is permitted to grow into the melt, while in the latter the seed is withdrawn at a rate that keeps the solid-liquid interface more or less in a constant position. Pull rates depend on the temperature gradient at the crystal-melt interface and can vary from 1 to 40 mmh^{-1}. The steeper the gradient the faster the growth rate and, hence, the faster the permissible rate of withdrawal.

Compounds that dissociate on melting cannot be grown by the vertical pulling technique unless means are provided for suppressing dissociation. A simple elegant means is provided by the liquid encapsulation technique in which the melt surface is covered by a floating layer of transparent liquid. The encapsulant, e.g. boric oxide, B_2O_3, a low melting- point glass, acts as a liquid seal provided that the inert gas pressure on its surface is greater than the dissociation pressure of the compound. As the crystal is pulled from the melt, a thin film of encapsulant adheres to its surface and suppresses dissociation. The technique is particularly useful for the growth of semiconductor and metal crystals; even highly dissociable compounds such as GaP (vapor pressure 35 bar at 1470 °C) can be grown successfully on a commercial scale by this technique (Mullin, 2001).

An example of the crucible method is one by Bridgman and Stockbarger (see Figure 6.25a). The melt is contained in a crucible with a conical bottom which is lowered slowly from a hot to a cold zone. The two sections of the furnace, which are kept at about 50 °C above and below the melting point of the substance, respectively, are isolated by a thermal shield. The bulk of the solidified material, apart from the tip of the cone where the nuclei gather, is a single crystal. This method has proved successful for many semiconductors materials and alkali halides.

The flame fusion technique (see Figure 6.25b) was originally devised in 1904 by Verneuil for the manufacture of artificial gemstones, such as corundum (white sapphire) and ruby. This method is now used for the mass production of jewels for watches and scientific instruments. A trickle of fine alumina powder plus traces of coloring oxides is fed at a controlled rate into an oxy-hydrogen flame. Fusion occurs and the molten droplets fall on to a ceramic collecting rod. A seed crystal cemented to the rod is fused in the flame and the rod is lowered at a rate that allows the top of the growing crystal (known as a bode) to remain just molten. Renewed interest has recently been shown in this method for the production of rubies for lasers.

Growth from vapor

Crystal growth from a vapor by direct condensation, without the intervention of the liquid phase, can be used to produce small strain-free single crystals of substances that sublime readily. Large single crystals cannot be grown by this method. A number of techniques are available. For example, a gas stream, such as N_2 or H_2S, may be passed over a sublimable substance, such as cadmium sulfide, in a heated container. The vapors then pass to another part of the apparatus, where they condense, in crystalline form, on a cold surface.

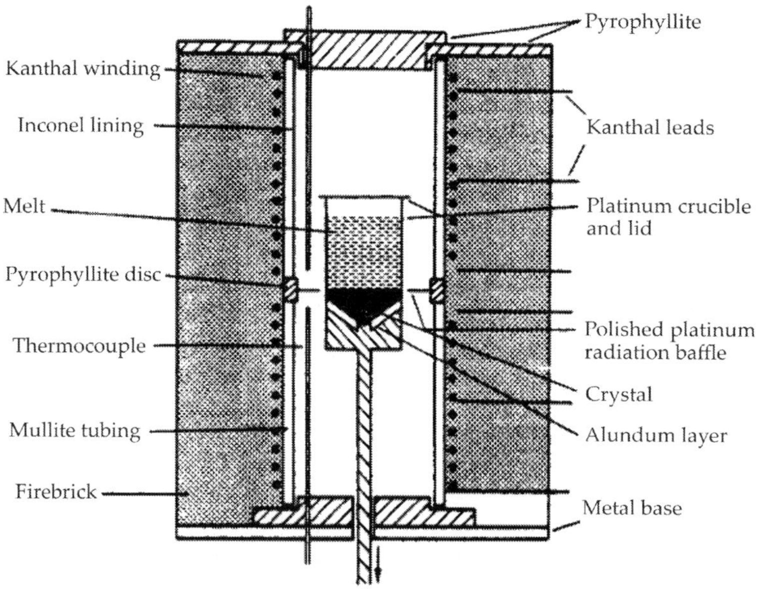

(a)

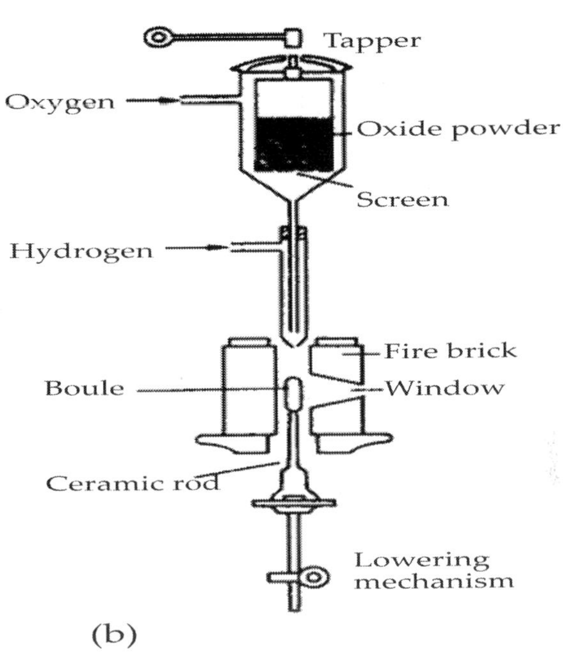

Figure 6.25: (a) The Bridgman-Stockbarger technique; (b) the Verneuil boule furnace (Mullin, 2001)

Sublimation in sealed tubes is also used for the preparation of single crystals of metals, such as zinc and cadmium, and non-metallic sulphides. A quantity of the material is placed at one end of the tube, along which a temperature gradient is maintained, so that sublimation occurs at the hot end and crystallization at the other. An electric furnace with a number of independently controlled windings is used to maintain the temperature gradient to give a rate of sublimation sufficiently slow for the grown crystals to be single and not polycrystalline.

CHAPTER 7

CONTINUOUS RECRYSTALLIZATION DURING AND AFTER LARGE STRAIN DEFORMATION

INTRODUCTION

A large number of low-angle boundaries are formed when a polycrystalline metal is deformed (or when the material is subsequently recovered by annealing at low temperatures. On annealing at elevated temperatures, such a microstructure will usually recrystallize discontinuously as discussed in chapter 6, this process being driven mainly by the energy stored in the low-angle boundaries.

However, after deformation to large strains, particularly at elevated temperatures, a microstructure consisting predominantly of high-angle grain boundaries may be formed. Minor boundary movements either during the deformation or on subsequent annealing may then result in a fine-grained microstructure consisting mainly of crystallites which are surrounded by high-angle boundaries. Such a microstructure is similar to that resulting from conventional recrystallization, but because no recognizable 'nucleation' and 'growth' of the recrystallized grains occurs, and the microstructure evolves relatively homogeneously throughout the material, the process can reasonably be classified as continuous recrystallization. It is worth emphasizing that terminology such as continuous or discontinuous recrystallization is purely phenomenological, referring only to the spatial and temporal heterogeneity of microstructural evolution, and implying no specific mechanism of recrystallization.

The occurrence of continuous recrystallization during high temperature deformation, which is known as geometric dynamic recrystallization, has been recognized for some time. However, recent research has shown that severe cold-working of a metal can also result in a microstructure which consists almost entirely of high-angle grain boundaries, and which, on annealing, may undergo continuous recrystallization. The microstructures which evolve after such low temperature processing, are often sub-micron grain (SMG) structures, and because such materials may have attractive mechanical properties, there has been extensive research recently in this field.

We consider that the phenomena discussed in this chapter can be explained almost entirely in terms of the fundamental processes discussed in earlier chapters, and that no unusual micro-scale mechanisms are involved. Nevertheless, this subject, which was briefly alluded to in the first edition of this book, now requires a separate chapter, in which the principles underlying the continuous recrystallization of metals deformed at either ambient or elevated temperatures are considered. In order to simplify discussion and to enable comparison with the conventional deformation processing discussed elsewhere in the book, we will mainly be concerned with microstructures produced by conventional rolling procedures (Rollett, Humphreys, Rohrer, & Hatherly, 2004).

MICROSTRUCTURAL STABILITY AFTER LARGE STRAINS

When a polycrystalline metal is deformed, the grain boundary area increases with increasing strain at a rate which depends on the mode of deformation. For example, during rolling, the grain thickness in the normal direction (H) is related to the strain (ε) and the initial grain size D_0 (figure 7.1) by the geometric relationship:

$$H = D^{\circ} \exp(-\varepsilon) \qquad (7.1)$$

During the plastic deformation, cells or sub-grains are formed, and after a moderate strain (ε~1) these generally do not change very much. Therefore, with increasing strain, the percentage of the boundaries which are of high angle (HAGB %) increases (Figure 7.1c). Additionally, new high-angle boundaries may form by grain fragmentation.

Although it is convenient to discuss the microstructures in terms of the fraction or percentage of high angle boundaries (HAGB %), it should be recognized that this is only a rather crude and arbitrary description, and that it is the overall grain boundary character distribution which is important. Nevertheless, for analytical purposes, it often convenient to use the simpler descriptions.

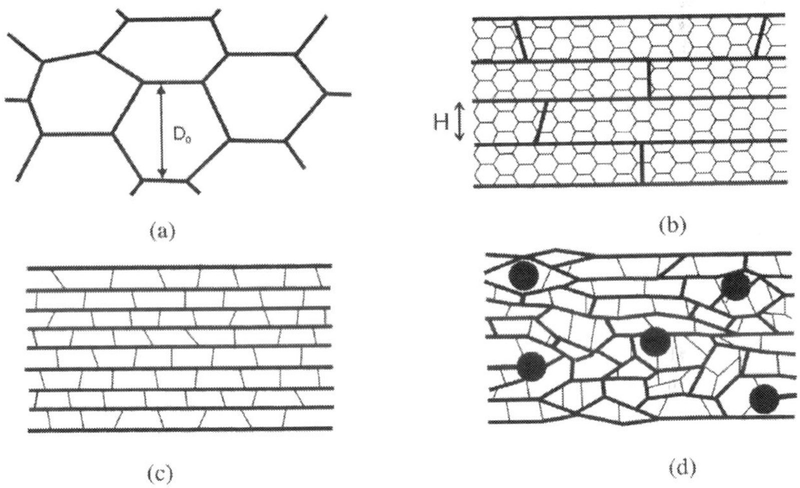

Figure 7.1: Schematic diagram of the development of microstructure with increasing strain; (a) initial grain structure, (b) moderate deformation, (c) large deformation, (d) effect of large second-phase particles on deformation (Rollett, Humphreys, Rohrer, & Hatherly, 2004).

The analysis of the stability of cellular microstructures showed that microstructures with a large fraction of low angle boundaries are intrinsically unstable with respect to discontinuous growth (i.e. discontinuous recrystallization). If the mean low angle boundary misorientation is increased, the boundary properties become more uniform, and the structure cannot recrystallize discontinuously.

During deformation, the mean boundary misorientation increases, and thus the microstructure become progressively more stable against discontinuous growth as strain increases. The criteria for the suppression of discontinuous recrystallization are therefore the conditions of strain, temperature and initial grain size, which result in a sufficient fraction of high angle boundary, (0.6–0.7) to ensure microstructural stability. Alternative criteria for continuous recrystallization can be formulated in terms of the impingement of corrugated high angle boundaries at large strains. However, purely geometric criteria are less rigorous, although both approaches give rather similar results.

DEFORMATION AT AMBIENT TEMPERATURES

The development of stable microstructures by large strain deformation

For simplicity, we consider only the case of deformation by cold rolling or plane strain compression. The size (D) of the cells or sub grains formed on deformation is reduced very little at strains above ~2 as shown in figure 7.2, and data from this figure is consistent with the following relationship for $\varepsilon>2$.

$$D=K\varepsilon^{-1} \qquad (7.2)$$

Where, K is constant.

The spacing (H) of the high angle boundaries in the normal direction (ND), of a material with initial grain size D_0, decreases with strain (Figure 7.1a–c) according to equation 7.1. Consequently, as shown in Figure 7.3, the HAGB spacing decreases more rapidly than the sub grain size with strain, and the fraction of boundaries which are of high angle therefore increases with strain. At a sufficiently large strain (), the fraction of high angle boundary becomes sufficient (~0.6–0.7) to make the microstructure stable against discontinuous growth (i.e. recrystallization). The precise rate of increase in the fraction of HAGB with strain and therefore the critical strain, is difficult to calculate precisely, as it depends on the shape and planarity of the boundaries, and will be increased by any new high angle boundaries created by fragmentation of the old grains.

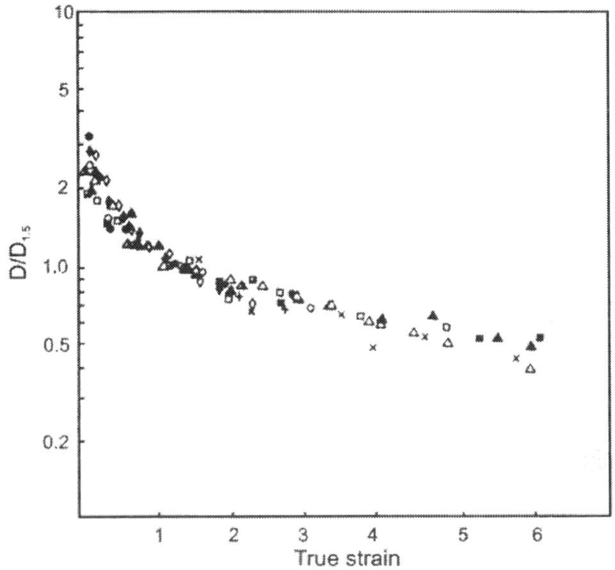

Figure 7.2: Average cell size as a function of strain for Al, Cu, Fe, Ni, Cr and Nb. The cell size is expressed as a fraction of $D_{1.5}$, the cell size at a strain of 1.5. Values of $D_{1.5}$ are Al~0.5 μm, Cu~0.3 μm, Ni~0.3 μm, Fe~0.3 μm, Cr~0.4 μm, Nb~0.2 mm (Rollett, Humphreys, Rohrer, & Hatherly, 2004).

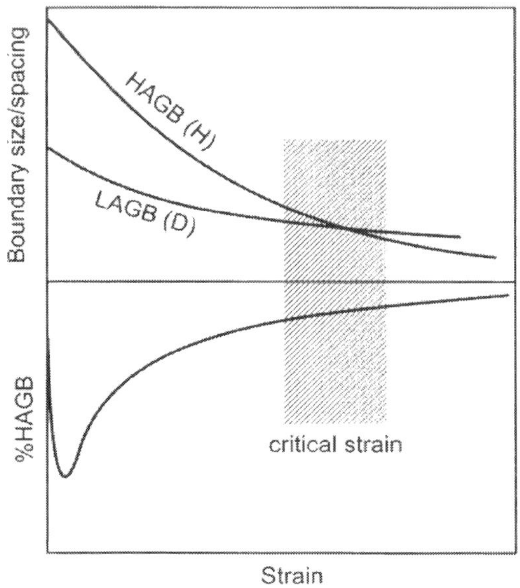

Figure 7.3: Schematic diagram of the effect of strain on the cell/sub grain size

(D), HAGB spacing (H) and %HAGB during deformation at ambient temperatures (Rollett, Humphreys, Rohrer, & Hatherly, 2004).

THE EFFECT OF THE INITIAL GRAIN SIZE

As shown by equation 7.1, a small initial grain size (D_0) will reduce the boundary spacing H to the critical value at a lower strain than will a larger initial grain size. The effect of strain and initial grain size on the spacing of high angle boundaries (H) and on the percentage of high angle boundaries in an Al–Fe–Mn alloy are shown in figure 7.4. The HAGB% initially decreases with increasing strain as low angle boundaries are formed during deformation, but then progressively increases as H becomes smaller. If the initial grain size is reduced, it may be seen that the HAGB% at high strains is larger, and the HAGB spacing (H) at a given strain, is smaller.

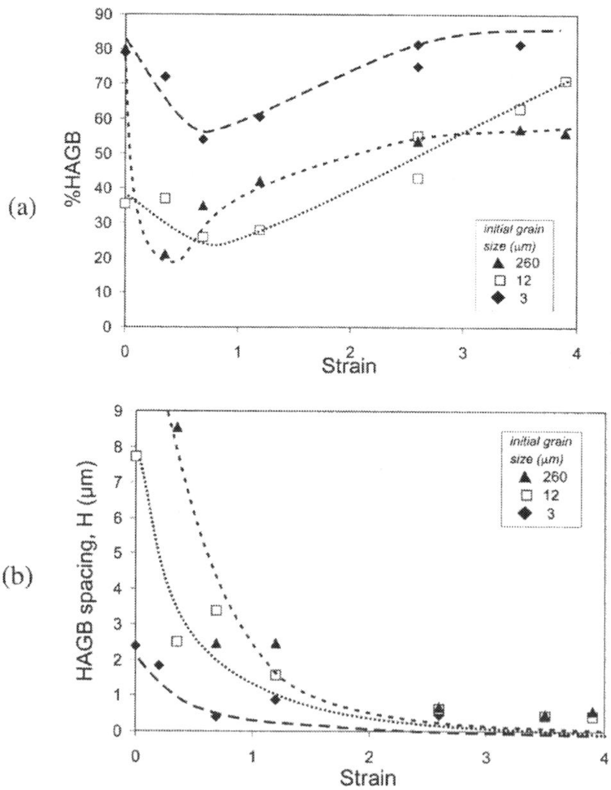

Figure 7.4: The effect of strain and initial grain size on the microstructure of an

Al–Fe– Mn alloy (AA8006); (a) The percentage of high angle boundaries, (b) The high angle boundary spacing (H) (Rollett, Humphreys, Rohrer, & Hatherly, 2004).

It is seen from Figure 7.4b that, at the largest strains, the HAGB separations for the materials of different initial grain size tend to converge. In table 7.1, the measured HAGB spacing (H) is compared with that predicted (HG) by the geometric reduction of the original grain size (equation 7.1) at a true strain of 2.6. The ratio HG/H, which will be unity if H is determined purely by geometrical factors, is seen to be greatly affected by the initial grain size. The large value of HG/H for the 260 μm material is consistent with extensive grain fragmentation in such a large-grained material. For the 12-mm material, the value of HG/H implies no grain fragmentation, which is consistent with previous results for small grain sizes. The 3 μm material is seen to have a grain spacing which is considerably larger than the geometric value. There are two factors which can contribute to this. At large strains as the deformation texture strengthens, some grain orientations converge, transforming high angle into lower angle boundaries. Statistical calculations of this effect for the materials of table 7.1, indicate that the loss of high angle grain boundaries in this way should be no more than ~15%, which is insufficient to account for the very small value of HG/H. The second effect which can cause this is the local dynamic grain growth which can occur when high angle boundaries approach closely. This effect is normally found only at elevated temperatures, but is clearly able to occur at ambient temperatures in a low-solute aluminum alloy under these conditions (Rollett, Humphreys, Rohrer, & Hatherly, 2004).

Table 7.1 The effect of strain and initial grain size on the spacing of high angle boundaries in the normal direction in Al–Fe–Mn alloy AA8006 cold rolled to a strain of 2.6 (Rollett, Humphreys, Rohrer, & Hatherly, 2004).

Initial grain size (μm)	Measured spacing (H) (μm)	Predicted spacing (H_G) (μm)	H_G/H
260	1.7	19.3	11.3
12	1.0	0.9	0.9
3	0.6	0.2	0.4

**rewrite this as a true table

THE EFFECT OF SECOND-PHASE PARTICLES

The presence of second phase particle in many alloys can be from the impurities of a substance deliberately added during the alloying process. Large (>1 μm) second-phase particles will increase the rate of formation of high angle boundaries by breaking up the planarity of the boundary structure and also by creating high angle boundaries associated with the large local lattice rotations close to the particles. An Electron Back Scatter Diffraction (EBSD) map demonstrating the effect of large particles on the microstructure is shown in Figure 7.5. Such particles therefore reduce , although it is difficult to quantify the effect.

For bimodal particles distribution such as that of Al-Si alloys, the situation becomes simpler due to the distinction of the particles' population. It was found out that even at very large (<5 μm) particles, Chan and Humpreys, 1984) reported that the recrystallization is mostly participated by the small particles, resulting in the microstructure that is almost the same as the microstructure of the one from an alloy with small particles.

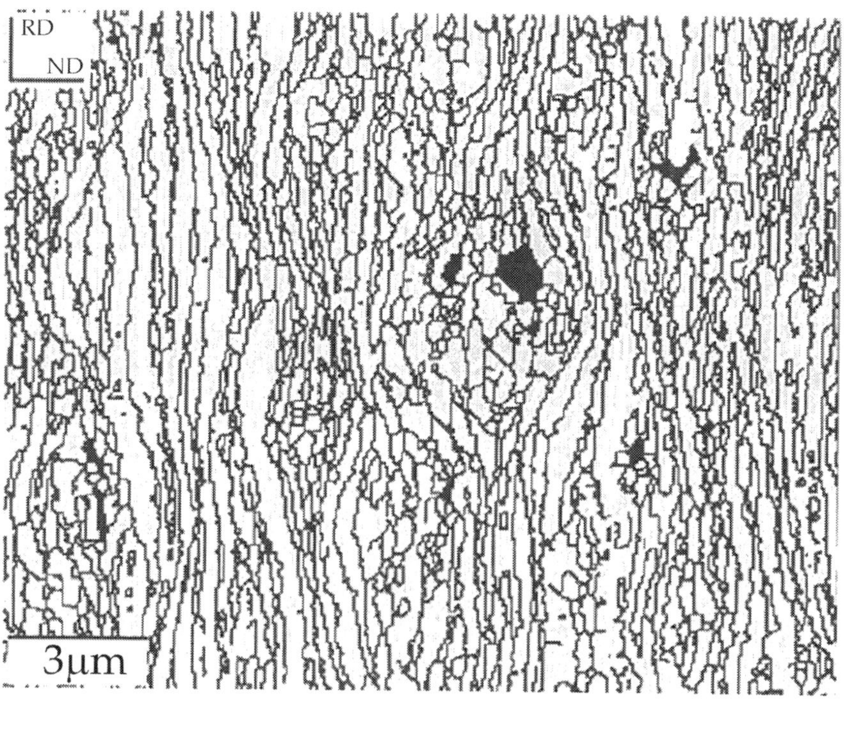

Figure 7.5: EBSD map showing the effect of large (>1 μm) second-phase particles on the microstructure of Al–Fe–Mn (AA8006) cold-rolled to a strain of 3.9. The particles are shown as black regions, HAGBs shown as black lines and LAGBs as white lines (Rollett, Humphreys, Rohrer, & Hatherly, 2004).

THE TRANSITION FROM DISCONTINUOUS TO CONTINUOUS RECRYSTALLIZATION

For the larger initial grain sizes and lower strains, normal discontinuous recrystallization occurred on subsequent annealing (Figure 7.6 a), but for larger strains and smaller grain sizes, a predominantly high angle boundary microstructure was formed at comparatively low annealing temperatures, with only minor boundary movements (Figure 7.6b), by a process of continuous recrystallization.

The transition between discontinuous and continuous recrystallization is best measured by monitoring the changes in HAGB content on annealing, and this is shown in Figure 7.7a and b. When continuous recrystallization occurs, there is a relatively sharp change in the HAGB% as the deformed substructure is consumed by the recrystallizing grains. However, when continuous recrystallization occurs, there is little change in the HAGB%. During annealing the grains/sub grains gradually become more equiaxed and larger (figure 7.7c). These changes are gradual, and quite different to those occurring during discontinuous recrystallization (Rollett, Humphreys, Rohrer, & Hatherly, 2004).

(a) 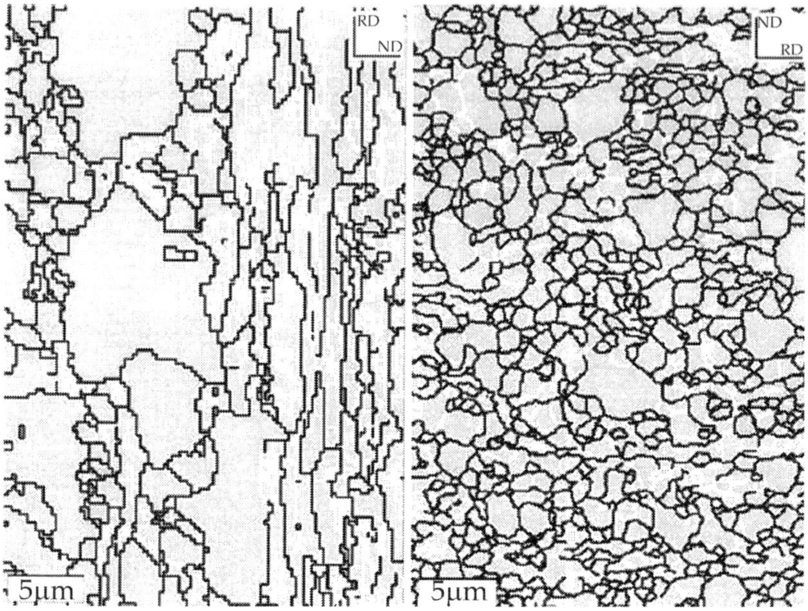 (b)

Figure 7.6: EBSD maps of Al–Fe–Mn (AA8006) annealed after cold rolling; (a) ε=0.69, T=250°C, showing discontinuous recrystallization; (b) ε=3.9, T=300°C, showing continuous recrystallization. HAGBs shown as black lines and LAGBs as white lines (Rollett, Humphreys, Rohrer, & Hatherly, 2004)

A major difference is that in discontinuous recrystallization, the strong deformation texture is generally replaced by a different texture, such as a strong cube texture. However, when continuous recrystallization occurs, the rolling texture is retained with little change, because there is little microstructural change on annealing following the deformation.

Figure 7.8 summarizes the effects of initial grain size and strain on the transition from discontinuous to continuous recrystallization in an AA8006 Al–Fe–Mn alloy.

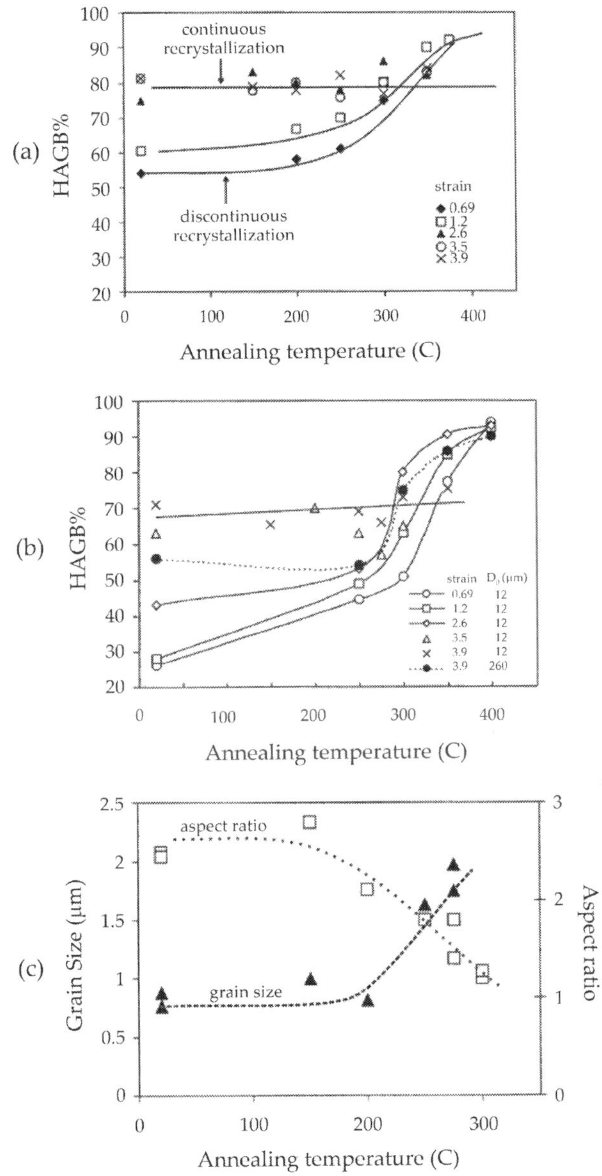

Figure 7.7: Microstructural changes during the annealing of AA8006, as a function of strain and initial grain size (D_0) (a) % HAGB for D_0=3 μm; (b) %HAGB for D_0=12 μm and 260 μm; (c) Grain/sub grain size and aspect ratio for D_0=12 μm, ε=3.9 (Rollett, Humphreys, Rohrer, & Hatherly, 2004)

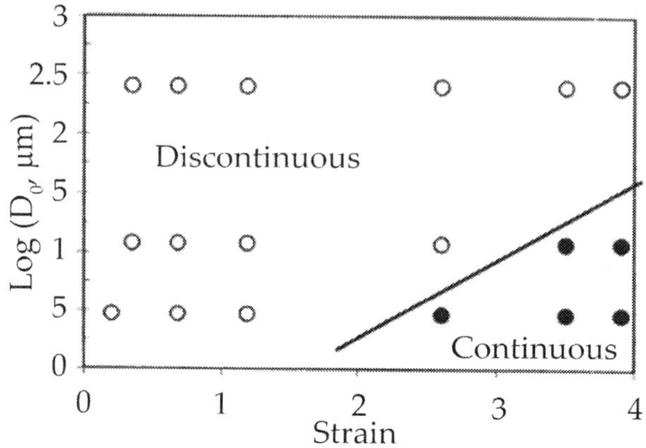

Figure 7.8: The effects of initial grain size and strain on the transition from discontinuous to continuous recrystallization in AA8006 (Rollett, Humphreys, Rohrer, & Hatherly, 2004)

THE MECHANISM OF CONTINUOUS RECRYSTALLIZATION IN ALUMINUM

The microstructural changes which occur when a highly deformed material undergoes continuous recrystallization are rather small, and occur incrementally as the annealing temperature is increased. As seen from figure 7.7c, there is no significant change in the fraction of high angle boundaries, but the deformed grains/sub grains become more equiaxed and slightly larger.

In its simplest form, the highly deformed microstructure is similar to that shown schematically in Figure 7.1c. It comprises lamellar high angle boundaries aligned parallel to the rolling plane, together with intersecting boundaries which are mainly of low angle. On annealing such a microstructure, it is thought that the energy is lowered by localized boundary migration as shown in Figure 7.9, and this can be considered to occur in two stages.

Collapse of the lamellar microstructure

The lamellar structure will tend to collapse due to the surface tension at the node points such as A, where the boundaries (of energy) aligned in the rolling plane, are pulled by the boundaries (of energy) aligned in the normal direction as shown in Figure 7.9b.

The critical condition for collapse of the structure is when the nodes A and A' touch. This depends on the grain length in the rolling direction (L) and the normal direction (N) and on the relative boundary energies. If then the critical aspect ratio (L/N) for impingement is ~2. However, the boundaries in the normal direction are usually of lower angle, and if for example (corresponding to boundaries in the normal direction of ~3°), the critical aspect ratio is ~4. A convincing vertex simulation of this process due to Bate and this has been shown to give good agreement with the annealing behavior of a highly deformed Al–3%Mg alloy. Additional break-up of the lamellar grain microstructure and production of high angle boundaries during the deformation by large second-phase particles (Figure 7.1d and 7.5) will increase , and lower the critical grain aspect ratio required for continuous recrystallization (Rollett, Humphreys, Rohrer, & Hatherly, 2004).

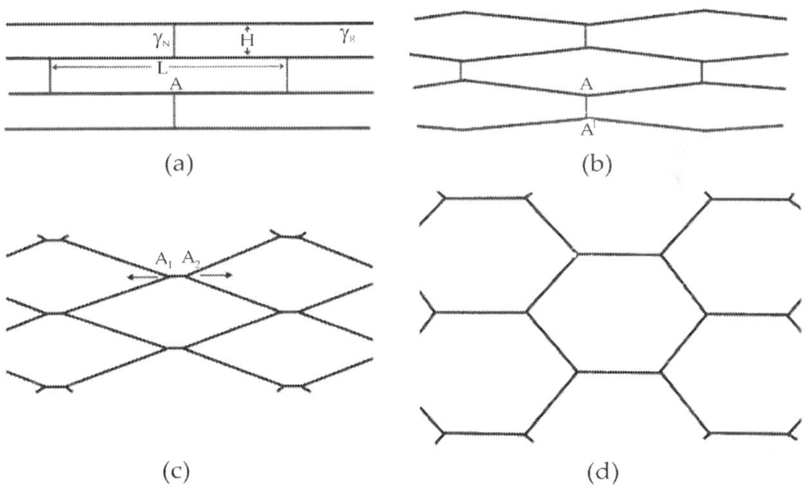

Figure 7.9: Schematic diagram showing the continuous recrystallization of a highly deformed lamellar microstructure; (a) Initial structure, (b) collapse of the lamellar boundaries, (c) spheroidization begins by Y-junction migration, (d) further spheroidization and growth (Rollett, Humphreys, Rohrer, & Hatherly, 2004).

Spheroidization and growth

When the nodes A and A' touch, node switching (Figure 7.9) will occur, and two new nodes A_1 and A_2 will form and be pulled apart by the boundary tensions as shown in figure 7.9c. Further spheroidization and growth will occur due to boundary tensions, as shown in figure 7.9 d, leading to a more equiaxed grain structure. Such a fine-grained microstructure will be unstable with respect to normal and perhaps abnormal grain growth (Rollett, Humphreys, Rohrer, & Hatherly, 2004).

DEFORMATION AT ELEVATED TEMPERATURES

Geometric dynamic recrystallization

As described in previous chapters, grain boundaries develop serrations during dynamic recovery, and the wavelength of these serrations is similar to the sub grain size. If the material is subjected to a large reduction in cross section, for example by hot rolling or hot compression, then the original grains become flattened, as shown schematically in figure 7.10. Because the sub grain size during high temperature deformation is almost independent of strain, the fraction of boundaries which are high angle, increases with strain, and eventually the size of the boundary serrations will become comparable with the grain thickness as shown schematically in Figure 7.10b.

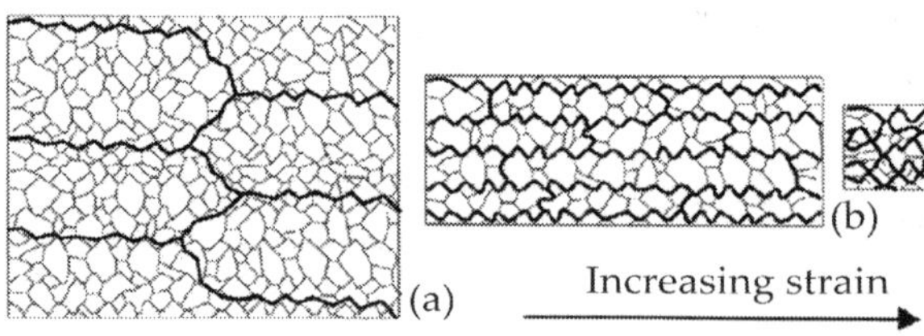

Figure 7.10: Geometric dynamic recrystallization. As deformation progresses, the serrated HAGBs (thick lines) become closer, although the sub grain size remains approximately constant. Eventually the HAGBs impinge, resulting in a microstructure of mainly high angle boundaries (Rollett, Humphreys, Rohrer, & Hatherly, 2004).

Interpenetration of the scalloped boundaries will occur, resulting in a microstructure of small equiaxed grains of a size comparable with the sub grain size (Figure 7.10c), and an example of this process in an Al–Mg–Fe alloy is shown in Figure 7.11. At lower strains (Figure 7.11a), the flattened old grains containing sub grains are seen, but at large strains (figure 7.11b) a microstructure of almost equiaxed grains has formed.

(a)

(b)

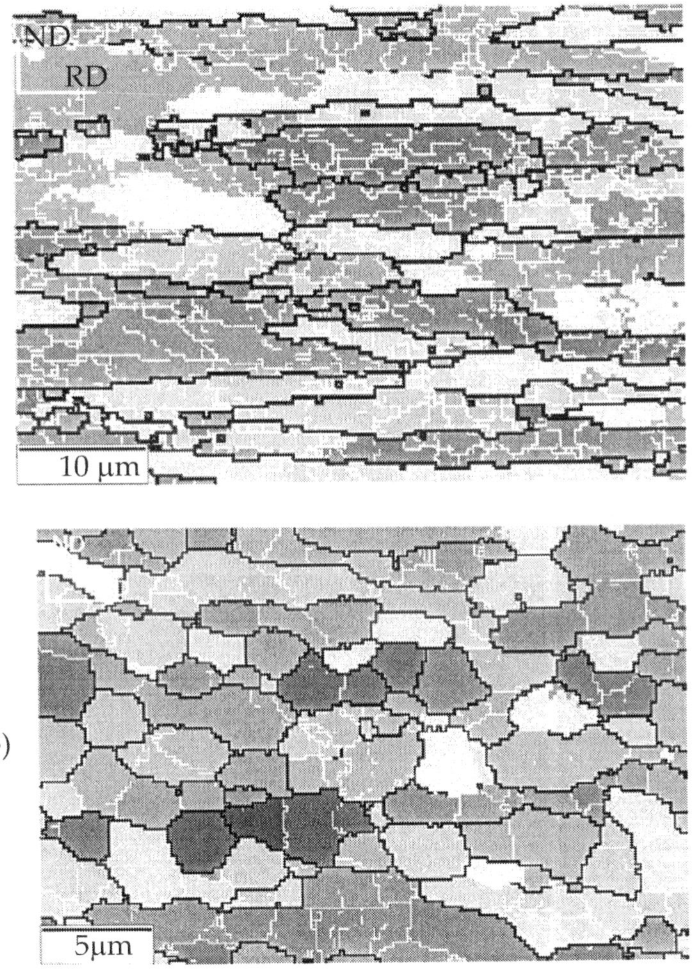

Figure 7.11: EBSD maps showing geometric dynamic recrystallization in an Al–3Mg– 0.2Fe alloy deformed at 350°C in plane strain compression; (a) ε=0.7, (b) ε=3. (HAGBs shown black and LAGBs white) (Rollett, Humphreys, Rohrer, & Hatherly, 2004).

An equiaxed microstructure with a large number of high angle boundaries therefore evolves without the operation of any new microscopic recrystallization mechanism and this process clearly differs from the discontinuous dynamic recrystallization discussed in. Microstructures of this type are commonly formed in aluminum and its alloys deformed to large strains. Humphreys (1982) showed that the origin of such microstructures was a process of grain impingement as discussed above, and this has been confirmed by later more extensive investigations and the term geometric dynamic recrystallization has been used to describe the phenomenon. The mechanical properties and texture resulting from this process are of interest for scientists.

One factor which often distinguishes geometric dynamic recrystallization from conventional discontinuous recrystallization is the crystallographic texture. Textures resulting from discontinuous recrystallization are often quite different from the deformation textures. However, during geometric dynamic recrystallization, there is little high angle boundary migration, and the texture remains largely unchanged (Rollett, Humphreys, Rohrer, & Hatherly, 2004).

THE CONDITIONS FOR GEOMETRIC DYNAMIC RECRYSTALLIZATION

Dynamic recrystallization is basically the recrystallization that occurs during deformation. The occurrence of geometric dynamic recrystallization will depend on both the original grain size (D_0) of the material and on the deformation conditions. If we assume that the condition for geometric dynamic recrystallization is that grain impingement occurs when the sub grain size (D) becomes equal to the grain thickness (H), then, using equation 7.1, the critical compressive strain () for the process is:

$$_{cr} = \ln\left(\frac{K_1 D_0}{D}\right) \qquad (7.3)$$

Where, K_1 is a constant of the order of unity. From the relationship between flow stress and sub grain size, equation 7.3 could be written as follows:

$$_{CR} = \ln(\ D_{\circ}) + K_2 \qquad (7.4)$$

And also, can be written by:

$$_{CR} = \ln\left(Z^{1/m} D_{\circ}\right) + K_3 \qquad (7.5)$$

When the specimen is deformed at high stresses (large Z), geometric dynamic recrystallization does not occur until large strains because the sub grain size is small, whereas at low stresses, geometric dynamic recrystallization occurs at smaller strains. The critical strain for geometric dynamic recrystallization is also seen from equation 7.3 to be reduced if the original grain size is small. As such, the grain size decreases as the applied stress is increased.

It should be noted that the condition for geometric dynamic recrystallization may be derived either from geometric conditions as above, or from the theory of the stability of cellular microstructures (Rollett, Humphreys, Rohrer, & Hatherly, 2004).

THE GRAIN SIZE RESULTING FROM GEOMETRIC DYNAMIC RECRYSTALLIZATION

The considerations above predict that geometric dynamic recrystallization should occur when the high angle boundary separation is similar to the sub grain size, and that this should result in a grain size which is approximately equal to the sub grain size. The question arises as to what happens if deformation is continued to larger strains, so that the HAGB spacing might become smaller than the sub grain size. In fact, it has been shown that this does not happen, even at very large strains, as seen in Figure 7.12, in which the HAGB separation, the sub grain size and the HAGB spacing given by equation 7.1 are shown.

It is seen that although the HAGB spacing initially decreases more rapidly than the theoretical value, when it reaches the sub grain size it decreases no further. This is because of dynamic grain growth. When the grain spacing is larger than the sub grain size, e.g. Figure 7.10a, large scale HAGB migration is prevented by the pinning effects of the substructure. However, if the HAGB spacing is smaller than the sub grain size, no such pinning is exerted, and the elongated grains will tend to spheroidize as shown schematically in Figure 7.9c, under the influence of their boundary tensions by lateral migration of the Y-junctions (A). As the grain spacing becomes similar to the sub grain size, substructure pinning will reoccur, and in this way a constant HAGB separation is achieved by dynamic equilibrium. It should also be noted that a similar process of dynamic grain growth under conditions where the grain size is similar to the sub grain size, is of importance during superplastic deformation, although in this situation the grain size is stabilized by a dispersion of second-phase particles (Rollett, Humphreys, Rohrer, & Hatherly, 2004).

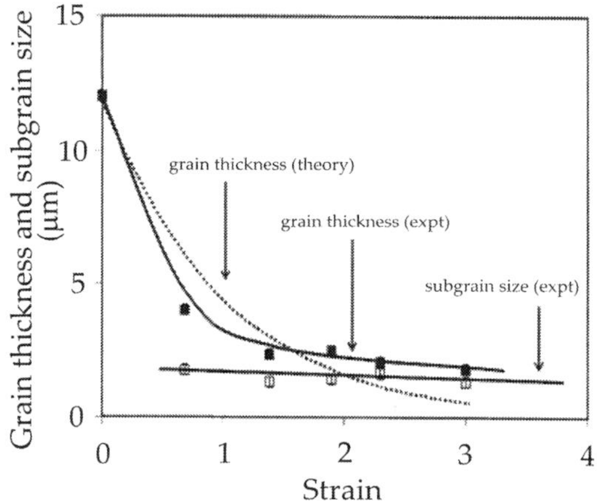

Figure 7.12: Effect of strain on HAGB spacing and sub grain size in Al–3Mg–0.2Fe alloy deformed at 350°C in plane strain compression (Rollett, Humphreys, Rohrer, & Hatherly, 2004).

THE STABILITY OF MICRON-GRAINED MICRO-STRUCTURES AGAINST GRAIN GROWTH

Although sub-micron-grained alloys may have impressive mechanical properties, they cannot be used at elevated temperatures unless the microstructures are stable. Here we are only concerned with the behavior of very fine-grained materials.

Single-phase alloys

Single-phase fine-grain microstructures have a large stored energy, due to the large area of grain boundary, and during high temperature annealing they are very susceptible to grain growth. Hayes et al. (2002) have measured the growth at 250°C of 0.5 mm grains produced in an Al–3%Mg alloy by severe deformation, and found normal grain growth, with an exponent of 2.6. As may be seen from Figure 7.13, the rate of grain growth at this temperature is very rapid. Concisely scientists proved that high temperature stability of the fine-grained microstructures produced by thermo-mechanical processing will be very dependent on the method of deformation, because of the effect that the processing route has on the texture.

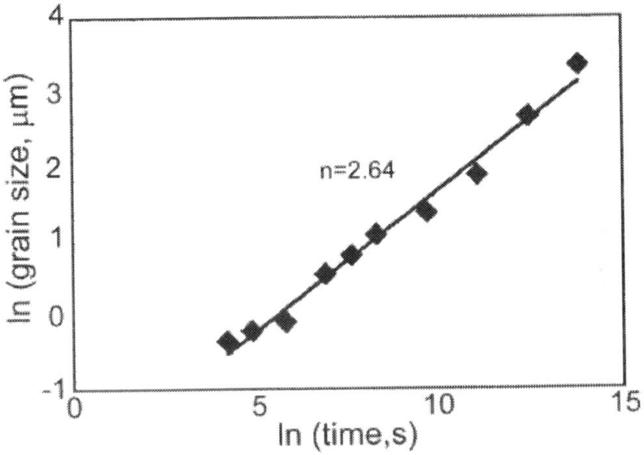

Figure 7.13: Annealing of a single-phase 0.5 mm-grained Al–3%Mg alloy at 250°C, showing normal grain growth (Rollett, Humphreys, Rohrer, & Hatherly, 2004).

Two-phase alloys

The most effective method of preventing normal or abnormal grain growth of a fine grained microstructure is with a dispersion of stable second-phase particles. Figure 7.14 shows the region relevant to micron-grained alloys, and which shows the predicted influence of the pinning particles (F_V/d) and the grain size (D) on the stability of fine grain structures. The figure explains the three types of grain growth behavior. If there are no or few second-phase particles, it is seen that normal grain growth is predicted to occur during high temperature annealing as discussed above. However, as F_V/d increase, normal grain growth will be prevented, but abnormal grain growth is increasingly likely for smaller grain sizes.

In an alloy such as the Al–Fe–Mn alloy discussed above, in which $F_V/d \sim 0.1\ \mu m^{-1}$, the 0.5 µm diameter grains are predicted to undergo abnormal grain growth after a small amount of normal grain growth is found. In order to maintain a very small grain structure at high temperatures, sufficient particles must be present to prevent abnormal grain growth, and this requires $F_V/d > 1.5\ \mu m^{-1}$ for 0.5 μm diameter grains. This represents very significant amounts of second-phase particles, e.g. a volume fraction of 0.1 of 60 nm particles, which may be compared with the level of $F_V/d \sim 0.1\ \mu m^{-1}$ which is typically required to prevent discontinuous recrystallization during conventional processing, and is rarely achievable in conventional industrial alloys. However, interest in the superplastic deformation of micron-grained alloys has stimulated the development of special alloys containing sufficiently large quantities of stable intermetallic particles to maintain microstructural stability at elevated temperatures (Rollett, Humphreys, Rohrer, & Hatherly, 2004).

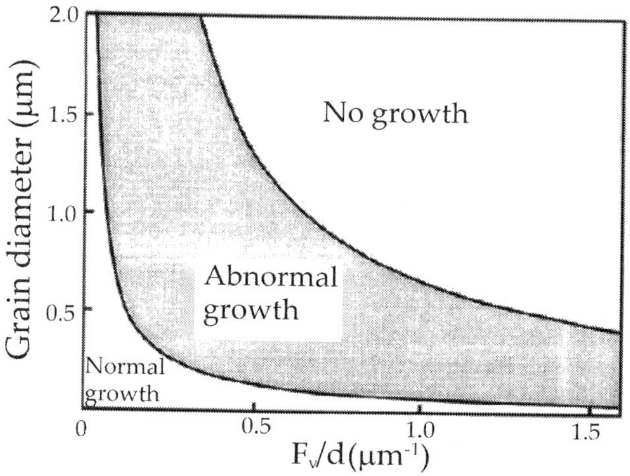

Figure 7.14: The effect of second-phase particles on normal and abnormal grain growth of micron-grained alloys (Rollett, Humphreys, Rohrer, & Hatherly, 2004).

CHAPTER 8

RECRYSTALLIZATION TEXTURES

INTRODUCTION

The recrystallization textures that develop when deformed metals are annealed have been the subject of extensive study. There are two reasons for this: first, and most important practically, these textures are largely responsible for the directionality of properties observed in many finished products and second, their origin is a source of much scientific interest. With respect to the latter, and despite frequent claims to the contrary, present understanding of the origin of recrystallization textures is little more than qualitative in nature. There are many reasons for this and most of them have already been discussed in earlier chapters. They are related to the nature of the nucleation event, the orientation dependence of the rate of nucleation in in-homogeneities of various type and orientation environment, and to the nature, energy and mobility of the boundaries between grains of various orientations. The determination of micro-textures by EBSD which enables a correlation to be made between local orientation and micro-structure is now enabling some of these questions to be answered (Rollett, Humphreys, Rohrer, & Hatherly, 2004).

There are several important matters should be taken in consideration when discussing texture evolution as follow:

- The effect of subsequent grain growth. The development of an annealing texture does not cease when recrystallization is complete. It continues during grain growth and the final texture is not necessarily representative of the texture present when primary recrystallization is complete. Much of the immense amount of material available in the literature refers to textures that have been altered by growth.
- The effect of strain. Much of the data, and particularly that for fcc metals, refers to the annealing of very heavily rolled materials, e.g. 95–99% reduction by rolling. Such extreme deformation is not generally typical of industrial practice.
- The starting texture. Although a great deal of attention has been given to the relationship between the deformation texture and the recrystallization texture, there is increasing evidence that the orientation of the material before deformation plays a significant role in determining the recrystallization texture.
- The effect of purity. The purity of a particular metal may strongly influence the recrystallization texture. In many cases very small amounts of second elements may, by affecting the boundary mobility, change the annealing texture completely, e.g. 0.03% phosphorus in copper (Rollett, Humphreys, Rohrer, & Hatherly, 2004).

THE NATURE OF RECRYSTALLIZATION TEXTURES

In this section we describe the nature of the textures formed on annealing, and compare them with the deformation textures.

Recrystallization textures in fcc metals

The range of recrystallization textures seen in fcc metals and alloys is much wider and more complex than is the case for the corresponding deformation textures, and in many cases the origin of the textures is still a matter of debate.

Copper and its alloys

The rolling texture of pure copper consists basically of a tube in orientation space which includes as major components the orientations {123}<634>(S) and {112}<111> (Copper), together with a less significant {110}<112> (Brass) component. In highly alloyed metals of low stacking fault energy such as 70:30 α-brass, the Brass orientation is dominant and there is additionally a less important {110}<001> (Goss) component. The intermediate alloys have textures in which there is a more or less gradual transition from one extreme to the other. A similar, simple transition is not found in the recrystallized metals.

In the case of heavily rolled copper, and other fcc metals and alloys of medium to high stacking fault energy, the recrystallization texture consists of a strong cube texture {001}<100> (Figure 8.1). The strong cube texture resembles a single crystal with subgrains which may contain a minor amount of material with different orientation. If the stacking fault energy is sufficiently low, minor components that correspond to the twins of this orientation are also present. The strong cube texture of copper is eliminated by quite small amounts of many alloying elements, e.g. Al 5%, Be 1%, Cd 0.2%, Mg 1.5%, Ni 4.2%, P 0.03%, Sb 0.3%, Sn 1%, Zn 4%. It will be recognized that for the most part these alloying elements reduce the stacking fault energy and change the nature of the deformation texture. A similar change in deformation texture occurs if copper is rolled at low temperatures (-196˚C) and as would be expected, the cube texture is then absent after annealing (Rollett, Humphreys, Rohrer, & Hatherly, 2004).

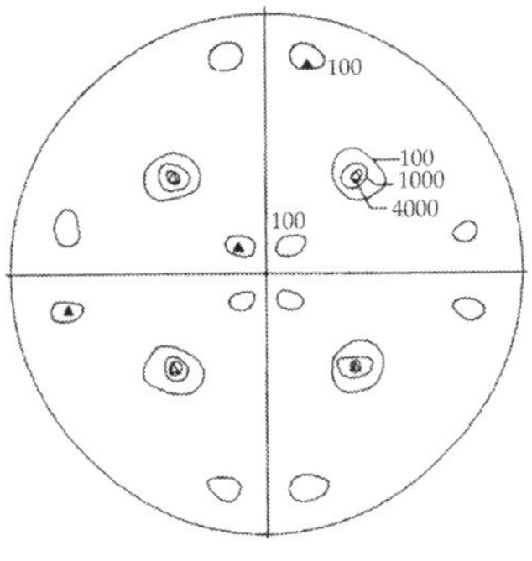

Figure 8.1: 111 Pole figure of 97% cold rolled copper, annealed at 200˚C (intensity in arbitrary units). Open symbols indicate the ideal 'cube' orientation

{001} <100> and filled symbols {122} <212> the 'twin' orientation (Rollett, Humphreys, Rohrer, & Hatherly, 2004).

LOW STACKING FAULT ENERGY COPPER ALLOYS

There have been many investigations of copper–zinc alloys. The recrystallization texture of 70:30 brass (figure 8.2) is usually reported to be {236} <385>, but it will be shown later that this can be significantly altered by variation of the annealing temperature. The strong cube texture of copper is lost rapidly as the alloy content increases and in the 3% zinc alloy is very weak. Similarly the {236} <385> texture is not a major component in alloys containing less than 10% zinc. In the intermediate range of 3–15%Zn a series of complex orientations is developed. These annealing textures have been investigated extensively by many scientists. Their results for this and several other alloy systems are summarized in figure 8.3. It is found that stacking fault energy is the dominant parameter and if the results are normalized with respect to stacking fault energy then consistent behavior is observed for a range of alloys. In figure 8.3 the texture components are plotted as a function of the reduced stacking fault energy, .

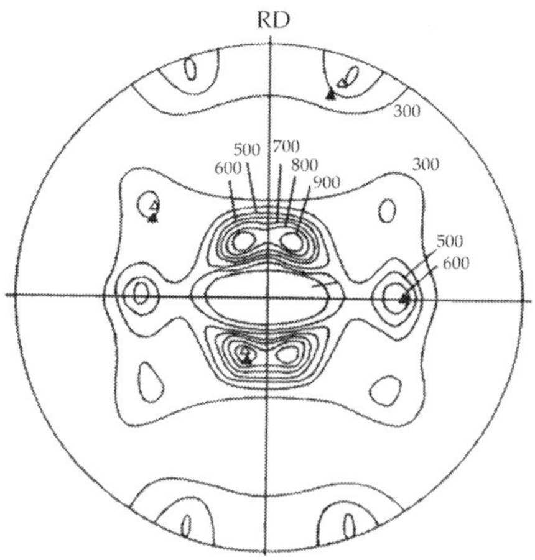

Figure 8.2: 111 Pole figure of 95% cold rolled 70:30 brass, annealed at 340°C, (intensity in arbitrary units). Opens symbols indicate ideal {225}<374> and filled symbols {113}<211> orientations (Rollett, Humphreys, Rohrer, & Hatherly, 2004).

There is a marked change in the nature of the recrystallization texture at ~3.10^{-3} corresponding to ~8%Zn, 3%Al or 2%Ge. At about this value the {236}<385> component begins to develop in the alloys and thereafter rises rapidly to a near maximum value at ~2.2x10^{-3}. At lower values, the cube texture is the strongest component although its strength is significant only for low alloy contents. Typical Orientation Distribution Functions (ODFs) for the system Cu–Zn are given in Figure 8.4, and the volume fractions of the various texture components are shown in figure 8.5. The Euler angles for these and other recrystallization texture components are given in Table 12.1.

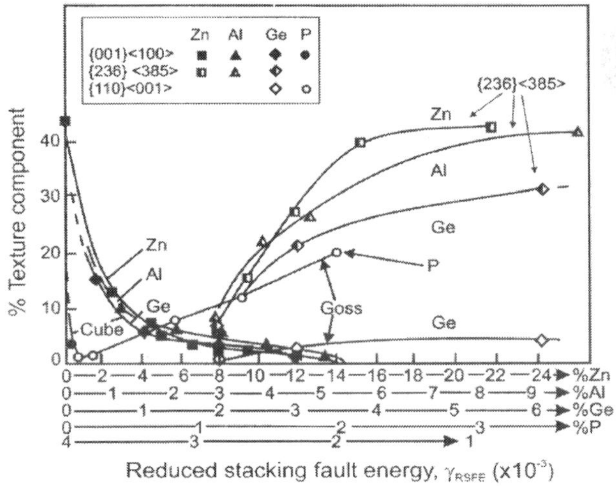

Figure 8.3: Volume fraction of major texture components ({100}<001>; {236}<385>; {110}<001>) in recrystallized copper based alloys, plotted as a function of the reduced stacking fault energy, (Rollett, Humphreys, Rohrer, & Hatherly, 2004).

Table 8.1: Components of recrystallization textures in fcc metals

Notation	Miller indices	Euler angles		
		φ_1	Φ	φ_2
Cube	{001} < 100 >	0	0	0
-	{236} < 385 >	79	31	33
Goss (G)	{011} < 100 >	0	45	0
S	{123} < 634 >	59	37	63
P	{011} < 122 >	70	45	0
Q	{013} < 231 >	58	18	0
R (aluminium)	{124} < 211 >	57	29	63

**rewrite this as a true table

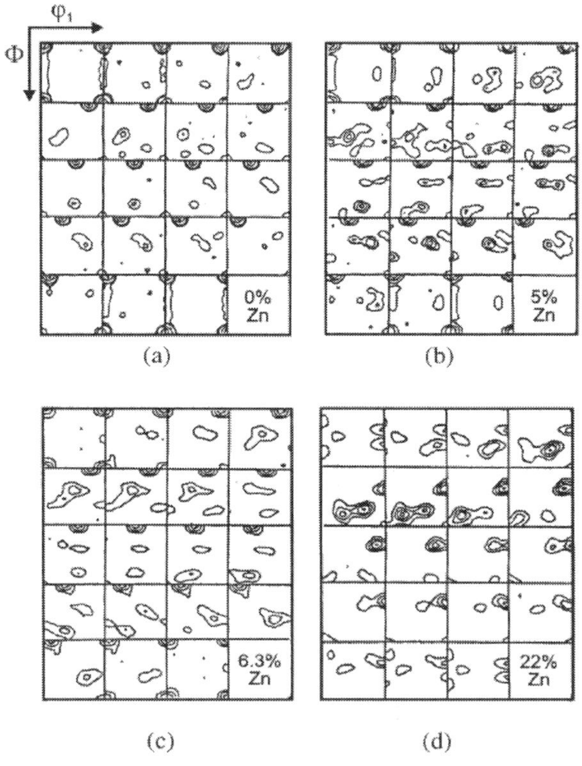

Figure 8.4: Recrystallization textures of rolled and annealed copper–zinc alloys. (a) Cu, (b) 5%Zn, (c) 6.3%Zn, (d) 22%Zn (Rollett, Humphreys, Rohrer, & Hatherly, 2004).

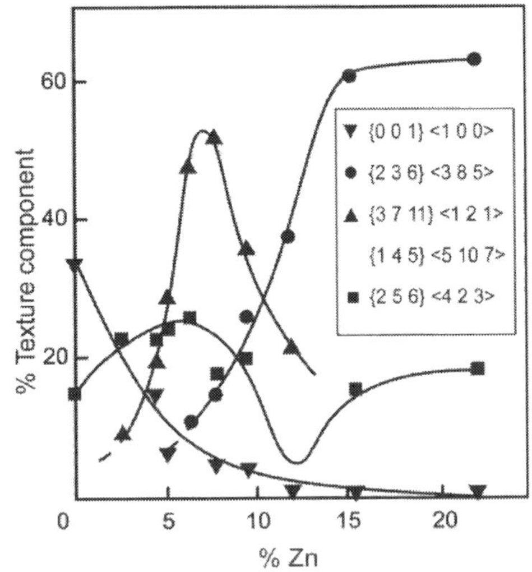

Figure 8.5. Volume percentage of texture components in recrystallized copper–zinc alloys (Rollett, Humphreys, Rohrer, & Hatherly, 2004).

COPPER–MANGANESE ALLOYS

Unlike most elements, the addition of manganese to copper does not lower the stacking fault energy and, increasing manganese content leads to increasing Brass and Goss deformation textures. On recrystallization, Engler (2001a) has shown that the cube texture disappears as the Mn level increases and α-fibre components such as Goss and {236}<385> dominate. However, the textures are rather weak except for very large rolling reductions.

Effect of processing variables

There have been many investigations of the effects of microstructure and processing variables such as prior grain size and texture, strain, time, temperature of annealing and rate of heating on these basic textures. It is not possible to review all of these and only a few specific examples are discussed. In the case of the cube texture it has been known for more than fifty years that this texture is reduced by:

- small rolling reductions (50%) with intermediate annealing,

- a large grain size at the beginning of the penultimate rolling stage, and
- a low final annealing temperature.

An example of the effects of prior grain size, rolling reduction and annealing temperature in 70:30 brass is found in the work of Duggan and Lee (1988). After annealing for 1 hr at 300°C (low temperature) the texture of material cold rolled 92% changed from the normal {236}<385> (Figure 8.6a) to {110}<110> (Figure 8.6b) as the prior grain size increased, but the magnitude of the change was strain dependent and non-existent at 97% reduction. Annealing at high temperatures (600_C) led to a {113}<332> texture for small grain sized material and high reductions (Figure 8.6c). After low reductions the {110}<112> and more general {110}<hkl> orientations were produced. The {110}<112> texture was present to higher strain levels as the grain size increased. It seems probable that the high temperature {113}<332> texture involved grain growth, and in the 97% rolled material, where grain growth was limited by the specimen thickness and the short annealing times, the texture was {236}<385>. For all grain sizes investigated (30–3000 mm) the texture of heavily rolled material was {236}<385> after annealing at 300°C, and {113}<332> for 600°C annealing. These results illustrate the importance of grain growth effects on annealing textures.

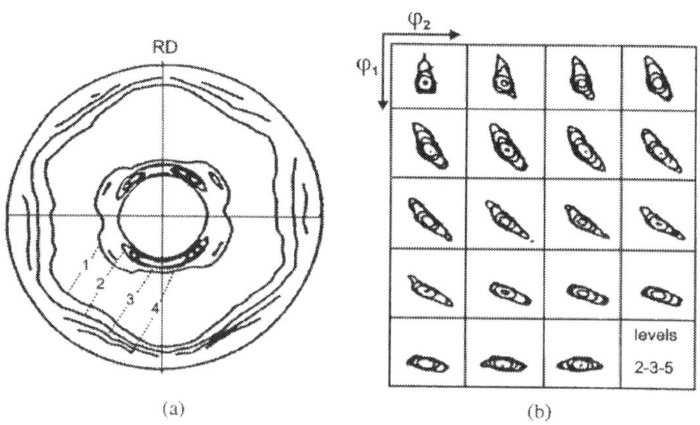

Figure 8.6: Recrystallization textures of interstitial-free steel after 80% prior reduction by cold rolling. (a) 110 pole figure, (b) ODF (Rollett, Humphreys, Rohrer, & Hatherly, 2004).

ODF data from Yeung et al. (1988) have confirmed the importance of annealing temperature. Specimens of 70:30 brass of 100 mm grain size, were cold rolled and annealed in a salt bath at 300°C and 900°C. The volume fractions of some of the main texture components measured from the ODF data are summarized in Table 8.2. Nucleation was found to occur in all of the various types of deformation in-homogeneities present in the microstructure after annealing at 900°C, and for 95% rolled material the volume fraction of the {236}<385> orientation was reduced from 0.35 at 300°C to ~0.1 at 900°C. The major components after the high temperature annealing were characteristic of the rolling texture, i.e. 32% by volume of a component with orientation between B and S and 16.8% of one that was developed by an ND rotation of the cube component. The changes were attributed to the absence of recovery during the rapid heating to the high annealing temperature, which resulted in recrystallization nucleating uniformly over all components of the deformation microstructure. When coupled with the rapid boundary migration rate associated with high temperature annealing, the retention of rolling texture components is to be expected (Rollett, Humphreys, Rohrer, & Hatherly, 2004).

Table 8.2. Volume percentages of some texture components in recrystallized 70:30 brass after the rolling reductions shown and subsequent annealing (Rollett, Humphreys, Rohrer, & Hatherly, 2004)

Component	Volume % of component after annealing					
	T = 300°C			T = 900°C		
	cr. 80%	cr. 87%	cr. 95%	cr. 80%	cr. 87%	cr. 95%
{001} < 100 >	2.4	3.1		0.5	3.3	
Brass–Goss					24.0	5.9
{236} < 385 >			35.2			9.4
ND rotated cube						16.8
Brass–S						31.8
Others	20.0	34.7	20.9	54.3	29.0	7.1
Random	77.6	62.2	43.9	45.2	43.7	29.0

**rewrite this table as a true Table..

Table 12.2 illustrates well, a little recognized aspect of recrystallization textures. It is commonly asserted that such textures are sharp and well defined, particularly after large strains. However, as shown in the table, the volume fraction of random and other minor components is high in all cases.

Fiber textures. The recrystallization textures of copper-based wires, and other products which have fiber textures, are normally similar to the corresponding deformation textures. In the case of copper, low temperature annealing of wires leads to fiber textures based on <100> or <112>, but high temperature annealing leads to mixed <111> to <112> type textures.

Single-phase aluminum alloys

When annealed after cold rolling, high purity aluminum develops the characteristic cube texture found in materials of high to medium stacking fault energy. As in the case of copper, the texture is strengthened by high rolling reductions and high annealing temperatures.

The presence of magnesium in solid solution reduces recovery, increases the strength of the alloy and leads to the formation of shear bands during cold rolling. These lead to a reduction in the strength of the normally dominant cube texture for two reasons.

The shear bands cut through the elongated cube bands which are responsible for the cube-oriented recrystallized grains and therefore suppress the formation of the cube texture.

Recrystallization nucleates in the shear bands, and texture components such as {011}<100> (Goss), {011}<122> (P) and {013}<231> (Q). The overall textures may be quite weak, suggesting that grains of more random orientation are also generated within the highly deformed shear bands.

In Al–3%Mg, the recrystallization texture is strongly affected by the amount of deformation as shown in figure 8.7. After a cold rolling reduction of 90%, the cube texture remains dominant after recrystallization (Figure 8.7a), but after a reduction of 97.5%, strong Q and Goss components are formed (Figure 8.7b). This behavior is rather similar to that of a solution treated Al–1.8%Cu alloy.

Because the small amounts of elements such as iron and silicon, which are present in aluminum of commercial purity, result in the formation of second-phase particles either before or during the recrystallization anneal, almost all commercial aluminum alloys are multiphase, and the amount of research on genuinely single-phase aluminum alloys is rather limited.

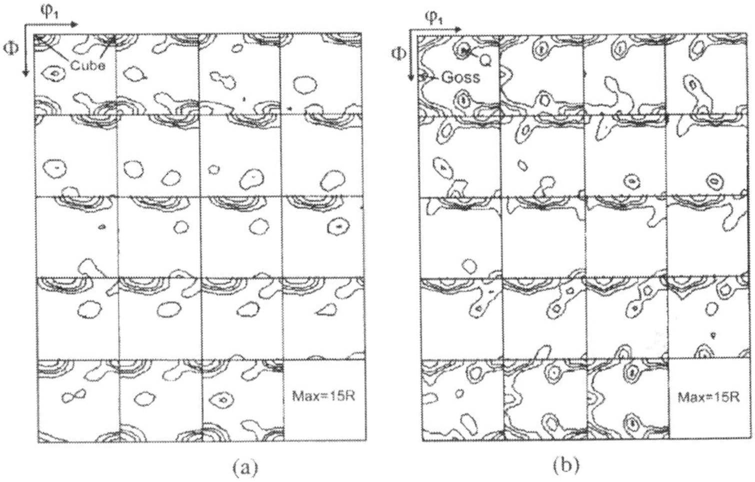

Figure 8.7: The effect of cold rolling reduction on the recrystallization textures of coarse-grained Al–3%Mg: (a) prior reduction 90%, (b) prior reduction 97.5% (Rollett, Humphreys, Rohrer, & Hatherly, 2004).

RECRYSTALLIZATION TEXTURES IN BCC METALS

Because annealed low carbon steel sheet is a product of major industrial importance the optimization of properties is of continuing interest. Emren et al. (1986) have made a detailed study of texture development in vacuum degassed, aluminum-killed and interstitial-free steels and in Armco iron. The extent of rolling, the time and temperature of annealing and the heating rate were all varied. The 110 pole figures for each material after 80% rolling and annealing at 700°C were basically similar, and that for interstitial-free steel is shown in Figure 8.8a; the corresponding ODF is shown in figure 8.8b. It is clear that the recrystallization texture is basically similar to the rolling texture and that the recrystallization texture can still be described by reference to the fibers used earlier for the rolled state, viz:

- The α-fiber, a partial fiber texture with a <110> fiber axis parallel to the rolling direction and intensity maxima at {001}<110>, {112}<110> and {111}<110>.

- The γ-fiber, a fairly complete fiber texture with the <111> fiber axis parallel to the sheet normal; the major components in this fiber texture have <110>, <112> and<123> aligned with the rolling direction.

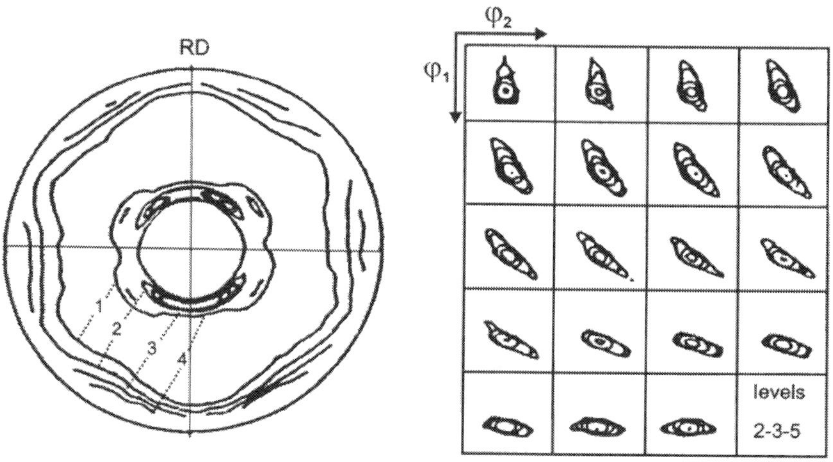

(a) (b)

Figure 8.8: Recrystallization textures of interstitial-free steel after 80% prior reduction by cold rolling. (a) 110 pole figure, (b) ODF (Rollett, Humphreys, Rohrer, & Hatherly, 2004).

A typical recrystallization texture plotted in the φ_2=45° section is shown in figure 8.9. Comparison with the corresponding deformation textures of figure 8.10 shows that the recrystallization orientations are generally close to the α and γ fibers, but that much of the α-fiber has been eliminated on recrystallization, especially in the range (001) to (112) . The γ-fiber is relatively unchanged.

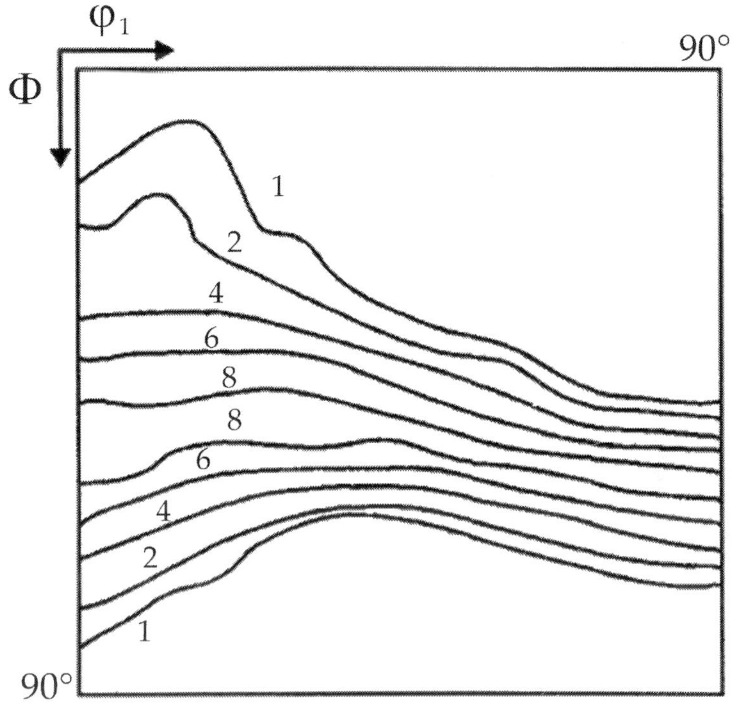

Figure 8.9: φ_2=45° section of ODF for recrystallized Interstitial-Free (IF)steel (Rollett, Humphreys, Rohrer, & Hatherly, 2004).

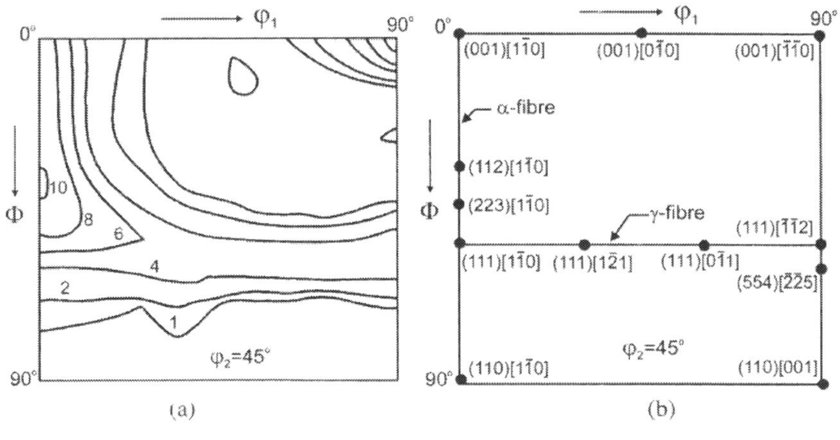

Figure 8.10: φ_2=45° section of ODF for cold rolled IF steel; (a) experimental data, (b) location of important texture components (Rollett, Humphreys, Rohrer, & Hatherly, 2004).

Examination of the development of texture during annealing at 700°C of a vacuum degassed steel (figure 8.11), shows a loss of orientations near {112} <110> and {001} <110> and an increase in the strength of the {111} <112> component. The {111} <110> component remains constant.

There is a significant difference in the nature of the recrystallization textures for bcc and fcc metals. The rolling textures for both can best be described by prominent fibers and in bcc metals these fibers are largely retained in the recrystallization texture. In contrast, the recrystallization textures of rolled fcc metals are better described by individual peak type components which are not always prominent in the deformation texture.

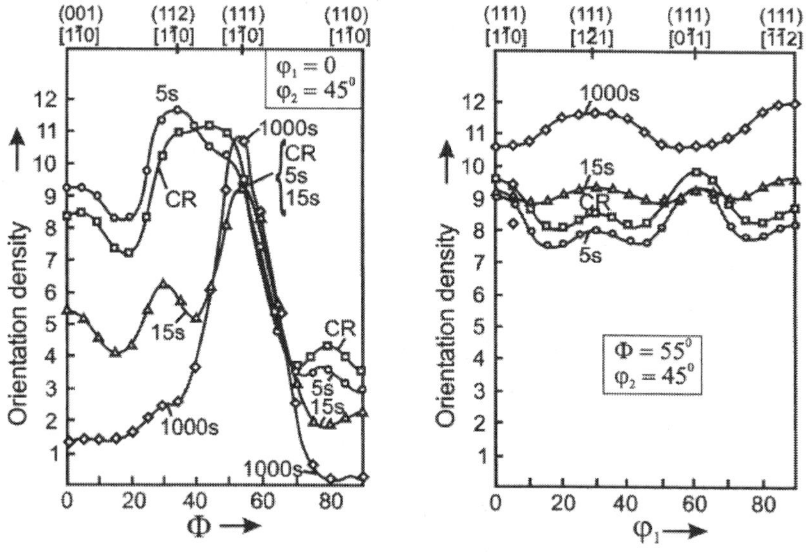

Figure 8.11: Orientation density f(g) along α- and γ fibers for different stages of recrystallization in 85%, cold rolled steel after annealing at 700°C. (a) Along α-fiber, (b) Along γ-fiber (Rollett, Humphreys, Rohrer, & Hatherly, 2004)

RECRYSTALLIZATION TEXTURES IN HEXAGONAL METALS

The early studies of recrystallization textures in hexagonal metals reported that, in general, the rolling textures were retained with some redistribution of intensity and this is largely true of the rather limited number of more recent studies. Orientation distribution function results from Inoue and Inakazu (1988) for rolled and recrystallized titanium showed that the major component after annealing within the α-phase field was {0225}<2110> irrespective of the extent of rolling or the annealing temperature. This orientation is related to the major component of the rolling texture, (1214) [] by a 30° [0001] rotation. However, the intensity increased considerably as the reduction or the annealing temperature was increased. When annealing was carried out at 1000°C, i.e. within the β-phase field, the texture was modified by variant selection associated with the phase transformation. Figure 8.12 shows selected sections of the ODFs for material recrystallized in the α and β temperature ranges (Rollett, Humphreys, Rohrer, & Hatherly, 2004).

RECRYSTALLIZATION TEXTURES IN TWO-PHASE ALLOYS

In addition to the size and spacing of the particles, the recrystallization textures that form when multi-phase alloys are annealed depend particularly on whether the second phase is present during deformation, or whether precipitation occurs on annealing. The particles may affect the recrystallization texture in two ways:

- Large (>1 μm) pre-existing particles are favored nucleation sites during recrystallization and the highly miss-oriented deformation zones associated with them lead to a wide range of nucleus orientations.
- A dispersion of closely spaced particles may pin boundaries, affecting not only the recrystallization kinetics, but also the final grain size and texture.

Table 8.3 illustrates collated data on the recrystallization textures of a number of alloys, and some selected examples which will be used to illustrate the effects of particles on texture.

Table 8.3: Major components of the recrystallization texture for some particle-containing alloys (Rollett, Humphreys, Rohrer, & Hatherly, 2004)

Material	Rolling reduction	Particles		Recrystallization texture
		$F_V \times 100$	Diameter (μm)	
Al (99.9965)	90	0	—	Cube
Fe–AlN	70	0.06	0.017	Rolling
Al–Al_2O_3	90	0.4	0.1	Deformation
Cu–SiO_2	70	0.5	0.23	Rolling + twins
Al–FeSi	90	0.5	0.2–7	Cube, rolling, random
Al–Si	90	0.8	2	Cube, rolling, random
Al–Ni	80	10.0	1	Rolling, random
Al–SiC	70	20.0	10	Random

** rewrite this as a true table

THE THEORY OF RECRYSTALLIZATION TEXTURES

Historical background

The nature of the recrystallization texture is determined primarily by two factors:

- The orientations of the new grains.
- The relative nucleation and growth rates of these grains.

The manner in which these factors operate to produce a unique recrystallization texture from a particular deformation texture and microstructure has been the subject of controversy for more than 50 years and for the whole of that time two major theories have been strongly advocated. It has been claimed that the recrystallization texture has its origin in either the preferred nucleation of grains with a particular orientation (oriented nucleation theory - ON) or the preferred growth of grains of specific orientations from a more randomly oriented array of nuclei (oriented growth theory - OG). In the ON model, it has been argued that a higher frequency of the special orientation randomly occurs, while the OG model argues that specially oriented grains have high mobility boundary and thus can migrate faster and grow to a larger size as compared to random orientations- both phenomenon contributes to the final recrsytallization texture. Most of the experimental work related to the controversy has been concerned with fcc metals and the early work was mainly carried out on copper and its single-phase alloys. The importance of the cube texture in the thermo-mechanical processing of commercial, multiphase aluminum alloys has resulted in an increased activity in this area in recent years (Rollett, Humphreys, Rohrer, & Hatherly, 2004).

It was pointed out that the nuclei from which the recrystallized grains originate are small regions which pre-exist in the deformed state, i.e. the orientations of the nuclei are already present in the deformed structure. With the exception of changes that may be introduced by twinning there is no way that other significantly different orientations can develop. To the early workers faced with a recrystallization texture that was apparently quite different to the deformation texture from which it developed this presented a major problem, and several models were proposed to explain how a new orientation might be generated. Most of these are no longer considered tenable and will not be discussed here; details of this early work are reviewed by Beck and Hu (1966).

The ON theory began with the assertion by Burgers and Louwerse (1931) that the preferred nuclei in lightly compressed aluminum single crystals consisted of crystal 'fragments' that were more heavily deformed than the bulk of the crystal. A formal crystallographic theory was devised which associated the 'fragments' with local lattice curvatures and involved a <112> rotation. Subsequently, Barrett (1940) found that the recrystallized grains in a single crystal specimen were related to the deformed matrix by a 45°<111> rotation. Results of this type led to speculation that the observed rotations were associated with a maximum growth rate, and as experimental methods improved, the value of the angular rotation was refined and the oriented growth theory was established. Considerable further evidence for the high mobility of boundaries with certain orientation relationships has been obtained (Rollett, Humphreys, Rohrer, & Hatherly, 2004).

An early alternative oriented nucleation theory was based on the inverse Rowland transformation (Verbraak and Burgers 1957, Verbraak 1958). This so called 'martensitic' theory relied on a formal crystallographic analysis (the Rowland transformation) which describes a mechanism capable of deriving two twin-related lattices of type {112} <111> from a single {001} <100> lattice. The martensitic model assumes that the two neighboring twin-related {112} <111> copper components of the rolling texture jointly undergo an inverse Rowland transformation to generate an {001} <100> cube texture nucleus. The necessary shear requires the presence of <112> partial dislocations and the analysis requires therefore that the stacking fault energy be sufficiently low to allow the formation of the partials. If these are absent, as in aluminum, the process cannot occur and the development of the cube texture in that metal would require a different mechanism. For this and other reasons the theory never attracted strong support and a later attempt by Verbraak (1975) to revive interest was not successful (Rollett, Humphreys, Rohrer, & Hatherly, 2004).

ORIENTED GROWTH

There are several difficulties in accepting that oriented growth due to specific crystallographic relationships between grains is solely responsible for the final texture, and these are outlined below.

The number of active rotation axes

Only a small number of the possible rotation axes are actually active and for any particular axis, rotation generally occurs only in one sense.

Many single crystal studies have been made in this area. Consider the orientations {112} <111>, {123} <634> and {110} <112> which form the basis of all fcc rolling textures. Single crystals of the first type retain that orientation during rolling to 80% reduction, and when recrystallized, orientations derived from 6 of the 8 possible 40° <111> rotations are found. {123} <634> crystals are a little less stable during rolling, but only one of the possible rotations occurs during recrystallization. The {110} <112> orientation is very stable during rolling and all 8 rotations occur on annealing, albeit with some scattering.

A similar situation is found for polycrystalline materials as can be illustrated by considering the cube texture, {001} <100>. Oriented growth theory maintains that this orientation is related to the {123} <634> (S) orientation present in the rolling texture of copper and aluminum by a near 40°<111> rotation. However only one of the eight possible rotations of each variant of S yields the cube texture and there is at present no way of predicting which.

These results can only be interpreted on the basis that the number of nuclei available is limited, i.e. oriented growth does not occur independently of oriented nucleation.

The precision of high mobility relationships

Although definite orientation relationships between deformed and recrystallized textures are often quoted and used to support the oriented growth theory, there are several factors that suggest that precise relationships do not exist during recrystallization.

- The spread in both the deformation and annealing textures is such that it is unrealistic to select a single orientation relationship to relate them. Figure 8.12 shows the orientations of recrystallized grains artificially nucleated in aluminum single crystal. The orientation of the deformed crystal is plotted in standard projection. It was asserted that the results demonstrated a 40°[] relationship with the deformed crystal although 'in most cases the scatter....around the ideal orientation was rather large'. The common [] directions are in fact up to 16° apart and [] for the annealed grains is spread over an angle of approximately 28°. Although there is clearly an orientation relationship based at or near <111>, involving a rotation of ~40°, the scatter is considerable (Rollett, Humphreys, Rohrer, & Hatherly, 2004).

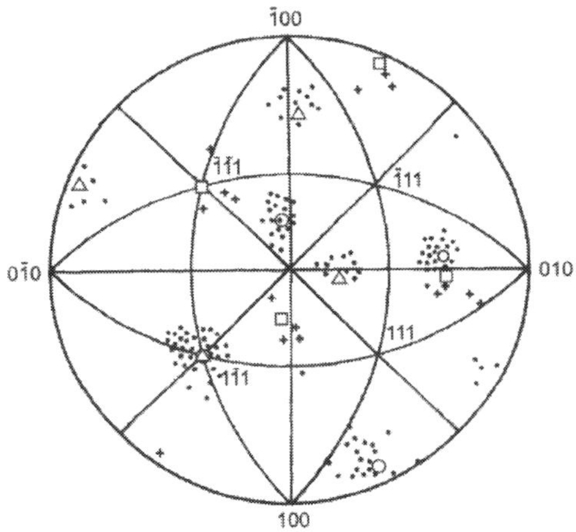

Figure 8.12: Orientation relationships between artificially nucleated

grains and the deformed matrix of aluminum crystals (Rollett, Humphreys, Rohrer, & Hatherly, 2004).

- Experiments on single crystals show that the high mobility of 40°<111> boundaries is restricted to tilt boundaries, and that 40°<111> twist boundaries do not have a high mobility. Thus, only a restricted set of these boundaries are likely to have high mobility. It should also be noted that the recrystallized grains in poly crystals are usually equiaxed, showing little evidence of the preferential migration of tilt boundaries.
- Although the reasons for the high mobility of a low-or CSL boundary can be broadly understood in the case of boundaries migrating in perfect material, it is difficult to see how the necessary ordered boundary structure can be maintained during recrystallization.
- Because of the rapid local orientation variations in the deformed microstructure, a precise relationship between a growing grain and the adjacent deformed matrix cannot be defined. During deformation a single grain breaks up into regions of different miss-orientation as cells or sub grains and other heterogeneities develop. The orientation of a deformed grain therefore varies over small distances. Such orientation changes have been measured in both single crystals and poly crystals.

In summary, we see that although there is clear evidence of oriented growth in single crystals and lightly deformed poly crystals where an orientation relationship between the recrystallizing grain and the deformed matrix may exist and be maintained during growth, it is now clear that during the recrystallization of highly deformed poly crystals, any orientation relationships will not only be very localized, but will be transient as a grain grows into new surroundings. Whether, during the later stages of growth, the existence of favorable orientation relationships over only part of the boundary of a growing grain will lead to a significantly faster overall growth rate for the grain remains to be clarified (Rollett, Humphreys, Rohrer, & Hatherly, 2004).

Oriented growth due to other factors

Although discussions of oriented growth have generally stressed the role of high mobility boundaries of specific character, there are other factors which may result in grains of some orientations growing faster or slower than others, and thereby influence the recrystallization texture.

Size. Grains of different orientations tend to originate at different type of site. If the recrystallization nuclei are large, e.g. those originating at cube bands; they will tend to grow more rapidly than others because of the smaller drag pressure from boundary curvature. This will be particularly important in a fine-grained, recrystallized microstructure.

Position. Recrystallizing grains adjacent to regions of high stored energy will recrystallize more rapidly than others. As stored energy is known to be orientation dependent, this can lead to the development of preferred orientations during recrystallization.

More general orientation relationships. The problems associated with ascribing high mobility to a precise orientation relationship such as 40° <111> were discussed above. However, there is mounting evidence that boundaries with more general orientation relationships may have higher or lower mobility's than average, and that these may make a substantial contribution to orientation growth effects, and we note the following:

- Boundaries of miss-orientation less than ~15° have low miss-orientations, as do 60° <111> (twin) boundaries. Juul Jensen (1995b) has pointed out that the effect of such low mobility boundaries on the growth of recrystallizing grains may be as significant as those of high mobility. This type of oriented growth is sometimes called 'orientation pinning' (Rollett, Humphreys, Rohrer, & Hatherly, 2004).
- Boundaries deviating by up to ~20° from 40°<111> have higher than average mobility.
- In a randomly oriented assembly of grains, large fraction (~0.35) of the high angle boundaries have miss-orientations within 20° of 40°<111>.

As discussed by Hutchinson and Ryde (1997), on the basis of the orientation dependence of both the frequency and the mobility of boundaries noted above, and the requirement of a successful recrystallizing grain to avoid a low mobility, it is to be expected that a substantial fraction of the boundaries of recrystallizing grains will inevitably have miss-orientations of 20°–40° about axes near <111>. Although much further work is required to substantiate the above ideas, a model of recrystallization growth based on more general considerations of boundary mobility and frequency overcomes many of the objections to the conventional oriented growth theory (Rollett, Humphreys, Rohrer, & Hatherly, 2004).

Oriented nucleation

For many years, it had realized that the nucleus of a recrystallized grain must have an orientation that is close to or identical with that of the volume element in the deformed structure from which it grew has existed. Attention was refocused on this by the development of electron microscope techniques that allowed orientation determinations from very small volumes of material and by advances in the theoretical analysis of texture development during deformation. Many of the developments in this field have been stimulated by the need to find the nuclei responsible for the cube texture. Although it is universally accepted that oriented nucleation occurs, many of the details of how nuclei of particular orientations develop remain a matter of debate. Such nuclei may be produced at heterogeneities such as deformation bands, or may be formed at specific components of the normal deformation texture, as discussed below (Rollett, Humphreys, Rohrer, & Hatherly, 2004).

Oriented nucleation at deformation/transition bands

The Dillamore–Katoh model

The benchmark work was the paper by Dillamore and Katoh (1974) who calculated the rotation paths during the compression of polycrystalline iron, and showed that a transition band centered on <411> would be formed when the rotation paths of neighboring volumes diverged. They argued that this would provide a very favorable site for the nucleation and growth of a recrystallized grain and pointed out that a <411> fiber texture was indeed prominent in the recrystallization texture of compressed iron. Inokuti and Doherty (1978) found <411> oriented nuclei in transition bands after annealing compressed iron samples. A second important result came from their analysis of rolling texture development in copper. In this case it was predicted that similar transition bands would form in which the central region of the curved lattice contained the cube orientation as shown in Figure 8.13. This shows that certain crystals will approach the cube orientation by rotation around the normal direction and then subsequently diverge by rotation about the rolling direction, eventually arriving at a stable orientation outside the plane of the figure. A small initial curvature, indicated by A-B-A will rapidly become larger, but the center point B, linking the two diverging orientations will continue to rotate into and remain at the cube orientation.

If nucleation were to occur at this site, which, because of its large orientation gradient is likely, then the cube texture should be found. The first experimental support for this was provided by Ridha and Hutchinson (1982) who found long narrow regions of precisely the predicted type in cold rolled copper. This observation has been confirmed many times since in both copper and aluminum by SAD and EBSD experiments. Somewhat related observations have also been made by the use of an orientation-sensitive deep etching technique and a typical microstructure is shown in Figure 8.14.

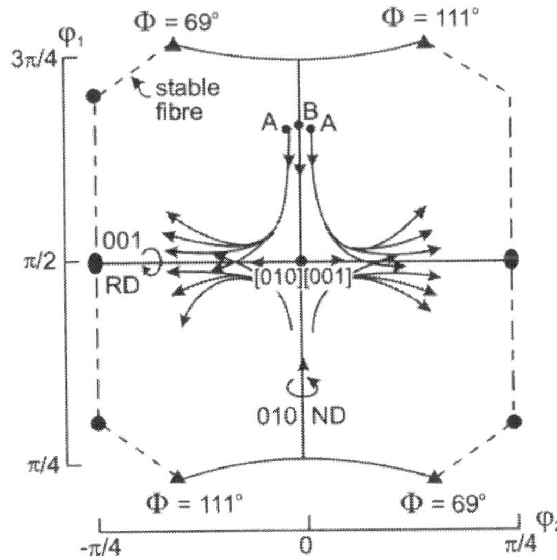

Figure 8.13: Section through orientation space at $\varphi = 90°$ showing schematically the slip rotations in the vicinity of the cube orientation in fcc metals (Rollett, Humphreys, Rohrer, & Hatherly, 2004)

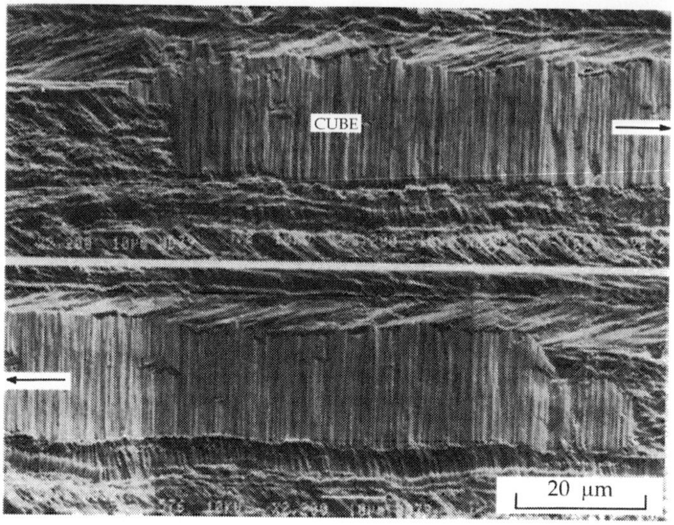

Figure 8.14: Long, thin, cube oriented-grain developing in 95% cold rolled copper after annealing for 100 sec at 140°C. Longitudinal section, cut at 45° to RD to expose cube oriented grains as {110} <001> (Rollett, Humphreys, Rohrer, & Hatherly, 2004).

CHAPTER 9

DESIGN OF CRYSTALLIZERS

INTRODUCTION

Several complex interrelating variables are responsible for regulation of the unit operation of crystallization, which can be defined as a concurrent heat and mass transfer process associated by a strong reliance on fluid and particle mechanics. It occurs in a multiphase, multicomponent system. It is attentive ***to*** particulate solids. The size and size distribution of these particulate solids are characterized by being incapable of unique definition and variable with time. The solids are suspended in a solution which has the ability of fluctuation between a named metastable equilibrium and a labile state, and also the solution composition can vary with time. Usually, profound influence on the nucleation and growth kinetics, the leading processes in this operation, can be done by simple traces of impurity in the system; a small number of parts per million may cause alteration of the crystalline product away from all recognition (Mullin, 2001).

Not surprisingly, crystallization process was once named an art rather than a science, but today to be called so, the fabulous advances that have taken place over the last three or more decades in both the theory and practice of the subject would be ignored (Mullin, 2001). Crystallization is the combination of the application of knowledge in particle formation and purification in a single operation – used to make a variety of products, from commodity chemicals to chiral pharmaceutical drugs,

For a commercial crystallizer product, factors like shape, size and its distribution, strength, and purity are important to the consumer, while the production rates and associated costs are important to the producer. The product specifications involving the production rate, mean size, and coefficient of variation are decided by a combination of process and market requirements. The crystallizer design that can meet the required product specification formulates the crystallizer type and size, along with the suitable operating conditions. In addition to these parameters, crystallizer design demands knowledge of system behavior and priority of functional duties to arrive at the best possible solution. The designer must also note the important point that crystallization by itself cannot be a final and isolated operation. It is normally an intermediate step in a process sequence. An interaction among these processes may have some influence which will not be considered here. The crystallization step may not be process controlling. Its performance will, however, have some influence on the performance of downstream operations, e.g., the separation systems (Tavare, 1995).

CRYSTAL SIZE DISTRIBUTION (CSD)

One of the earliest investigations aimed at studying the size distribution of crystals in a continuous crystallizer was made by Montillon and Badger (1927) with $Na_2SO_4 \cdot 10H_2O$ and $MgSO_4 \cdot 7H_2O$. Shortly afterwards McCabe (1929) analyzed the problem of crystal size distribution (CSD) and developed the *AL* law, making the following assumptions: (a) all crystals have the same shape; (b) they grow invariantly, i.e. both the growth rate and crystal size are self-governing; (c) super saturation is constant throughout the crystallizer; (d) no nucleation occurs; (e) no size classification occurs in the crystallizer; (f) the relative velocity between crystals and liquid remains constant. For a full account of the derivation of the AL law reference should be made to the original paper, but the main steps are briefly as follows.

The mass of one crystal of a chosen characteristic size L is given by , where is a volume shape factor and is the crystal density. The number of crystals, dN, of size L in a mass dM is thus . Assuming no nucleation, the number of seeds dN, of size L, is equal to the number of product crystals dN, of size L,, i.e.

$$\frac{dM_s}{\alpha\rho_c L_s^3} = \frac{dM_p}{\alpha\rho_c L_p^3} = \frac{dM_p}{\alpha\rho_c (L_s + \Delta L)^3}$$

Where, ΔL is the growth increment. Therefore

$$dM_p = \left(\frac{L_s + \Delta L}{L_s}\right)^3 dM_s$$

$$M_p = \int_0^{M_s} \left(1 + \frac{\Delta L}{L_s}\right)^3 dM_s \qquad (9.1)$$

Where, M_p is the product crystal yield obtained from an initial mass of seed crystals, M_s.

McCabe recognized that the ideal conditions assumed in this derivation are unlikely to be attained in a real crystallizer. Temperature and super saturation gradients are unavoidable; invariant growth is comparatively rare; different crystal faces usually grow at different rates; crystal growth rates may be dependent on solution velocity, in which case large crystals may grow faster than small; and so on. On the other hand, the derivation has an interesting feature; it does not require knowing of the dependence of growth on super saturation. In spite of its limitations, the ΔL law still offers an interesting, if oversimplified, approach to the development of a CSD (Mullin, 2001).

One use of the law is demonstrated by the following example. If a quantity of seed crystals of known CSD is added to a crystallizer, it is possible to estimate the CSD of the final product by the following procedure.

Find the value of ΔL compatible with the product yield M_p calculated from solubility data and the mass of added seeds M_s. This involves a trial and error evaluation of equation 9.1, which may be integrated graphically (plot M_s, versus $(1 + \Delta L/L_S)^3$; the area enclosed under $M_s = 0$ and M_s = mass of seeds added is equal to M_p).

Plot the integral curve of equation 9.1 (M_p versus M_s) using the correct value of ΔL.

Plot the CSD curve, e.g. from sieve test data, of the product crystals (M_p versus L_p) remembering that $L_p = L_s + \Delta L$. For comparison purposes, this plot may be made on the same diagram as the CSD curve of the seed crystals (M_s versus L_s).

The tedious trial and error solution of equation 9.1 to find the appropriate value of ΔL, as outlined in step 1, may be simplified by the use of a nomograph (*Figure 9.1*). For example, *Table 9.1* shows a seed sieve analysis based on 100 g and gives the values of $(1 + \Delta L/L_S)^3$ for a first trial using ΔL = 100 μm. A plot of these values against M_s, as explained in step 1, gives an area $M_p = 165$. If, for example, $M_p = 1.57M_S$, then the correct value of ΔL is found from the nomograph as follows. Draw a line from ΔL = 100 μm through $(1 + \Delta L/L_S)^3 = 1.65$ and extend it to the L_s scale (at 520μm). Now draw a line from this point on the L_s scale through $(1 + \Delta L/L_S)^3 = 1.57$ and extend it to the ΔL scale. The desired value of ΔL is found to be 90 μm. Column 4 in *Table 9.1* can now be completed and a plot of M_S versus $(1 + \Delta L/L_S)^3$ yields the correct value of $M_p = 157$. It is interesting to note that the focal point on the L_s scale is near the maximum of the size distribution curve of the seeds plotted on a differential basis (between 500 and 600 μm) (Mullin, 2001).

Table 9.1: Solution of the ΔL law (Mullin, 2001)

L_s	M_s	$(1 + \Delta L/L_s)^3$	
μm	g	$\Delta L = 100$ μm	$\Delta L = 90$ μm
1400	0	1.23	1.21
1180	0.1	1.28	1.25
1000	3.0	1.33	1.30
850	15.7	1.41	1.35
710	28.7	1.49	1.43
600	54.5	1.60	1.52
500	74.1	1.74	1.64
425	87.4	1.91	1.78
355	93.7	2.12	1.97
300	97.3	2.40	2.20
250	98.3	2.78	2.52
212	99.5	3.25	2.89
180	100	3.88	3.38

**rewrite this table as a true table

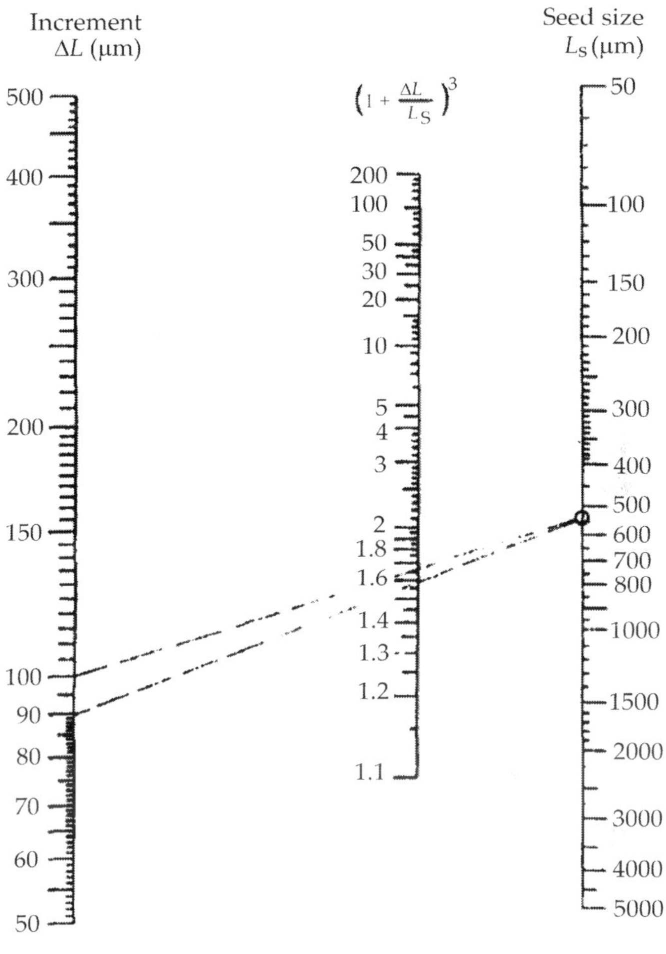

Figure 9.1: Nomograph to aid solution of the McCabe AL law (Mullin, 2001).

The postulation of the ΔL law, despite its many limitations, was a most significant step forward: it paved the way for a better understanding of the complex behavior of crystallizers.

The population balance

For a complete description of the crystal size distribution (CSD) in a continuously operated crystallizer it is necessary to quantify the nucleation and growth processes and to apply the three conservation laws of mass, energy and crystal population. The importance of the population balance, in which all particles must be accounted for, has been a central feature in the pioneering work of Randolph and Larson (1962, 1988).

The crystal population density, *n* (number of crystals per unit size per unit volume of system) is defined by

$$\lim_{\Delta L \to 0} \frac{\Delta N}{\Delta L} = \frac{dN}{dL} = n \tag{9.2}$$

Where, ΔN is the number of crystals in the size range ΔL per unit volume. The value of n depends on the value of L at which the interval dL is taken, i.e. n is a function of L (Figure 9.2a)T. he number of crystals in the size range L_1 to L_2 is thus given by

$$\Delta N = \int_{L_1}^{L_2} n \, dL \tag{9.3}$$

Application of the population balance is most easily demonstrated with reference to the case of a continuously operated MSMPR crystallizer (Figure 9.3) assuming (a) steady-state operation, (b) no crystals in the feed stream, (c) all crystals of the same shape, characterized by a chosen linear dimension L, (d) no break-down of crystals by attrition, and (e) crystal growth rate independent of crystal size.

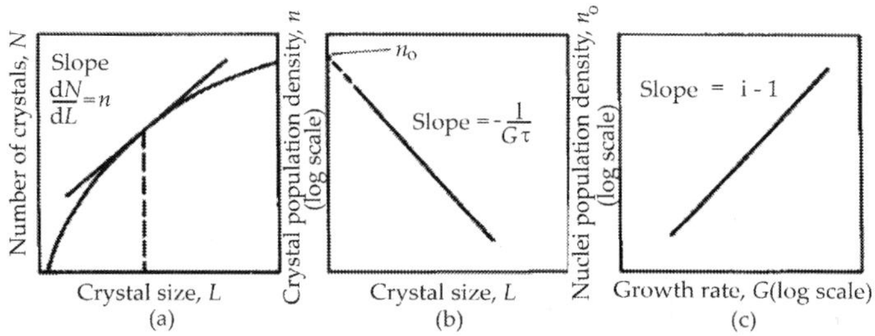

Figure 9.2: Population plots characterizing the crystal size distribution, and the nucleation and growth kinetics for a continuous MSMPR crystallizer (Mullin, 2001).

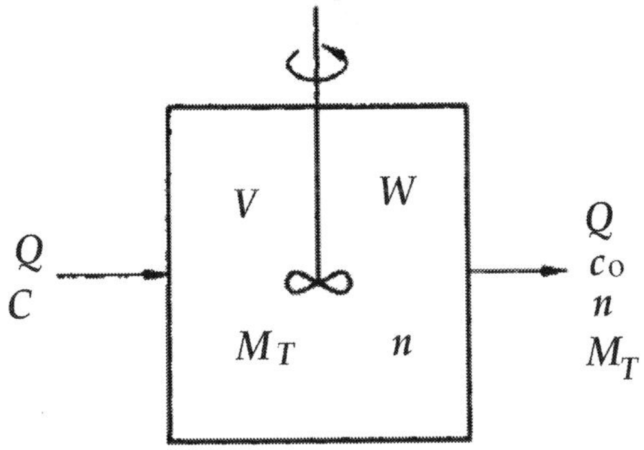

Figure 9.3: A continuous mixed-suspension, mixed-product removal (MSMPR) crystallizer (Mullin, 2001).

Alternative operating modes

Occasionally, crystallizers are operated in a connected series of well-mixed vessels, with the magma flowing from one stage to another. The most common aim of such series or 'cascade' operation is to reduce heat utilization, e.g. by dividing the overall temperature gradient over many stages and operating each stage at a lower temperature down the cascade. Additional, in cooling crystallization, encrustation problems may be considerably reduced owing to the smaller temperature drops required across the heat exchange elements.

Besides, there is the chance that cascade operation can have an effect on the product CSD, though this is a complex problem to evaluate. For instance, in a simple cascade of *k* vessels in which all act individually as basic MSMPR crystallizers operating at the same super saturation, at constant crystal growth rate, and with nucleation confined to the first stage only, the mean crystal size and product CSD leaving the kth stage may be represented by

$$L_k = \left(\frac{k+2}{3k}\right) L_1 \tag{9.4}$$

$$CV = 100/(k+3)^{1/2} \tag{9.5}$$

Equations 9.4 and 9.5 suggest that in this uncomplicated situation, which admittedly is hardly realistic from an industrial point of view, the mean crystal size and CSD would both decrease, e.g. from L_1 to 0.56 L_1 and from 50 to 41 *% CV,* at the end of a three-stage cascade, i.e. smaller crystals with a narrower size distribution (Mullin, 2001).

CLO, FD and CPR

The expectation of perfect MSMPR operation is unrealistic for industrial crystallizers, and CSD analyses based on the strict simplifying assumptions need to be modified for application to practical situations. The effect of size-dependent growth, for example, which introduces curvature into population density plots and makes their interpretation difficult, has already been discussed in the previous section. There are other conditions, such as size-dependent nucleation, crystal attrition, breakage, agglomeration, classification and so on, that can exert equally profound influences (Mullin, 2001).

It is appropriate, however, to make some comment here on a few common departures from MSMPR operation. Three basic types of continuously operated mixed-suspension crystallizer are shown in Figure 9.4. In the simple MSMPR unit (Figure 9.4a), where the outflow is of identical composition to that in the perfectly mixed vessel, the crystal and liquid residence times are equal:

$$\tau_c (= W / P) = \tau_l (= V / Q) \tag{9.6}$$

Where, Q is the feedstock rate ($m^3\ h^{-1}$) to a vessel of working volume V (m^3) containing a suspended mass of crystals W (kg). P is the crystal production rate (kg h^{-1}).

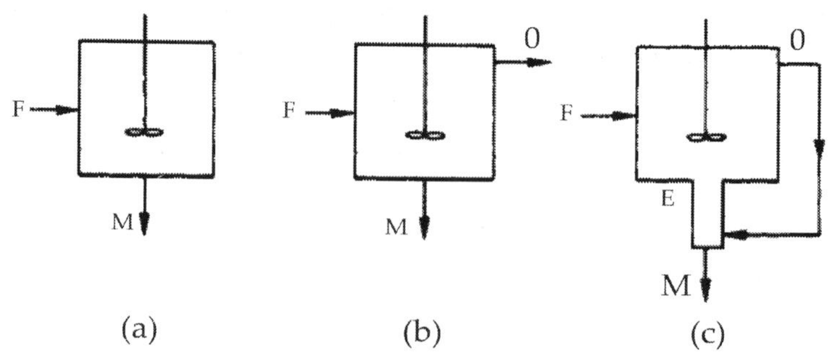

Figure 9.4: Three basic types of continuously operated mixed suspension crys-

tallizer, (a) mixed product removal (MSMPR), (b) clear liquid overflow (CLO), (c) CLO with recycle through an elutriating leg (classified product removal MSCPR). F =feed, E = elutriating leg, M = magma outlet, 0 = overflow (Mullin, 2001).

By allowing some clear liquid to overflow from the crystallizer (Figure 9.7b) the crystal residence time can be increased and rendered independent of the liquid residence time, i.e. . If required, the liquid overflow can be recycled back into the crystallizer through an elutriating leg, thus imposing some classifying action on the product crystals leaving the crystallizer (Figure 9.7c) .This mode of operation is generally referred to as mixed-suspension and classified-product removal (MSCPR) (Mullin, 2001).

Clear liquid overflow (CLO), also known as *double draw-off (DDO),* increases the residence time of crystals in the growth zone and increases the magma density. This mode of operation also allows excess fine crystals to be removed from the system. Most crystallizers suffer from excessive nucleation, and to produce reasonably large crystals the excess fines must be removed. In terms of mass, the small crystals may appear insignificant; the cumulative masses of crystals in a mixed suspension varies as the fourth power of their size so, for example, crystals up to 1/2 the mean product size constitute only $(1/2)^4$, one-sixteenth, of the total mass in suspension. In terms of numbers, however, these small crystals in a crystallizer exert a dominant effect, often outnumbering the larger crystals by > 1000 : 1.

Fines destruction (FD) can be accomplished by recirculating the liquid overflow externally through a fines trap, e.g. a hydro-cyclone or steam-heated dissolver, or internally. Figure 9.5 shows the effect of fines removal. The small-sized crystals undergo a very rapid decay up to the cut size of the fines trap (Figure 9.5a). Thereafter the decay proceeds at the rate appropriate to mixed product removal operation, and extrapolation of the population density line back to zero yields a lower *effective* density of nuclei appropriate to the crystallizer. The mean product size is increased by fines destruction (Figure 9.5b). A rough estimate of the rate of fines dissolution can be made by making the assumption that the Sherwood number = 2.

Classified product removal (CPR) is frequently adopted in crystallizer practice, e.g. with an elutriator or hydro-cyclone. As Randolph (1970) has pointed out, however, although this practice may produce a narrower CSD it also leads to a smaller product mean size (Figure 9.6). The cases of mixed and classified product removal are compared on the population density plot in Figure 9.6a and the CSD implications are shown in Figure 9.6b (Mullin, 2001).

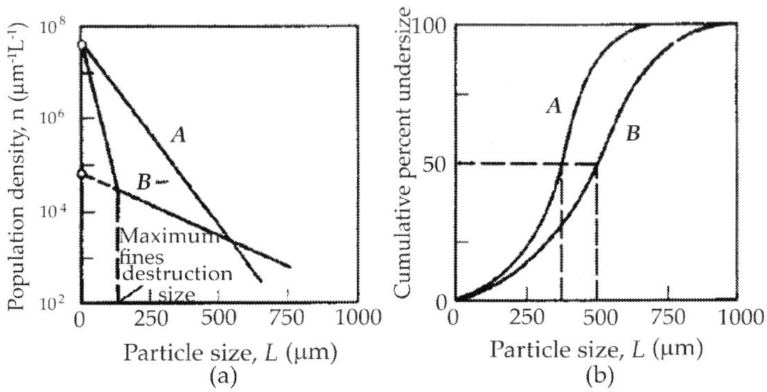

Figure 9.5: The effect of fines removal on (a) the population density and (b) the crystal size distribution. A, no fines destruction; B, fines destruction (Mullin, 2001)

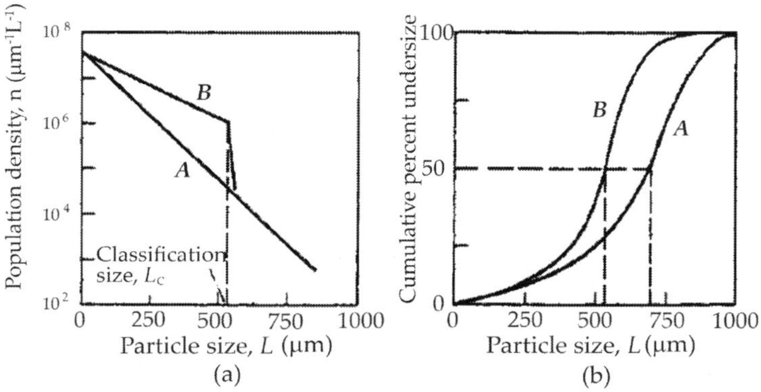

Figure 9.6: The effect of classified product removal on (a) the population density and (b) the crystal size distribution. A, mixed product removal; B, classified product removal (Mullin, 2001).

Unsteady-state operation

Continuous crystallizers commonly develop unsteady-state behavior.

Feed rate or composition changes, temperature fluctuations, etc., will manifest themselves as transient disturbances in the product CSD until the system regains its original, or adjusts to a new, steady-state mode. Even though operating conditions ostensibly remain constant, however, a crystallizer can exhibit unstable behavior and develop periodic changes in crystal production rate and CSD, although it has often been suggested that a continuous crystallizer should be self-stabilizing. For example, an increase of the super saturation would lead to a higher nucleation rate, but as this would give a total crystal surface area increase, owing to the growth of the nuclei, the super saturation would then decrease and the nucleation rate would fall to its former value.

In practice, however, the dynamic behavior of a continuous crystallizer is not so simple. The above stabilizing effect is subject to a considerable time lag, because newly formed nuclei have no appreciable surface area for a long time. Before stabilization can occur, therefore, large numbers of nuclei might be formed which will later reduce the super saturation below its steady-state value. The resulting slow nucleation will lead to a decrease of the total crystal surface area, below its steady-state value, and this in turn will cause an increase of the super saturation above its steady-state value, and so on.

The result of above sequence of events will be the occurrence of limit cycles, and mathematical models of continuous crystallizers with perfect and imperfect mixing, with and without product classification have been developed to show how periodic changes of super saturation, solids content, crystal size and production rate can readily occur. Periodic behavior is most pronounced at the beginning of the crystallization process, and in most cases the fluctuations are subsequently damped. This means that under favorable conditions a steady state is usually reached (Figure 9.7a). However, under some operating conditions the damping is not effective and the cyclic behavior may continue for a considerable time (Figure 9.7b). Indeed, in some cases steady state may not be achieved at all.

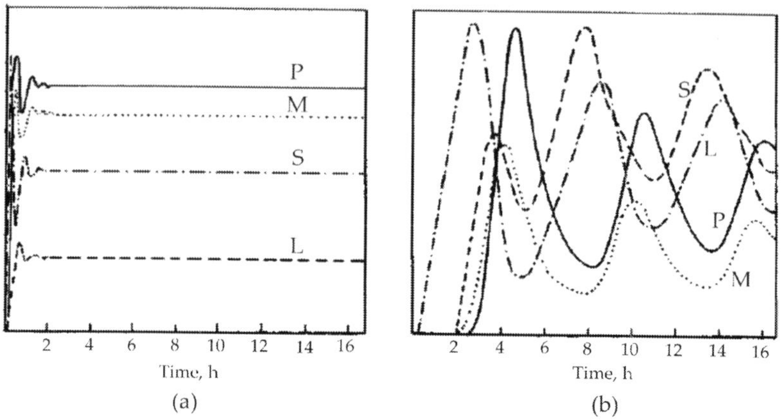

Figure 9.7: Dynamic behavior of a mixed suspension crystallizer showing (a) mixed product removal ~ short period damped cycles; (b) classified product removal via elutriating leg - long period slowly damped cycles. P, production rate; M, magma density; S, super saturation; L, median crystal size (various scales) (Mullin, 2001).

The stability of a crystallizing system tends to increase with increasing crystal growth rate and magma density, and with decreasing nucleation 'order', minimum product size, and the relative amount of crystals withdrawn. Continuous seeding has the same effect as decreasing the effective nucleation order, and should lead to a stabilization of the system.

In general terms, clear liquid overflow (CLO) tends to reduce cycling behavior while classified product removal (CPR) increases it. A common form of unintentional CPR is the use of a badly designed product outlet which causes segregated magma removal. Fines destruction (FD) can be used to control the population density of nuclei in the crystallizer. Much attention is being paid at the present time to ways in which on-line CSD measurement can be reliably used to assist the transmission of information to control systems to allow rapid adjustments to be made in temperature, feed rate, FD rate, etc. to stabilize the system (Mullin, 2001).

Figure 9.8a shows schematically a mixed suspension crystallizer being operated with FD and CPR: the R-Z model of Randolph and Larson (1988). Figure 9.8b shows an idealized population density plot for the crystals suspended within the crystallizer. The CSDs in the three size ranges divided by the cut sizes Lf and L, may be expressed by

$$n = n_0 \exp(-RL/G\tau) \qquad 0 < L < L_f \qquad (9.7)$$

$$n = n_0 A \exp(-L/G\tau) \qquad L_f < L < L_c \qquad (9.8)$$

$$n = n_0 B \exp(-ZL/G\tau) \qquad L_c < L < \infty \qquad (9.9)$$

**use equation editor in writing equations 9.7 – 9.9

Where, A and B are exponential functions relating to R and Z, respectively. R, the ratio of the total fines removal rate to the flow through the crystallizer, expresses the decrease in residence time of the fine crystals (< *Lf*), i.e. $\tau_f = T/R$, where τ is the overall crystal mean residence time. The ratio of the classified product removal rate to the total flow, *Z*, expresses the increase in removal rate for the large crystals (> *Lc*), i.e. $\tau c = \tau/Z$ (Mullin, 2001).

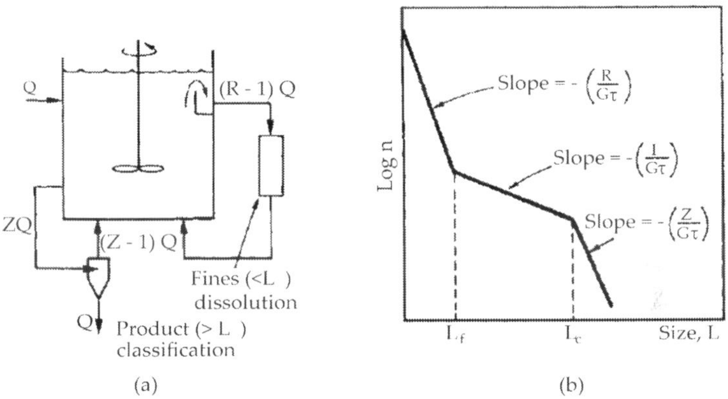

Figure 9.8: Mixed suspension crystallizer operating with fines destruction and classified product removal: (a) the R-Z model, (b) population density plot (Mullin, 2001)

CRYSTALLIZER SELECTION

Batch versus continuous crystallization

The design of a crystallizer involves decision-making processes at several stages and requires a fund of reliable information and experience to make scientific and engineering judgments. For a designer, a systematic analysis and comparative study of several alternative possibilities would be desirable prior to a selection of process and type of equipment. The method of super saturation generation, mode of operation, basic principles associated with suspension handling, and economic analysis appear to be major considerations in most crystallization processes. The solubility-temperature variation of the solute in the solvent is of prime importance in the selection of a crystallizer. For solutes that show large solubility variations with temperature, yielding appreciable quantities of crystals on cooling, the choice of equipment generally lies between a simple cooling and a vacuum (flash-cooling) crystallizer. For solutes that show small solubility variations with temperature, an evaporating crystallizer are normally used (Tavare, 1995).

It is difficult to suggest a suitable mode, given the choice between batch and continuous, for a given system. Continuous crystallizers are generally suitable for large production rates and are cost efficient from operating and labor viewpoints. The required crystal production rate and the supply of feed liquid to the plant are generally among the deciding factors (see Table 9.2) in the choice between a batch and a continuous unit. There is no general rule, as such, for a specific choice. The product specifications (in terms of required shape, size, and size distribution of crystals) also have an influence on the type of the crystallizer selected. The ideal conditions for crystallizer operation are to control the nucleation level and to grow crystals in a satisfactory environment, and can be achieved in practice by several methods.

Table 9.2: Batch versus Continuous Crystallizer (Tavare, 1995)

Item	Continuous	Batch
Capacity	Usually large	Generally small
Productivity/working volume	High	Low
Equipment	Elaborate	Simple
Equipment cost	High	Cheap
Flexibility	Not so flexible	Flexible
Operating costs	Low	High
Operation	Generally steady	Dynamic
Labor demand	Low	High
Product quality	Uniform, improved with crystallizer modifications	Variable, improved with programmed operation
Product characteristics	Generally small mean size with large CV	Possibility of large mean size with small CV
Utilization of mother liquor	Good	Good (?)
Problems of encrustation	Likely	May be avoided
Product	Bulk commodity	Fine, speciality chemicals

**rewrite this table into a real table

The circulating liquid type of crystallizer is designed so that super saturation of the liquid is induced in one part of the apparatus and then relieved in another part containing a crystal bed, the latter being maintained in a fluidized suspension by the upward flowing liquid. If the circulating liquid is substantially free from solids, the most highly super saturated part of the liquid makes first contact with the crystals in the lower part of the suspension and becomes depleted as it moves upward through the bed. In circulating liquid, crystallizers can be operated on the circulating magma principle. The design intention and the distinguishing features of the circulating magma crystallizer are that the liquid flow rates are chosen so that the suspension of crystals in the mother liquid is circulated throughout the system, and therefore through the super saturation zone.

Both types of crystallizers are used to produce a wide variety of chemicals with product sizes ranging from fine to coarse. Once a particular class of crystallizer is decided on, the choice of a specific unit depends on such considerations as initial and operating costs, space availability, type and size of crystals required, physical characteristics of mother liquid and crystal slurry, need for corrosion resistance, sterile operation, environmental protection, safety, and so on. To minimize encrustation problems, all liquid mixing zones in the crystallizer unit should be properly designed. Both temperature and composition at the point of mixing should be assigned to avoid excessive nucleation (Tavare, 1995).

Design strategies

Information for design

It is useful to know what sort of information a design engineer generally needs in order to specify an industrial crystallizer. A detailed description of the product should be given, including its full chemical name and formula, specifying if a hydrate is required or not. A realistic purity specification should be laid down. The production rate should be given, both as kgh^{-1} and ton $year^{-1}$ or some other suitable long-term rate, making generous allowances for maintenance and other shutdown periods.

A reliable solubility curve of the product in the actual liquids should be established. Solubility data reported in the literature usually refer to pure solutes and solvents, and these may be quite inapplicable to the industrial case. If impurities are known to be present in the working liquids, the relevant solubility data should be measured independently.

Measurements of the metastable zone width should be made in a manner relevant to the expected working conditions, if possible using the actual liquids to be processed, but most importantly in the presence of crystalline phase.

The size and size distribution of the required crystalline product should be specified, but it is important to remember that the more rigid the size specification the more difficult the crystallizer design becomes. An actual sample of the desired crystals, if available, is usually very helpful. Single-size specifications, such as 'about 450 μm', have no useful meaning, but upper and lower size limits may be quoted. For example, a specification such as '> 80% to be retained between 600 and 300 μm sieves' would be more helpful. Alternatively, the desired median size, MS, and coefficient of variation, CV, may be designated (Mullin, 2001).

The following liquid and solids data are also useful:

Liquid

- feedstock analysis and pH;
- density of the feedstock at the feed temperature;
- boiling-point elevation of the saturated solution;
- density and viscosity of the saturated solution over the working range of temperature.

Crystals

- true density of the substance;
- settling velocity of crystals of the desired size and shape in saturated solutions over a range of temperature;
- crystal growth and nucleation rates as functions of super saturation and temperature;
- heat of crystallization.

Design procedures

Over the past 30 years or more, several approaches have been done headed for the mathematical modeling of crystallizer behavior and the development of design strategies. In spite of their seeming differences, all the methods occupy shared theoretical ground, and also they suffer from a shared weakness in that they symbolize only idealized cases. The basic equations should only be used as qualitative guides to design. Usually, uncritical application of set formulae to common industrial situations can result in very deceptive outcomes. No design equation is any more reliable than the data fed into it, but there is a particular problem with crystallization data which are in very short supply and of extremely variable quality. Great care should be taken, when attempting to measure reliable design data, to use experimental methods of appropriate scale and mode of operation in relation to the envisaged large-scale equipment, and if at all possible to use actual process liquids (Mullin, 2001).

The objectives of this section is to simply outline some typical initial approaches to design which in practice will need to be tailored for specific circumstances and data availability. The design of any crystallizer does not have a single 'correct' answer; however the goal is to reach a rational combination of design values within an identified framework, which should involve downstream requirement as well.

Amongst the first quantities that may have to be specified, or calculated, in a design procedure are the crystal production rate, a clearly defined 'mean' product crystal size, the crystallizer working volume, liquid circulation rate, magma density, mass of crystals in the suspension and the residence times of both crystals and liquid. Several of these quantities are interrelated and will not need separate calculation. Some are directly linked to the effective overall crystal growth rate, a sensitive system-specific quantity.

The following brief examples demonstrate the manipulation of some of the key relationships. In an actual design procedure the result of each calculation step should be assessed as to its reasonability, either in the light of experience with similar systems, or within restrictions set by existing equipment and operating procedures (Mullin, 2001).

CONTINUOUS CRYSTALLIZATION

Applying vacuum technology opened the possibility to apply crystallization apart from the atmospheric pressure and herewith the opportunity to precisely design the crystallization processes to the respective phase systems and the material properties. Today, these vacuum crystallization processes have become the common standard in the industrial continuous single mass crystallization from solutions.

While the unit operation evaporation, that is, the mass transfer from the liquid phase to the vapor phase, still possesses a direct connection with vacuum techniques, the connection of today's "single mass crystallization from solution" with vacuum techniques is only indirect. The techniques of vacuum cooling and vacuum evaporation are only the mostly used means for inducing the crystallization process.

The reason for the dominant position of vacuum crystallization over classical surface cooling crystallization is the considerably reduced inclination to form incrustations. Vacuum crystallization is used in the low vacuum field down to 1 mbar. There are also applications in the overpressure field, although with increasing pressure the number of applications is reduced. In vacuum crystallization, one can find all the classical process control options used in the more familiar vacuum evaporation processes. However, an important difference to evaporation is the fact that the separation process is not concluded with the crystallization step. The suspension formed still has to be separated into crystal mass and mother liquid. Crystallization is therefore always associated with a mechanical separation process; better separation leads greater purity of the crystallized masses (Beckmann, 2013).

The separation process is all the more complete, the more compactly the crystals are formed and the coarser they are when they become separated. The degree of separation also depends on the separation technology. Centrifugation is preferred, as the adhering mother liquid is less and washing is also less complicated. As centrifugation with pusher-type or worm/screen centrifuges is possible from a mean crystal size of approximately 0.2mm, the success and cost effectiveness of the crystallization substance separation process are also determined by the crystal size obtained and the crystal size distribution (CSD). These parameters also influence the suitability of storage, freedom from dust, and strew ability of the product. For example, ammonium sulfate, which is used as a straight fertilizer, has to be sufficiently coarsely crystallized so that when it is strewn it does not remain on the leaves of the plants. Other crystallized products have to be crystallized as finely as possible in order to positively influence their dissolving behavior. An important difference between vacuum evaporation and vacuum crystallization is the fact that in vacuum crystallization the supersaturated solution tends to induce incrustations on all solid surfaces. The design and layout of a vacuum crystallizer has to be adapted to this, otherwise the cost effectiveness of the process will be jeopardized by the lack of availability. Not all evaporator types can therefore be used in vacuum crystallization. Furthermore, the requirements placed on the granulometry and purity of the crystallized masses are also determined by consumer and market customs, and frequently also by the requirements of the follow-up processes (Beckmann, 2013).

Concept and Design of Continuous Crystallizers

Importance of Secondary Nucleation

Provided that primary nucleation is reliably avoided by working within the metastable zone, the granulometry of the crystallized masses produced depends on the control of secondary nucleation. Secondary nucleation is therefore the most important control factor for the granulometry of crystallization processes.

The smaller the secondary nucleation, the more massive the individual crystal can become. For improved understanding of this process, an example is explained graphically in Figure 9.9. Here the super saturation is set proportional to the production quantity.

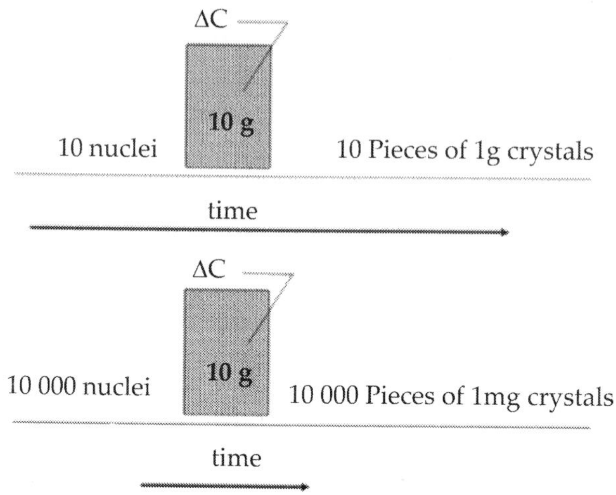

Figure 9.9: Influence of number of nuclei formed on final particle size (Beckmann, 2013).

For a produced quantity of 10 g, 10 crystals with a mass of 1 g each are produced from 10 crystal nuclei, while 10,000 crystal nuclei will lead to 10,000 crystals with a mass of only 1mg each. As the super saturation level cannot be increased above the metastable zone – otherwise the number of nuclei would increase spontaneously – it takes approximately 20 times longer (see time arrow) to incorporate the production quantity of 10 g in 10 crystals of 1 g each than to create 10,000 crystals of only 1 mg each.

In the case of the uncontrollable spontaneous nucleation with its tremendously higher nuclei production instead of secondary nucleation, the distribution of the production capacity to all of these nuclei will consequently lead to extremely fine crystals, difficult to separate from the impure mother liquid; that is, the targets for granulometry and purity will be entirely missed (Beckmann, 2013).

On this basis, it is possible to deduce the following two principles for the design and layout of crystallizers:

At no location within a crystallizer may the super saturation reach or exceed the limit of the metastable zone in order that primary nucleation is avoided and only secondary nucleation can occur during the crystallization process.

The metastable zone width must, however, be exploited to a large extent in order that the crystal growth rate available is sufficient.

Control of Super saturation

These two principles are the most important command variables in mass crystallization from solutions. Crystallizers in which these principles are taken into consideration will certainly function correctly. Building on this, various basic types of crystallizers have been developed with which the entire field of demands placed on grain size distributions and the mean crystal sizes can be fulfilled as shown in Figure 9.10.

Each of these basic types is aligned to a certain particle size range. The forced circulation (FC) crystallizer is used for the smaller particle sizes up to 0.8mm, the draft tube baffle (DTB) crystallizer for the coarser particles up to 2.5mm, and the Oslo-type crystallizer for even larger particle sizes. The volume of the crystallizers increases in Figure 9.10 from (c) to (a) reflecting the need to spend more retention time for crystal growth, if coarser crystals have to be produced. In order to allow the crystals become coarser by crystal growth during longer retention times, one has – in parallel – to carefully reduce the effective particle generation rate (secondary nucleation, attrition, and breakage) before the crystal growth can lead to larger particles.

The first to avoid in all these crystallizers is the primary (spontaneous) nucleation as only the secondary nucleation can be used to steer the nucleation rate via the energy input and thereby the particle sizes that shall be produced. Therefore, one aspect that they all have in common is the manner in which the level of super saturation is controlled and kept within the metastable zone width to ensure the avoidance of the primary nucleation. This method, too, is independent of the respective crystallization process; hence, all of these basic designs can be used for all crystallization processes. How that is realized in the crystallizers exhibited in Figure 9.10 is described below.

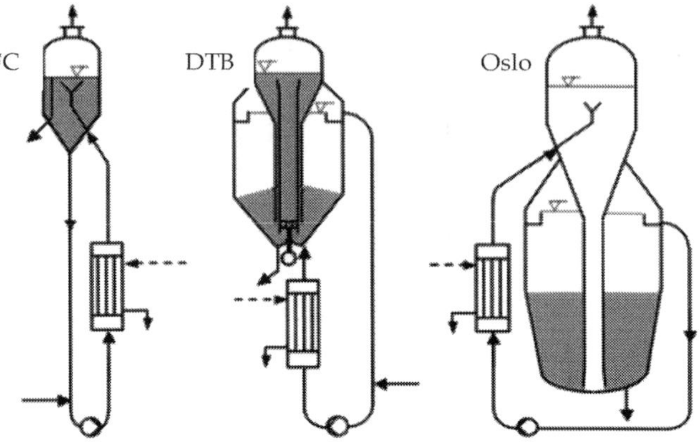

Figure 9.10: Types of crystallizers: (a) FC crystallizer, (b) DTB crystallizer, and (c) Oslo crystallizer (Beckmann, 2013).

Figure 9.11 shows the functional principle of this control method, using vacuum cooling crystallization as an example. Figure 9.11a shows a sketch of a FC crystallizer used for vacuum cooling crystallization. This is the simplest form of basic crystallizer design that has an external circulating line. Due to the recirculation, the crystallized mass is suspended and the level of super saturation adjusted at the same time. Figure 9.11b shows a sketch of a solubility system, whose clear temperature dependence is typical of vacuum cooling crystallization. The width of the (assumed) metastable zone is marked here in gray. The processes within the crystallizer in Figure 9.11a are reproduced in the solubility system in Figure 9.11b and the locations and processes allocated to one another in each case by means of figures (Beckmann, 2013).

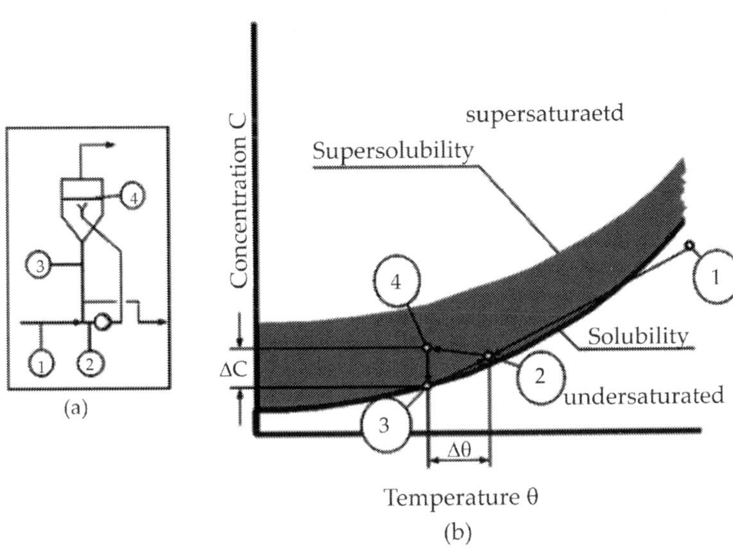

Figure 9.11: (a) Control of the tip super saturation. (b) The concentration at different points in the crystallizer is represented in the solubility diagram (Beckmann, 2013).

The solution (3) in the crystallizer is mixed with the feed solution (1). In the case of vacuum cooling crystallization, this feed solution is hotter and more concentrated than the mother liquid. Thus, a mixed solution (2), is created that is also hotter and more concentrated. Depending on the position of the mixture line (always straight connection) in the respective solubility system, it is possible – as already indicated – that super saturation is produced in the solution as a result of the mixing. The mixed solution is then conveyed to the liquid level by the recirculation pump. Above the level of the solution, the desired saturated vapor pressure of the solvent is maintained, which is controlled by means of a condensation device with downstream vacuum pump. When emitted from the central funnel of the recirculation line, the superheated solution now begins to boil (4), as it is superheated with respect to the vapor pressure above the level of the solution. As a consequence, the solution starts to boil adiabatically, dropping its temperature down to the temperature that corresponds to the saturated steam pressure in the vapor chamber. Due to the cooling process and the simultaneous removal of solvent, super saturation is formed, as shown in the solubility system by the section between points (3) and (4). The super saturation Δc created is dissipated on the suspended crystals by means of crystal growth, and the next cycle begins when point t is reached.

In this example, it is assumed that de-supersaturation is complete when point (3) is again reached. For this case, it is clear that the level of the super saturation produced at the level of the solution depends on the recirculation flow rate. High recirculation flow rates reduce the super saturation produced there (dilution), while low recirculation flow rates increase it. The recirculation flow rate, adjusted to the production output, is therefore the most important design parameter in industrial crystallizers. Where the production outputs are the same, this parameter is equal for all crystallizer designs. The required recirculation flow rate depends on the metastable zone width. If this is not known, it has to be determined beforehand by means of measurements. In practice, half of the metastable zone width is used for determining the required recirculation flow rate. Therefore,

$$\Delta c = 0.5\Delta c_{met}, \qquad \frac{dV}{dt} = \frac{dP/dt}{\Delta c},$$

Where, dV/dt is the recirculation flow rate in m^3/h, dP/dt is the production output in kg/h, and Δc_{met} is the metastable zone width in kg/m^3 or g/l. As a result, the performance of the crystallizer laid out in this way is restricted to dP/dt if primary nucleation is to be excluded. As the metastable zone widths usually have a spread of just a few g/l, large recirculation flow rates are necessary: The recirculation flow rate for a permitted super saturation of maximum 1 g/l is already 1000m^3/h for only 1 t/h crystal production rate. These high recirculation rates are best achieved by axial flow pumps.

For the purpose of simplification, the assumption has been made in Figure 9.11 that the time for a recirculation process is sufficient to reduce the super saturation down to negligible levels. This reduction in super saturation is, however, a function of the mean mass deposition rate and the growing crystal surface area present

$$\frac{dm}{dt} = k_g A \Delta c^g \propto -\frac{d(\Delta c)}{dt}$$

Here dm/dt denotes the mean mass deposition rate, A is the surface of the suspended crystals, Δc is the super saturation, and kg is the proportionality constant. From this relationship, it is evident that the de-supersaturation rate -d(Δc)/dt depends on the crystal surface available. Where the crystal growth rate is slow or the crystal surface available is small, it is possible that the time for a recirculation cycle is not sufficient to allow negligible super saturation to occur before the new super saturation is produced. This is shown in Figure 9.12. The selected example is again vacuum cooling crystallization. After the creation of super saturation (1) at the level of the solution, the super saturation is degraded on the crystals that have been carried along. Through the cyclical sequence of the generation of super saturation and de-supersaturation (as a result of crystal growth), the saw tooth curve shown, which is typical of this process, is created.

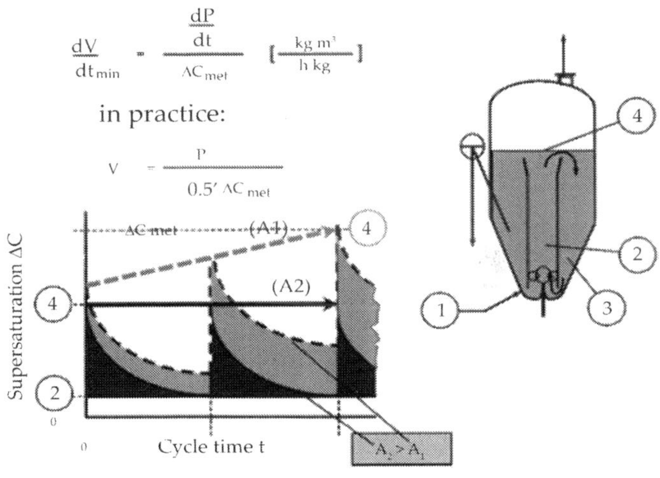

Figure 9.12: De-supersaturation as a function of crystal surface area (Beckmann, 2013).

The smaller the active crystal surface available, the slower the mass deposition rates dm/dt, and the larger the super saturation remaining after each recirculation cycle. The point (4), in Figures 9.11 and 9.12 moves up, if crystal growth does not de-supersaturate to . As this residual supersaturation is added to the newly created super saturation, it is certainly possible that the metastable zone width will be exceeded and primary nucleation will occur. In addition to the "correct" recirculation flow rate, when the crystallizer is being designed attention must also be paid to the sufficient availability of crystal surface (function of suspension density), in order that primary nucleation can be reliably prevented.

This is mainly of importance where the mass balance resulting from the crystallization process would produce values that were too low for the suspension density. In this case, special measures have to be taken, as explained below. In general, suspension densities of between 15 and 25% by mass are sufficient in order to achieve an adequate desupersaturation rate.

With these measures and specifications relating to the level of super saturation and suspended crystallized mass, it is generally possible to obtain sufficiently granular (coarse enough) crystallized masses, so that separation with centrifuges can be performed (Beckmann, 2013).

Adjustment of the Granulometry – Mean Crystal Size and Crystal Size Distribution

At this point, however, after having avoided the formation of spontaneous nucleation in these types of crystallizers, the question arises as to which measures remain in order to design the crystallizers for producing coarser crystals. The solution is clearly shown in the two kinetic relationships for crystal growth rate dm/dt and secondary nucleation rate B^0, which describe the crystallization process, though the specifications described above mean that the majority of the variables are now fixed by agreement. These fixed variables are circled in Figure 9.13. The remaining means available for controlling the granulometry are the proportionality constant kg and the specific energy input ε.

$$\frac{dm}{dt} = \textcircled{k_g} * A * \Delta C^m = - \frac{d(\Delta C)}{dt}$$

$$B_0 = k_N * \textcircled{\varepsilon^r} * m_T^l * \Delta C^n$$

ΔC	=	$0.4 - 0.6\ \Delta C_{met}$
m_T	=	10 - 30 wt% crystals in suspension
k_G	=	f(T,...) Target: T high

ε^r = specific energy input

T = crystallization temperature

Figure 9.13: Importance of energy input (Beckmann, 2013).

As described before, secondary nucleation is a function of the mechanical energy input (Figure 9.13). This is introduced by the circulation pump. The blades of the impeller pass this on to the suspension in the form of kinetic energy, whereby the acceleration process causes direct collisions between the blades of the impeller and the suspended crystals. The strength of these collisions increases with the increasing mass of the individual crystal, with the increasing difference in the specific densities of the crystal and mother liquid (mass inertia effect), and with the tip speed (circumferential velocity) of the impeller, while the quantity of collisions increases with the suspension density m_T.

In addition to the direct collisions of the crystals with the rotor blades, there are also consequential collisions between the crystals and the walls of the crystallizer or its components, or even collisions between the crystals themselves. The energies involved in these collisions are lower and the resulting secondary nucleation of secondary importance; however, they are also a function of the energy input via the circulation pump. The relationship of these attrition mechanisms to one another may be imagined as

$$\frac{\text{Crystal}}{\text{Crystal}} \Big/ \frac{\text{Crystal}}{\text{Wall}} \Big/ \frac{\text{Crystal}}{\text{Impeller}} = 1/10/1000$$

The largest source of attrition and nucleation is the collisions between the suspended particles and the blades of the circulation pump.

The less energy that has to be expended for the recirculation process, the fewer crystal nuclei will be produced. The nucleation frequency is therefore the key to producing different crystal sizes (compare Figure 9.8). The designs of crystallizers also differ from one another mainly with respect to the specific amount of energy ε, which has to be expended for the recirculation process. For the different types of crystallizers, these specific energy inputs vary from around 10 W/kg for the FC types down to 0.1–1 W/kg for the DTB types and to below 0.1 W/kg for the fluidized bed crystallizers (Beckmann, 2013).

VARIOUS CONTINUOUS CRYSTALLIZERS

Within every model category there are various designs, which are adapted to the specific individual applications.

FC Group

Figure 9.14 shows the different designs of the simple FC group; depending on the crystallization process such FC crystallizers can be used for (vacuum cooling crystallization or vacuum evaporative crystallization), one can find the typical FC (4), and (5) with and without a tube and shell heat exchanger. These are the simple stirred tank (1), the draft tube (DT) crystallizer (2), and the FC crystallizer (4), (5). The suspension is recirculated by stirrers or circulation pumps and the super saturation created within the suspension. The stirred tank (1) is suitable for vacuum cooling crystallization. The disadvantage of this design is the absence of directed flow recirculation, as only a simple, but inexpensive stirrer is used. This means that the maximum super saturation is not fully under control, which is why this design should only be used for systems with large metastable zones.

The draft tube crystallizer (2) (the difference of the DTB crystallizer to the DT crystallizer is the baffle for the clear liquid overflow to the external recirculation) is equipped with a draft tube propeller pump instead of a stirrer. This provides full control of the maximum super saturation. This design is used in vacuum cooling crystallization for larger production outputs. Also, the so-called horizontal crystallizer (3) is used exclusively for vacuum cooling crystallization. It features several crystallizer stages that are connected in series in a single vacuum vessel in order to save space and costs. Such multistage vacuum cooling crystallization is found in applications in which the vapors are colder than the condensation possibilities allow and which for this reason initially have to be raised to a sufficiently high temperature and pressure level by means of vapor recompression (thermo compression). The more cooling stages can be carried out, the less energy is required to compress the vapors to the respective condensation conditions (Figure 9.14). (The worst-case scenario with the highest consumption of steam is the single-stage design, in which the entire quantity of vapors has to be increased in temperature and pressure to the condensation level by means of thermo compression.)

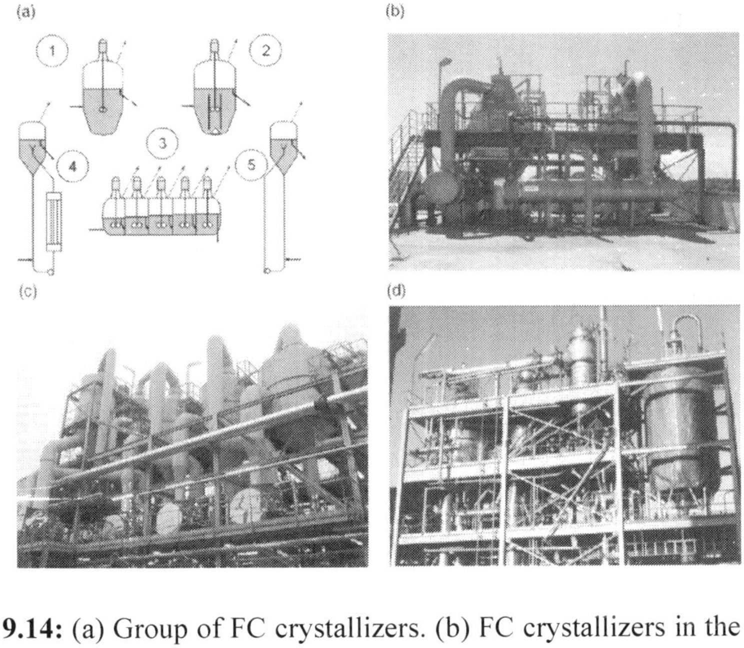

Figure 9.14: (a) Group of FC crystallizers. (b) FC crystallizers in the food industries. (c) FC crystallizers in vinasse concentration (K_2SO_4 crystallization). (d) FC crystallizers (vacuum evaporation) for Na_2SO_4 (Beckmann, 2013).

The disadvantage of this design is the unfavorable geometry of the chambers for the suspension process. Sedimentation is unavoidable over the long term, which results in incrustations. The maximum realizable operation periods are approximately 1 week. Its use is therefore limited today to companies where production stops running at weekends. If longer operating times are planned, a few draft tube crystallizers are connected in sequence today instead of the horizontal crystallizer and if necessary the condensation carried out on cooling brines, which are re-cooled using ammonia cooling units (Beckmann, 2013).

Use of the FC crystallizer (4) is preferred for evaporative crystallization. The directed and controlled recirculation is carried out with an axial flow pump (elbow type). This design can also be used for vacuum cooling crystallization, so that then the heat exchanger is no longer required (5). The use of this version is recommended if due to the lack of barometric cover in the draft tube crystallizer the incoming solution would boil out of the feed-in point.

In these simple crystallizers, the crystal content of the suspension can be determined directly from the mass balance. Higher suspension densities can only be adjusted if parallel to the suspension discharge clear liquid is also removed. This is necessary when the mass balance shows suspension densities (crystal surface area) that are not sufficient for the de-supersaturation. In this case, it is possible to install clarification surfaces even in these simple crystallizers. If this is carried out with the draft tube crystallizer, the result is a typical crystallizer of the DTB group as already pointed out above (Beckmann, 2013).

DTB Group

The DTB crystallizers (Figure 9.15), whose eponymous representative has already been introduced, show three different representatives. In the subcategory DP crystallizer, the propeller of the circulation pump is divided into two sections. The section located within the draft tube creates an upstream, while the section found outside the draft tube, which is bordered by the partition of the clarification section, creates a downstream. The external and internal sections are adapted to one another and ideally produce exactly the same degree of recirculation. In this manner, it is possible to use very low tip speeds for the desired recirculation flow rate.

In the Messo turbulence crystallizer, a secondary circulation loop driven by the primary circulation loop is created in the outer suspension section. This secondary circulation loop serves the purpose of sorting out the coarser crystals from the inner draft tube recirculation process, concentrating them in an outer vortex and thereby removing these coarser crystals from the influence of the circulation pump for the purpose of further growth. In this way, the product crystals can be pre-classified. With all DTB crystallizers, this classification during removal, which serves the purpose of improving the uniformity of the particle size distribution, can be achieved by adding an elutriation leg. Typical products that are produced by means of this category of crystallizers are ammonium sulfate and potassium chloride, which must be readily strewable as compound fertilizers or which are added as already coarse particles to mixed fertilizers.

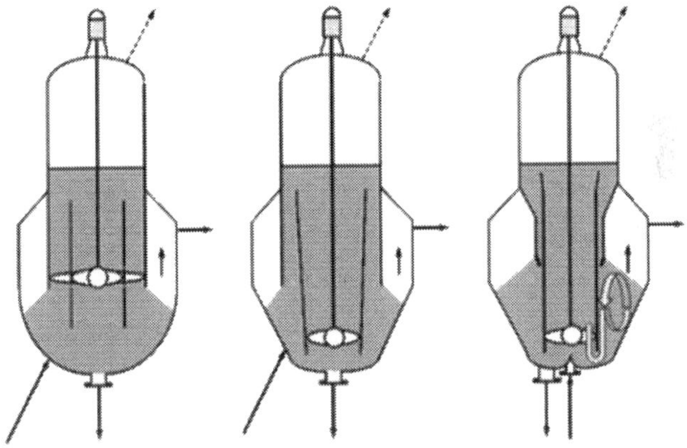

Figure 9.15: DTB group crystallizers (Beckmann, 2013).

Group of Fluidized Bed Crystallizers

The best-known representative of the group of fluidized bed crystallizers is the "Oslo crystallizer". Today, there are two known versions (Figure 9.16). The original design was developed in the 1920s by a firm called Krystal A/S in Oslo, which explains the name. This design is susceptible to malfunctions in the case of products that tend to produce incrustations, as falling crusts can block the annular gap at the entrance to the fluidized bed. In the case of the crystallization of sodium chloride, the incrustations in the evaporation chamber of the "Krystal" crystallizer become so thick after just 3 days that they fall down and block this annular gap. The fluidized bed collapses and the blockage can only be removed by thorough cleaning that takes several days without any production at all. The more modern design ("Messo") has been specially designed to overcome this problem. By reversing the direction of flow in the inflow area of the evaporation section, in this modified design the solution that has been superheated by the heat exchanger and is therefore under saturated is passed over the conical surface that was previously susceptible to incrustations, before the super saturation is created by means of the boiling process. As a result, no incrustations can form on the conical surface and the possibility of malfunctions is excluded. In this manner, it is possible to achieve operation periods of several weeks (Beckmann, 2013).

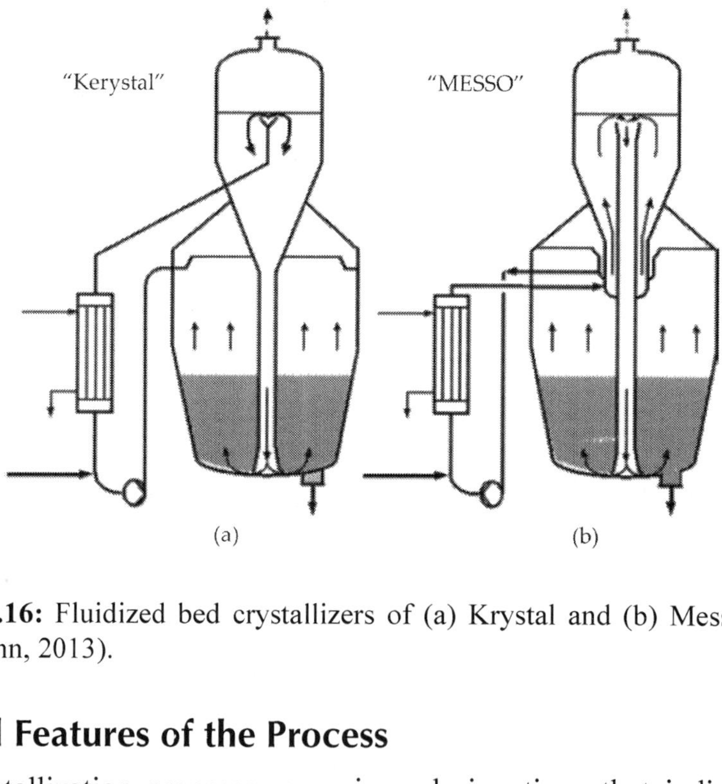

Figure 9.16: Fluidized bed crystallizers of (a) Krystal and (b) Messo types (Beckmann, 2013).

Special Features of the Process

The crystallization processes are given designations that indicate the method of creating the super saturation. Super saturation can be created in solutions by various measures: vacuum evaporation crystallization, for example, when vacuum evaporation is the means used to create the super saturation, or cooling crystallization (vacuum cooling crystallization or surface cooling crystallization) if the super saturation is created by cooling. The principle is shown in Figure 9.17. In the case of solubility functions that are not dependent on the temperature – or only slightly so – solvent removal is selected, while in cases where there is considerable temperature dependence the chosen process is the cooling method, which is mostly more favorable with respect to the amount of energy used. In addition, there are other important factors that influence the selection of the process. The most important of these are presented below (Beckmann, 2013).

Surface Cooling Crystallization

Continuous surface cooling crystallization is not a modern process. In this method, the super saturation is created using the surface of a heat exchanger. As soon as this surface is covered by incrustations, the production output is directly affected.

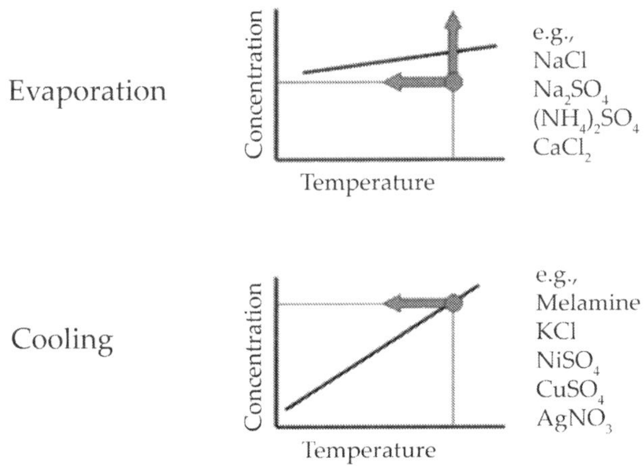

Figure 9.17: Typical examples to generate super saturation (Beckmann, 2013)

The probability of this is very high, as the highest super saturation in the crystallizer takes place directly on the surface of the heat exchanger. The operating times between the requisite cleaning periods are therefore significantly shorter than those of vacuum crystallization processes. The tendency to develop incrustations can be reduced somewhat by means of large (and expensive) heat exchanger surfaces. As a result, the super saturation directly on the cooling surface is reduced in size and the rate of incrustation growth slowed.

Surface cooling crystallization can be used for discontinuous processes without the need for any further measure, as the incrustations are dissolved again at the beginning of each new batch. In continuous processes today, surface cooling crystallization is only used where the vacuum methods become uneconomical, for example, when the vapor pressures of the solvent are too low. This is mainly the case with solvents with a high boiling point, for example, phenol. For aqueous solutions today, surface cooling crystallization is only used if the boiling point elevation is so great that vacuum crystallization is uneconomical due to the pressures being too low (Beckmann, 2013).

Vacuum Cooling Crystallization

Due to the disadvantage mentioned above, vacuum cooling crystallization is the preferred cooling method for the continuous process. Cooling is carried out on a liquid surface whose process function cannot be disturbed by incrustations. This makes it possible to achieve long operation periods. Vacuum cooling only becomes uneconomical if cooling down to very low temperatures is required. However, as long as the vapor temperatures are not below the freezing point of the solvent concerned, vacuum cooling – for example, through the use of cold water or coolants with high boiling point elevations – is always the more economical alternative due to the longer service life in comparison to surface cooling crystallization. Frequently, processes are found in which the existing surface cooling crystallization at low temperatures could be replaced by vacuum cooling crystallization at a higher temperature. This may be the case when the upstream process task is only the balance-related ejection of impurities and does not relate to the process yield. Instead of cooling a small quantity down to low temperatures, in these cases a larger quantity can also be treated at moderate temperatures and then the more favorable vacuum cooling crystallization is used (Beckmann, 2013).

Vacuum Evaporation Crystallization

Unlike vacuum cooling crystallization, this method does not depend on the concentration and temperature of the feed solution. Additional heat can be introduced by means of a heat exchanger and thus under saturated solutions treated in a crystallizer. Furthermore, the concentration factor of the mother liquid can be selected; that is, the quantity of solvent to be evaporated is adjustable to the requirements of the mass balance. The mass balance is based on the corresponding solubility system and must ensure that only the desired substance is crystallized out of the mother liquid. The permitted concentration of impurities in the mother liquid is to be geared on the one hand to the properties of the solubility system. On the other hand, the impurities of the product resulting from the adhering mother liquids have to be taken into account. As is the case with vacuum cooling crystallization, in evaporative crystallization there are no particular incrustation problems. Only in conjunction with the relatively rare inverse solubility do super saturations occur on the surface of the heat exchanger. This is the case, for example, during the crystallization of gypsum. This is remedied by means of seeding and limiting the temperature differences using the heat exchanger, as in surface cooling crystallization. Higher flow speeds and higher suspension densities can also help. Overall, the problems should be considered to be small. Particular attention should be paid to the falling incrustations from the evaporation chamber, as these can block the heat exchanger from below. Such incrustations of the evaporator section can be reduced by polishing the vessel walls. The energy carrier used in evaporative crystallization is steam or recompressed vapors. In order to save operating costs, the plants used in evaporative crystallization are frequently designed as multiple-effect evaporator systems. Here the first stage works near to atmospheric pressure and the last stage under such a vacuum that the vapors can be condensed on cooling water without any additional measure (Beckmann, 2013).

CHAPTER 10

PRECIPITATION

INTRODUCTION

Precipitation is a widely used industrial process. It is also a very popular laboratory technique, especially in analytical chemistry, and the literature on this aspect of the subject is voluminous. Precipitation plays an important role not only in chemistry but also in metallurgy, geology, physiology, and other sciences. In the industrial field, the manufacture of photographic chemicals, pharmaceuticals, paints and pigments, polymers and plastics utilizes the principles of precipitation. Regarding the production of ultrafine powders, precipitation may be well thought-out as a beneficial substitute to conventional crystallization followed by milling (Mullin, 2001).

The term precipitation very often refers to nothing more than fast crystallization, although sometimes it also implies an irreversible process, e.g. many precipitates are virtually insoluble substances produced by a chemical reaction, whereas the products of most conventional crystallization processes can usually be re-dissolved if the original conditions of temperature and solution concentration are restored.

Another distinguishing feature of precipitation processes is that they are generally initiated at high super saturation, resulting in fast nucleation and the consequent creation of large numbers of very small primary crystals. Although precipitation, like all crystallization processes, consists of three basic steps (super saturation, nucleation and growth) two subsequent secondary steps usually have a profound effect on the final crystalline product. The first is agglomeration, which generally occurs soon after nucleation, and the second is ageing, a term used to cover all irreversible changes that take place in a precipitate after its formation.

A common method for producing a precipitate is to mix two reacting solutions together as quickly as possible, but the analysis of this apparently simple operation is exceedingly complex. Primary nucleation does not necessarily commence as soon as the reactants are mixed, unless the level of super saturation is very high, and the mixing stage may be followed by an appreciable time lag before the first crystals can be detected. This event, which is often referred to as the end of the induction period, depends on the super saturation, temperature, efficiency of mixing, intensity of agitation, presence of impurities, and so on. However, an experimentally measured induction period is not identical with the critical nucleation time since critical nuclei have to grow before they can be detected. Primary nucleation, both homogeneous and heterogeneous, together with growth of the nuclei, may continue for some time until sufficient crystal surface area has been created to cause a rapid de-supersaturation. This event, which will be referred to as the end of the latent period, can be virtually identical with the experimental induction period at high super saturations. Secondary nucleation is probably a rare event in precipitating systems (Mullin, 2001).

Some indication of the particular nucleation and/or growth mechanisms predominating during the early stages of precipitation may be gained by plotting measured induction periods as a function of (1) $(\log S)^{-2}$, (2) $(\log s)^{-1}$ and (3) $\log(S^{-1})$. A straight line obtained from any of these plots can give an indication of the predominant mechanism. For example, (1) suggests either homogeneous nucleation or primary nucleation followed by diffusion controlled growth, (2) indicates poly nuclear growth, and (3) screw-dislocation growth. Further support for such mechanisms, however, must be sought by evaluating a relevant physical property from the slope of such functions (Mullin, 2001).

PRECIPITATION FROM SOLUTION BY MIXING TWO STREAMS

Devices and Mixing Schemes

The precipitation by mixing two streams, during a solvent or pH shift, or by mixing two salts can be carried out in a number of arrangements. The most commonly used, though not optimal, configuration is the semi-batch arrangement using a vessel containing the first component and adding, feeding, the second stream under a more or less vigorous stirring (Figure 10.1, geometry D). At the point of addition, the super saturation is relatively high, leading to moderately small crystals. Feeding both streams into the mother liquid, the local super saturations can be decreased (geometries C, B, and A). On the other hand, mixing the two streams in a nozzle, the local super saturations can be increased substantially (geometry E).

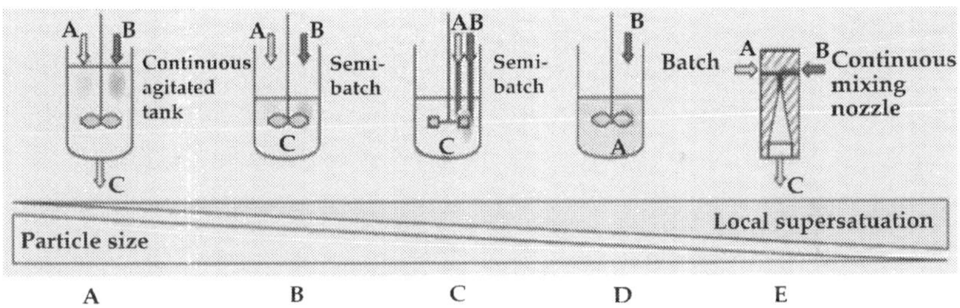

Figure 10.1. Geometries for the addition of components in a drowning-out or pH-shift precipitation. Depending on the local super saturation, either large crystals (left) or smaller particles (right) are obtained. The local super saturation is a function of intensity of mixing (Beckmann, 2013).

The super saturation in a precipitation is highly dispersed: it varies along the axis of the mixing and with time. For the mixing schemes B and E depicted in Figure 10.1, the profile of the super saturation is shown in Figure 10.2. The T-mixer has a very pronounced maximum in the super saturation, leading to locally very high super saturations and thus small particles, while in the semi-batch arrangement; the super saturations are lower, leading to larger particles. The lower super saturation also entails a longer induction time, and a longer presiding super saturation.

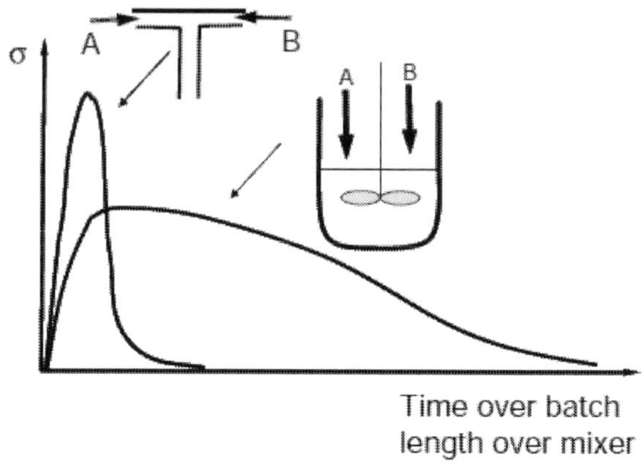

Figure 10.2: Profile of the super saturation as a function of time for the mixing schemes B and E depicted in Figure 10.1 (Beckmann, 2013).

The geometry of the T-mixer, for example, as shown in Figure 10.1, can be improved to increase the efficiency of mixing by modifying the geometry of the mixing chamber and the outlet. Various geometries have been proposed. The closed geometry of the T mixer bears the danger of clogging. Free-flowing impinging jet geometries might be used; Figure 10.3 shows two of the more commonly discussed geometries.

The efficiency of mixing can be increased by subjecting the mixing zone to a sonication or by using an Ultra-Turrax.

It should be emphasized that though the geometry of mixing will influence the super saturation and nucleation rate and consequently the particle size; the absolute particle size obtained will highly depend on the system crystallized (Beckmann, 2013).

SEMI-BATCH PRECIPITATIONS

The most commonly used mode of operation for precipitations in the life science and specialty chemicals field is the semi-batch operation, the fed-batch (geometry D in Figure 10.1). One of the two solutions is present in a stirred tank vessel and the other solution is added. The goal of these precipitations is often not the (fine) particle size but rather the fact that no alternative and more gentle crystallization technique is available.

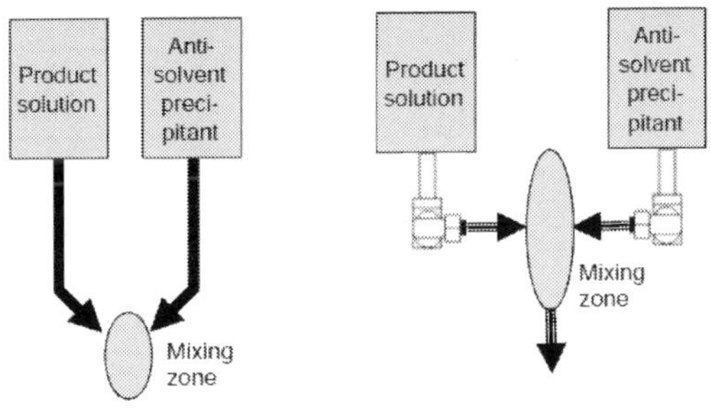

Figure 10.3: (a and b) Design examples of impinging jet mixing devices, where the two streams are not confined (Beckmann, 2013).

Two modes of operation are possible for drowning-out and pH-shift operations: either the anti-solvent is added to the solution of the target molecule (the classical operation) or the solution of the target molecule is added to the anti-solvent (also called inverse drowning-out). The path in the solubility diagram is schematically shown in Figure 10.4. Depending on the nature of the droplets, these contain either solution or solvent; the diffusion process leads to relatively low super saturation in the bulk of the mother phase for the classical drowning-out and relatively high super saturations in the droplets containing the target molecule in the case of the inverse process.

Products of the two modes of operation are shown in Figure 10.5. If the anti-solvent is added to the solution of the active, relatively well-developed particles are obtained. The size will depend on the speed of the dosing; however, the effects of variations in the dosing scheme are mostly minute.

If the solution is dosed into the anti-solvent, the local super saturations are far higher, leading to smaller primary particles. The size is on the order of 10 μm. Note also that the habit changes to elongated particles. It is also evident from Figure 10.5 that the primary particles are heavily agglomerated into larger agglomerates with a size on the order of the primary particles from the classical drowning-out.

PRECIPITATIONS USING SUPERCRITICAL FLUIDS

Several techniques have been proposed using supercritical fluids for precipitations. The main advantage is the lower viscosity of supercritical fluids compared to classical fluids, which facilitates mixing and thus higher super saturations can be achieved. Due to its ready availability, negligible toxicity, and convenient critical point of 32 °C and 100 bar, carbon dioxide is used in most cases as a supercritical fluid.

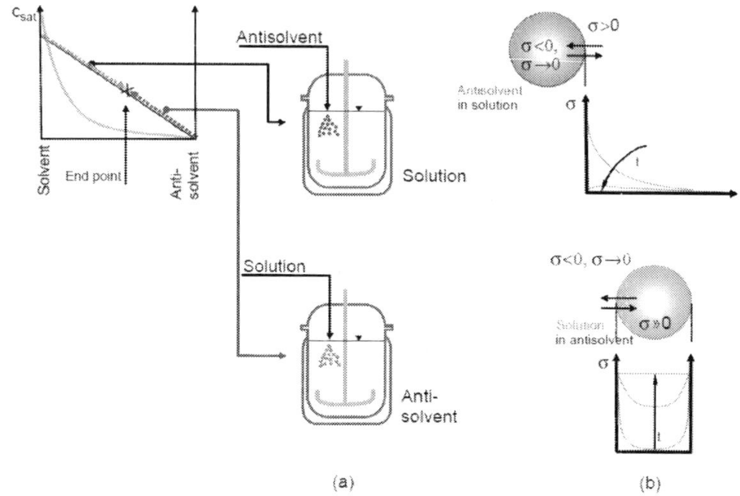

Figure 10.4: (a) Path of a classical and an inverse drowning-out crystallization in the solubility diagram and (b) mixing of solution and anti-solvent on the droplet level. The diffusion of anti-solvent from the droplets of anti-solvent into the bulk of the solvent- phase is much slower, entailing much lower super saturations, that the diffusion of anti-solvent from a bulk-phase of anti-solvent into the small droplets of solution, top and bottom right (Beckmann, 2013).

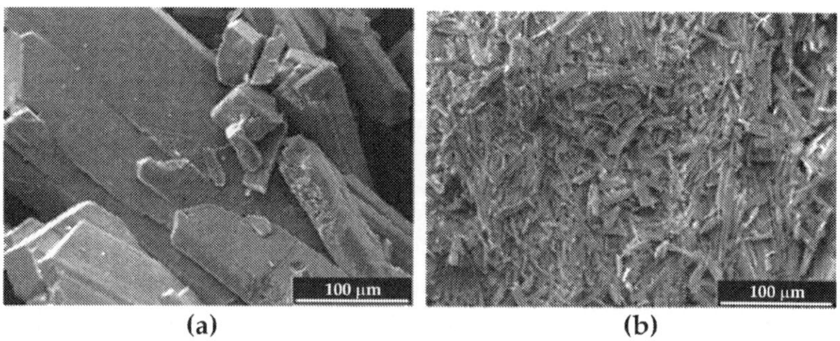

Figure 10.5: Comparison of the product from a drowning-out precipitation using methanol as anti-solvent. (a) The anti-solvent is dosed into the solution of the active, leading to large crystals. (b) The addition of the solution of the active into an anti-solvent leads to small particles. Note that the small primary particles are heavily agglomerated (Beckmann, 2013).

The supercritical or near-critical carbon dioxide is used as a solvent, as the solubility of most organic compounds is substantial near the critical point and decreases readily with decreasing pressure. On the other hand, near-critical carbon dioxide can also be used as anti-solvent, as the critical fluid can dissolve to substantial amounts in organic solvents. At the same time, the solubility of organic compounds in the organic solvent is substantially decreased.

In the Rapid Expansion of Supercritical Solution (RESS) process, the target compound is dissolved in the supercritical solvent, that is, in supercritical carbon dioxide. Due to the a polar nature of the solvent, the solubility is limited; nevertheless, the pressures of up to 300 bar allow for a solution of sometimes >2% at 50 °C. These solutions are rapidly expanded to subcritical conditions, that is, often ambient pressures. The cooling due to the adiabatic expansion limits the pressure ratio. As mentioned earlier, the solubility decreases with falling pressure, leading to a super saturation and subsequently nucleation and growth of crystals. The expansion itself can be carried out very fast; thus, very small crystalline particles are formed. In the case of benzoic acid, a mean particle size on the order of 100nm can be achieved (Figure 10.6).

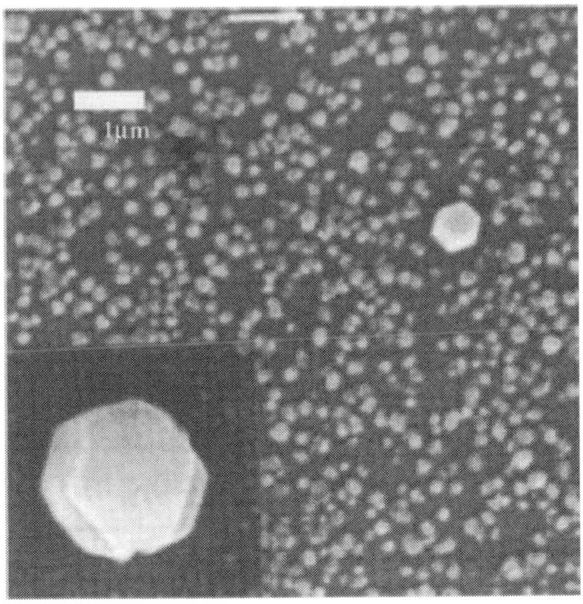

Figure 10.6: Particles of benzoic acid precipitated using the RESS process with supercritical carbon dioxide as solvent (Beckmann, 2013).

Organic solvents are known to dissolve considerable amounts of carbon dioxide when pressurized to near- or supercritical conditions. Concurrently, the solubility of an organic solute in the solvent will decrease accordingly. When the organic solution is rapidly pressurized (Figure 10.7), the precipitation of fine particles occurs. The precipitations can be regarded as an anti-solvent crystallization (Figure 10.8). It is noted that carbon dioxide behaves nearly similar to hexane regarding the modification of solubility and anti-solvent an organic solute in the solvent will decrease accordingly. When the organic solution is rapidly pressurized (Figure 10.7), the precipitation of fine particles occurs.

The precipitations can be regarded as an anti-solvent crystallization (Figure 10.8). It is noted that carbon dioxide behaves nearly similar to hexane regarding the modification of solubility and anti-solvent (Beckmann, 2013).

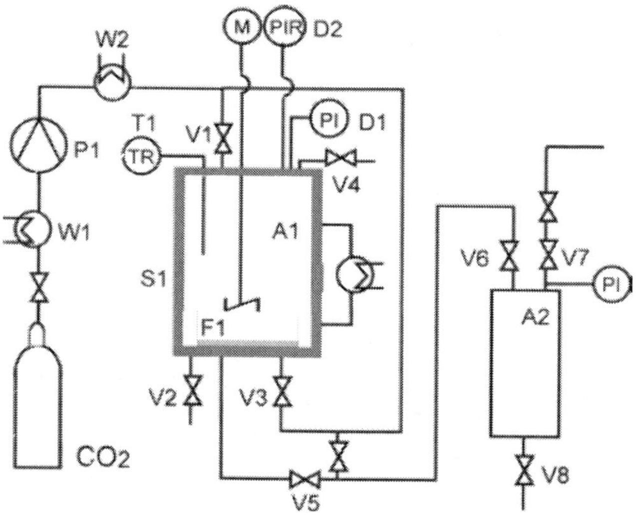

Figure 10.7: Process of an anti-solvent precipitation of a solution of an organic solute in an organic solvent using near-critical or supercritical carbon dioxide. A solution of critical carbon dioxide in the active or solution of active in an organic solvent is prepared and rapidly expanded into a second vessel. The precipitated solid is collected in the latter (Beckmann, 2013).

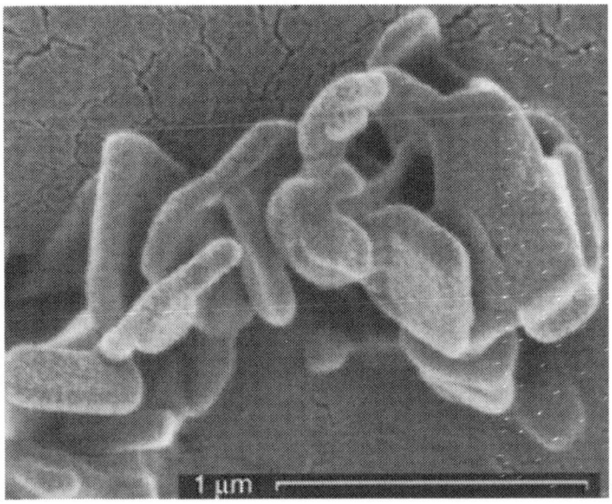

Figure 10.8: Crystals of an active compound crystallized by an anti-solvent crystallization using supercritical carbon dioxide. In the case shown, an inverse drowning-out mixing scheme was used, leading to small particles (Beckmann, 2013).

INDUSTRIAL APPLICATIONS

There are many examples of industrial precipitation practice; calcium carbonate will be studied in the following sections.

Precipitated calcium carbonate is widely used industrially, with different applications usually demanding different physical and granulometric properties. Three different crystallographic forms of $CaCO_3$ can be precipitated (calcite, aragonite and vaterite) depending on the super saturation, temperature, presence of trace impurities, etc (Mullin, 2001).

Calcium sulfate can crystallize from aqueous solution as the dehydrate (gypsum), hemihydrate (α and β forms) or the anhydrous salt (anhydrite: α, β and ‹insoluble› forms), many of which can coexist in contact with solution in metastable equilibrium. The solubility of all forms of calcium sulfate decrease with increasing temperature (gypsum only above 32 °C) so $CaSO_4$ scale commonly forms on heat-transfer surfaces. Simulations of calcium sulfate precipitation as encountered in the phosphoric acid wet process have included studies of nucleation and growth kinetics and the effects of trace impurities.

The precipitation of alumina trihydrate (Bayer hydrate) from caustic (sodium aluminate) solution is an important step in the production of alumina from bauxite by the Bayer process. The reaction below

$$2NaAlO_2 + 4H_2O \rightarrow Al_2O_3 \cdot 3H_2O + 2NaOH$$

is normally carried out in the presence of alumina trihydrate seeds to induce the formation of the appropriate species. Agglomeration plays an important part in the development of the final product.

Magnesium is recovered from seawater and brines on a large scale worldwide by the precipitation of its hydroxide. Some processes use lime as the precipitant while others use calcined dolomite (dolime) which reacts with $MgC1_2$ according to

$$CaO \cdot MgO + MgCl_2 + 2H_2O \rightarrow 2Mg(OH)_2 + CaCl_2$$

Photographic emulsions consist of dispersions of silver halide crystals, around 1 μm in size, in a protective colloid such as gelatin. The light sensitivity of the finished photographic emulsion is to a large extent dependent on the crystal size distribution. Emulsions of high light sensitivity are based on silver bromide containing small percentages of silver iodide. Most commercial photographic emulsions are made by adding silver nitrate solution slowly to an agitated vessel containing an aqueous solution of gelatin containing excess halide. After precipitation of the silver halide, the dispersion may be held at a given temperature to permit ripening (Mullin, 2001).

Many pigments, such as Prussian blue, iron oxide reds and yellows, and the chromate pigments, are manufactured by precipitation processes. The most important group of synthetic organic pigment are the azo group produced by the diazotization of an aromatic amine followed by coupling the soluble diazo compound with a suitable agent to form an azo linkage. This latter step is the precipitation stage:

$$Ar{-}N^+{=}N + Ar'H \rightarrow Ar{-}N{\equiv}N{-}Ar' + H^+$$ Where, Ar and Ar' represent different aryl groups. The color of a pigment depends on the physical form in which it is precipitated and this, in turn, depends on the kinetics of the coupling reaction and the subsequent nucleation and growth steps. The important variables include concentration, temperature, pH, time and presence of impurities and/or additives. Stringent particle

size and size distribution limits around 0.2 μm are usually specified.

Crystalline sucrose develops interesting physicochemical properties when its specific surface exceeds about 2000 $cm^2\ g^{-1}$, which corresponds roughly to a crystal size of around 20 pm. Sucrose crystals of this size and smaller are of importance for icing sugar and for use as seed material in large-scale sugar boiling practice. Such small crystals are normally produced by commination, e.g. in hammer mills, or in the latter case by ball-milling in the presence of a liquid such as isopropanol. Precipitation, however, would appear to offer an alternative procedure with potential advantages. Precipitation of sucrose from aqueous solution may be done through the addition of a water-miscible organic liquid (Mullin, 2001).

Precipitation is increasingly used for the production of colloidal nanoparticles in many fields such as pharmacy and cosmetics. Plasari, Grisoni and Villermaux (1997) described a wide-ranging study on the production of ethylcellulose nanoparticles from solution in ethanol by drowning-out with water. The ethylcellulose concentration in the ethanolic solution was discovered to be the most significant process parameter regulating the distribution of the product particle size. In contradiction of all other expectations, other parameters such as type of agitator and speed, so1vent:non-solvent ratio and temperature had practically no effect. Moreover, the particle size distribution was insensitive to scale-up and a distinct double population of single nanoparticles (< 100 nm) and aggregates (> 200 nm) was obtained in all runs. A comparison was made between the experimental results and predictions made from classical nucleation theories and aggregative growth models.

The recovery of proteins from solution is frequently done by using salting-out, as their solubility is significantly declined by increasing the ionic strength. Ammonium and sodium sulfates are most commonly used for this purpose. It is of critical importance to control the pH: a protein molecule may be represented as a dipolar ion and pH controls the overall net charge on the molecule. A protein exhibits its minimum solubility at the isoelectric point (zero net charge) and for a pure protein this can be very sharp, the solubility increasing greatly even at 0.5pH on either side of the minimum.

Until comparatively recently, the only reported stable inorganic hydrosols were primarily sols of elements such as gold, sulfur, selenium, etc. and compounds such as silica, lead iodate, silver halides, etc. On the other hand, a significant attention is now given to the preparation of mono dispersed hydrous metal oxides, which are chemically considerably more complex than other crystalline or stoichiometrically well-defined materials and are of interest as potential catalysts. Examples include the hydrous oxides of chromium and aluminum (spheres) and copper and iron (polyhedra) with particle sizes $< 1\ \mu m$. One manufacturing procedure consists of ageing aqueous solutions of metal salts at elevated temperatures under controlled conditions of concentration, pH and temperature for several hours in the presence of complexing anions, notably sulfates or phosphate (Mullin, 2001).

REFERENCES

1. Mersmann, A. (2001). Crystallization technology handbook. Marcel Dekker, Inc.
2. Beckmann, W. (2013). Crystallization: Basic concepts and industrial applications. John Wiley & Sons.
3. Borchardt-Ott , W. (2011). Crystallography: An introduction. Springer Science & Business Media.
4. Commission, B. P. (2000). British Pharmacopoeia 2000. London: The Stationary office.
5. Drobny, j. G. (2012). Polymers for electricity and electronics. New Jersey: John Wiley & Sons.
6. Glebovsky, V. (2015). Recrystallization in Materials Processing. InTech.
7. Jones, A. G. (2002). Crystallization process systems. Butterworth-Heinemann.
8. Mullin, J. W. (2001). Crystallization. Butterworth-Heinemann.

9. Nalluri, B. N., Krishna, M. R., Rao, T., & Crooks, P. A. (2012). Effect of Recrystallization on the Pharmaceutical Properties of Valsartan for Improved Therapeutic Efficacy. Journal of Applied Pharmaceutical Science,2(10), 126.
10. Rollett, A., Humphreys, F. J., Rohrer, G. S., & Hatherly, M. (2004). Recrystallization and related annealing phenomena. Elsevier.
11. Tavare, N. S. (1995). Batch Crystallizer: Industrial Crystallization. Springer.
12. Yadhav, A. (2007). Significance of recrystalization on pharmaceutical dosage form processing. Latest Reviews, 3(3).

Integrated Control of Cereal Mildews: Monitoring the Pathogen

Proceedings of a Seminar in the Community Programme of Coordinated Research on Energy in Agriculture held in Freising-Weihenstephan, Federal Republic of Germany, 4–6 November, 1986

Sponsored by the Commission of the European Communities, Directorate-General for Agriculture, Coordination of Agricultural Research

edited by

M.S. WOLFE and E. LIMPERT

Plant Breeding Institute, Trumpington, U.K.

organized by

G. FISCHBECK and E. LIMPERT

Technische Universität München, F.R.G.

1987 **MARTINUS NIJHOFF PUBLISHERS**
a member of the KLUWER ACADEMIC PUBLISHERS GROUP
DORDRECHT / BOSTON / LANCASTER
for
THE COMMISSION OF THE EUROPEAN COMMUNITIES

Distributors

for the United States and Canada: Kluwer Academic Publishers, P.O. Box 358, Accord Station, Hingham, MA 02018-0358, USA
for the UK and Ireland: Kluwer Academic Publishers, MTP Press Limited, Falcon House, Queen Square, Lancaster LA1 1RN, UK
for all other countries: Kluwer Academic Publishers Group, Distribution Center, P.O. Box 322, 3300 AH Dordrecht, The Netherlands

Library of Congress Cataloging in Publication Data

Integrated control of cereal mildews.

(Advances in agricultural biotechnology ; 22)
1. Barley powdery mildew fungus--Integrated control--Europe--Congresses. 2. Barley powdery mildew disease--Europe--Congresses. 3. Wheat powdery mildew fungus--Integrated control--Europe--Congresses. 4. Wheat powdery mildew disease--Europe--Congresses.
I. Wolfe, M. S. II. Limpert, E., 1948- .
III. Commission of the European Communities. IV. Series.
SB608.B2I58 1987 633.1'04943 87-24854

ISBN 90-247-3626-9
ISBN 90-247-2790-1 (series)
EUR 10842 EN

Book information

Publication arranged by: Commission of the European Communities, Directorate-General Telecommunications, Information Industries and Innovation

Copyright/legal notice

© 1987 by ECSC, EEC, EAEC, Brussels-Luxembourg.

All rights reserved. No part of this publication may be reproduced, stored in a retrieval system, or transmitted in any form or by any means, mechanical, photocopying, recording, or otherwise, without prior written permission of the publishers,
Martinus Nijhoff Publishers, P.O. Box 163, 3300 AD Dordrecht, The Netherlands.

Neither the Commission of the European Communities nor any person acting on its behalf is responsible for the use which might be made of the following information.

PRINTED IN THE NETHERLANDS

FOREWORD

This book collects the proceedings of a workshop on "Integrated control to reduce the damage caused by cereal mildews" which was held at Freising-Weihenstephan (FRG) on 4-6 November 1986. Among the 27 participants there were scientists and farm advisers from 15 countries.

Powdery mildew is the most damaging disease of barley, a crop grown on approximately 8.5 millions ha in 1984 in EUR-10 with a total production of 44 millions tons for EUR-10.

The disease is controlled by the use of resistant varieties and by application of fungicides. The latter are expensive and can be harmful to the environment. Therefore, there is a need to develop and improve alternative control techniques.

This workshop, which focussed mainly on barley mildew, comprised 4 sessions dealing with the epidemiology of virulence of barley mildew in Europe, the methods of virulence analysis, the status of resistance gene analysis in various countries and the regional differentiation of virulence and fungicide resistance. A particular emphasis was given to the terminology for resistance and virulence.

The overall objective of the EC agricultural research programme is that of bringing together the considerable resources of scientific knowledge within the Community and stimulating research in subject areas which are vital to guide and support the reorientation of the common agricultural policy.

Vittoria Alliata
Agricultural Research Programme
on Energy in Agriculture

EDITOR'S NOTE

The members of the Workshop were extremely grateful to the Commission of the European Communities, and particularly to Dr. V. Alliata, for help in the organisation and in providing funds and advice, to Prof. Fischbeck and his colleagues in the Technical University of Munich for providing excellent facilities, and to Dr. I.M. Smith, Director-General of EPPO, for his constructive involvement. The editors also wish to thank Mrs K.L. Parr for typing the Proceedings.

M.S. Wolfe
E. Limpert

CONTENTS

CEC Workshop on
INTEGRATED CONTROL TO REDUCE THE DAMAGE CAUSED BY CEREAL MILDEWS
4-6 November 1986, Weihenstephan

RECOMMENDATIONS

Powdery mildew of cereals is an extremely damaging disease in most of Europe. It is caused by a pathogenic fungus which is extremely variable and spread by wind over large distances.

There are three principal ways of controlling the disease: by the use of resistant cereal varieties, by the application of fungicides, or by the integrated use of both measures. The use of fungicides alone has some major disadvantages: increased cost of production, the risk of unwanted environmental effects, and loss of their initial high effectiveness due to adaptation of the pathogen. The use of resistant varieties is therefore likely to gain further importance since it is cheap and environmentally safe. However, genetic resistance may also lose its effectiveness because of genetic changes in the pathogen population.

The widespread use of particular resistances and fungicides, pathogen adaptability and the ease of its dissemination, all encourage the rapid emergence of new and damaging pathogen races throughout Europe.

In an attempt to solve this problem, national programmes have been set up individually in the form of virulence surveys to improve the efficiency of breeding programmes and to ensure effective use of varieties and fungicides.

As the results of the Workshop have confirmed, uncoordinated national efforts are inadequate. The Workshop emphasised the basic need for European surveys of the distribution of virulence and fungicide resistance similar to the survey carried out at Weihenstephan with standardized methods. For further progress a coordinated European approach is essential in order to:

i. Define resistance characters of cereal varieties
ii. Determine distribution of host resistances and fungicide use in different European countries
iii. Detect new virulences and fungicide resistances in the pathogen population wherever they occur in Europe
iv. Monitor the changing composition of the pathogen population over the whole continent.

To this end the Workshop recommends that:

1. National programmes which form the cornerstones of the system should be maintained and extended
2. National programmes should be linked by harmonizing aims, methods and terminology, by exchanging information and materials, and by organizing regular meetings of experts
3. Barley mildew should be used as a model for the establishment of a coordinated information system between European countries which in principle could be extended to all wind-borne diseases of cultivated crops
4. The Workshop considers that these goals can be achieved only by establishing a coordinator to continue the European survey, supported by a European Technical Committee

The Workshop notes and thanks the CEC Directorate-General for Agriculture for its support and recognizes the existence of the EPPO panel on "Strategies for Deployment of Resistance Genes".

The Workshop invites these organizations to consider providing further support for the urgent implementation of these recommendation.

JOINT PROPOSALS

During the course of the Workshop, a number of joint working proposals were agreed. These are summarised below:

1. To determine the changing nature of selection of the pathogen population and to facilitate the interpretation of the pathogen surveys, it is necessary to obtain reliable and precise information on the distribution of the use of varieties and fungicides throughout Europe. Members of the Workshop were encouraged to obtain such information from appropriate sources including breeders and seed companies, extension and plant protection services, and manufacturers.

2. A proposal by J.H. Jørgensen for a common designation of varietal resistances to barley mildew was accepted with slight modifications by the Workshop (see Jørgensen, this volume).

3. An integrated list of barley varieties with their identified resistances designated by these common symbols will be compiled by M.S. Wolfe and J.H. Jørgensen and distributed widely.

4. The different methods of monitoring were compared in order to improve interpretation of different surveys and to facilitate common evaluation. In this context, it was agreed that the Workshop should respond to an offer from Denmark of the supply of barley lines isogenic for mildew resistance genes in the Pallas background, available from Dr. O. Stølen. Sets of these lines could provide an excellent means of integrating data obtained by the different methods using other varieties with the same resistance genes.

5. Several groups reported apparent shifts to increased pathogenicity on ml-o varieties. In order to verify these observations, E. Schwarzbach proposed that each group should carry out a blind test of the candidate isolates in comparison with control isolates on a standard set of 30 differential varieties; the data would then be compiled and distributed.

6. The Workshop agreed that future efforts should be integrated through a European Co-ordinator, supported by a European Technical Committee. The Co-ordinator and Committee would consider establishing an annual survey, analogous to the Cereal Pathogen Virulence Survey in the UK. These Proceedings might be considered to be the first report of such a European survey. Data for a second European Cereal Mildew Survey would then be needed by December 1, 1987, and a second Workshop would be envisaged for November, 1988.

JOINT PROPOSAL 2: DESIGNATIONS OF BARLEY POWDERY MILDEW RESISTANCE AND VIRULENCE IN EUROPE

J. Helms Jørgensen
Risø National Laboratory, DK-4000 Roskilde, Denmark

ABSTRACT

The designations of virulences in barley powdery mildew in Europe are derived from the designations of the corresponding resistances, and those of virulence genes are derived from those of the resistance genes. In order to simplify the information flow between national virulence surveys in Europe a common series of designations for resistances (and virulences) is proposed (Table 1) and recommended.

The national 'Lists of cultivars' in several European countries list the source(s) or designation(s) of powdery mildew resistance(s) for each barley cultivar. The designations used in five countries are listed in Table 1 together with the commonly used names for the respective resistances, and the gene(s) usually present. In most cases the designation is a code related to the name of the donor of resistance, e.g., code Ar refers to 'Arabische' and MC to 'Monte Cristo'. In the Netherlands the codes are mostly derived from the name of a variety, e.g., code e refers to 'Emir' and m to 'Minerva'. In Great Britain BMR (Barley Mildew Resistance) followed by the appropriate number (and letter) are used as codes.

For virulence of powdery mildew the symbol V (virulence) is commonly used followed by the code for the corresponding resistance, e.g., V_{AR} or V_{Ar}, and V_{MC} or V_{Mc} for 'Arabische' and 'Monte Cristo' virulence, respectively. In Great Britain the codes are BMV (Barley Mildew Virulence) followed by the appropriate number (and letter), e.g., BMV 5 and BMV 8. If virulences refer to individual resistance genes the symbol V is followed by the gene symbol, e.g., V_{a9} or Va9 or V(Ml-a9) for virulence corresponding to resistance gene Ml-a9.

This heterogeneity in designations (codes) used in Europe for powdery mildew resistances in barley, and, therefore, also for virulences in the powdery mildew fungus, was discussed in the Workshop. The following two conclusions were reached:

1. A common series of designations for resistances in spring and winter barley cultivars (and for virulences) was proposed and recommended (Table 1). A two-letter code is used with a capital and a small letter (or two capitals) derived from the common name of the source of

resistance, e.g., Ar for 'Arabische' or MC for 'Monte Cristo'. When it becomes known that a new source has a resistance gene in common with a named source, the code of the latter is used and, eventually, with a new code for additional resistance derived from the new source. For example, the code for resistance from 'Multan', comprising gene *Ml-a7* the main gene in 'Lyallpur' (code Ly) and one or more additional resistance genes, will thus be Ly, Mu.

2. The symbols used for powdery mildew resistance genes (and the corresponding virulence genes) are fairly common in Europe. They differ mainly in style, e.g., *Ml-a* = *Mla* = *Mla1* = *Ml-*a_1 and *Ml- (Ab)* = *Ml-Ab* = *MlAb*. In the present context there is no need for a common designation.

REFERENCES

Doll, H. and Jensen, H.P. 1986. Localization of powdery mildew resistance gene *Ml-ra* on barley chromosome 5. Hereditas *105*, 61-65.

Christensen, E. 1986. Specifikke meldugresistensgener i byg (in Danish). Thesis. Royal Vet. and Agric. Univ. Copenhagen. Risø-I-Report.

Søgård, B. and Jørgensen, J.H. 1987. Supplementary list No. 1 (to masterlist of barley genes): Genes for reaction to *Erysiphe graminis hordei* (powdery mildew). Barley Genetics Newsletter *16* (in press).

DESIGNATIONS OF BARLEY POWDERY MILDEW RESISTANCES IN SOME EUROPEAN COUNTRIES (from national 'List of Varieties').

Common name of resistance	Resistance gene(s) usually present[1]	Designations (codes) of resistances in: AU	D	DK	GB	NL	Designation recommended
Abessinian(EP72)	Ml-(Ab)	-	-	Ab	BMR6c	E^{11}	Ab
Algerian	Ml-a	Al	AL	Al	BMR7	-	Al
Arabische	Ml-al2 [=Mlas]	Ar	AR	Ar	BMR5	e	Ar
Laevigatum	Ml-(La) [=Mlv]	La	La	La	BMR4	m	La
Long Glumes	Ml-a7, Ml-(LG2)	-	-	LG	-	-	Ly, LG
Kwan	Ml-k	Ly	LY	Ly	BMR6a	l	Kw
Lyallpur	Ml-a7				BMR6b		Ly
Ml-o (L 100) / (EP79)	ml-o	mlo	ml-o	Ml-o	BMR9	L / E^{1}	Mlo
Monte Cristo	Ml-a9	Mo	MO	MC	BMR8	MC	MC
Multan	Ml-a7, Ml-(Mu2)	-	-	Mu	-	-	Ly, Mu
Ricardo	Ml-a3	-	-	Ri	-	-	Ri
Rupee	Ml-al3, Ml-(Ru3)	Ru	RU	Ru	BMR10	-	Ru
Spontaneum	Ml-a6, Ml-al4	Sp	SP	Sp	BMR3	-	Sp
Turkish	Ml-a(3), Ml-(Tu2)[3]	-	-	Ty	-	-	Ri, Tu
Weihenstephan	Ml-g, Ml-(CP)	CP	CP	We	BMR2	w	We
Hauters	Ml-h	-	HA	We	BMR1a	-	Ha
(Ragusa)	Ml-ra[2]	-	-	We	BMR1b	-	Ra
Weihenst./Spont.	Ml-g, Ml-a6, Ml-al4	-	-	-	-	ip	-
Wong	Ml-(Wo)	-	WO	-	BMR11	-	Wo
Unknown 1	unknown	u1	u1	?	-	div.	u
Unknown 2		u2	u2				
None	none	-	keine	ingen	BMR0	-	-
Slow mildewing	-	s	-	-	-	-	-

[1] Gene designations used in Denmark and [] in England. Gene symbols with parentheses, e.g., Ml-(Ab), are tentative (cf. Søgård and Jørgensen, 1987).

[2] Gene Ml-ra = Ml-(41/145), Doll and Jensen (1981).

[3] Christensen (1986).

JOINT PROPOSAL 5: SHIFTS TO INCREASED PATHOGENICITY ON ML-O VARIETIES

Compiled by E. Schwarzbach
Agricultural University of Vienna, Vienna, Austria

ABSTRACT

Suspect ml-o virulent isolates were tested in different laboratories on a coded set of varieties. Subsequent analysis of the decoded data revealed that several isolates from different countries had some virulence for ml-o, but less than that of isolates previously selected on ml-o in the laboratory.

INTRODUCTION

The ml-o resistance gene is now widely used in barley breeding programmes and commercial production because it is the only resistance gene so far known against which virulent races have not developed under field selection. Monitoring for the possible occurrence of virulence is thus of primary importance for barley breeding and agriculture in Europe.

During the course of the workshop, there were reports from Austria, Denmark, Poland and the UK of mildew infection on crops of ml-o varieties. A joint experiment was therefore undertaken to confirm whether or not the pathogen was adapting to ml-o under field conditions.

Three replicates of a set of barley varieties including some with ml-o and others with different resistance genes were distributed to each of the participants. The seed packets were coded randomly so that the varieties could not be identified. The participants were asked to inoculate their suspect ml-o virulent isolates on to the test varieties, using guidelines agreed during the workshop:

1. Uniform inoculation in a settling tower on to horizontally exposed primary leaves of the test varieties. This is necessary because of the quantitative nature of the ml-o resistance.
2. Inoculum density not to exceed 300 spores/cm^2 of leaf area because plants with ml-o tend to become infected under heavy inoculum load.
3. Incubation at less than 18°C because the Ml-La resistance in some of the test varieties does not operate at high temperatures.
4. Evaluation in terms of successful infections per cm^2 of leaf area.
5. Results to be sent to Dr. Schwarzbach for analysis and compilation.

EVALUATION OF DATA

For various reasons, the guidelines were not followed exactly by all participants and so all of the evaluation was done on a relative basis. The variety with the highest level of infection was set at 100%. This was usually Carina, but in the UK this variety was moderately resistant to most cultures and so Golden Promise, combined with inoculum density, was used instead. On the basis of spores applied per cm^2 and colony numbers produced, the data from all participants were then comparable.

Data from Triumph were inconsistent, due to the multiline character of the variety, and were therefore discarded. Isolates with too many missing data were also excluded from the final analysis.

RESULTS

The results are summarised in Table 1; data from Denmark were excluded because of inconsistency between replicates and the low infectivity of the inoculum. Data from the other countries were mostly consistent between replicates. The nature of the data does not permit a satisfactory test of statistical significance, but the summarised results were clear enough for the following conclusions:

1. Most of the isolates had no appreciable ml-o virulence and were similar to those isolated before the introduction of ml-o varieties.
2. Some of the isolates, for example, B272 from the UK, AT10 and AP20 from Austria, and APE and SAL from Poland, exhibited an obviously increased level of ml-o virulence.
3. The ml-o virulent isolates differed in their spectra of virulence suggesting independent origins.
4. The three ml-o varieties differed in their level of resistance to ml-o virulent isolates. Apex was the least resistant, followed by Atem, with Alexis the most resistant, indicating additional genes with minor effects in the resistance background of Atem and Alexis.
5. None of the field isolates were as virulent as the laboratory-derived isolate HL-3 and its sub-culture AP6. This suggests that the pathogen adapts gradually to ml-o and that higher levels of virulence are likely to be found in field isolates during the coming years.

FUTURE WORK

In order to obtain more significant data on further adaptation of the pathogen to ml-o in the field, it would be helpful if suspect isolates could be tested simutaneously under identical conditions in a single laboratory well suited to such experiments.

TABLE 1 Summary of data on ml-o virulence from all experiments. Values are colony numbers relative to those on susceptible standards.

Isolate	Origin	Carina	Aramir	Koral	Varunda	Apex	Atem	Alexis	comment
GE-3	D	100	5	0	55	0	0	0	Old isolates with no apparent
C/1	UK	0	0	0	1	0	0	0	ml-o virulence and isolated
C/21	UK	40	1	0	50	1	1	0	around 1972. No ml-o varieties
C/12	UK	0	100	0	10	4	1	0	grown at that time.
B266	UK	26	38	0	49	1	2	0	Recent field isolates with no
B270	UK	48	1	0	52	1	2	0	or very low level of ml-o
B271	UK	50	99	0	55	2	2	0	virulence.
B274	UK	0	0	0	58	2	2	0	
B269	UK	52	100	0	56	1	3	0	
B267	UK	58	100	0	59	2	3	0	
B268	UK	36	86	0	51	2	4	0	
B273	UK	20	82	0	69	3	3	0	
B275	UK	76	0	0	57	7	1	0	Recent field isolates showing
B272	UK	0	0	0	61	8	4	0	some degree of ml-o virulence.
VA14	A	86	100	0	81	6	6	3	
AP15	A	43	100	0	58	14	2	1	
AT10	A	100	73	0	95	12	14	0	
RU10	A	100	76	88	36	14	0	0	
AP20	A	100	75	0	43	24	1	3	
APE	PL	88	61	100	83	13	3	0	
SAL	PL	63	78	100	99	24	5	0	
HL-3	D	100	1	0	41	33	6	1	Isolates developed from GE-3 by
HL-3	A	100	0	0	32	54	0	10	48 generations of mass screening
AP-6	A	100	0	0	16	39	0	9	for ml-o virulence

TABLE 2 Decoded numbers of test varieties used.

1	Carina	9	Triumph	17	Koral
2	Koral	10	Varunda	18	Alexis
3	Triumph	11	Aramir	19	Atem
4	Alexis	12	Carina	20	Apex
5	Varunda	13	Atem	21	Triumph
6	Atem	14	Koral	22	Varunda
7	Aramir	15	Alexis	23	Aramir
8	Apex	16	Apex	24	Carina

DISTRIBUTION OF VIRULENCE AND OF FUNGICIDE RESISTANCE IN THE EUROPEAN BARLEY MILDEW POPULATION*

E. Limpert and G. Fischbeck

Technische Universität München, Lehrstuhl für Pflanzenbau und Pflanzenzüchtung, 8050 Freising-Weihenstephan, Deutschland (BRD)

ABSTRACT

For monitoring populations of wind-dispersed pathogens like *Erysiphe graminis* f.sp. *hordei*, investigations within and between regions are desirable. Representative random samples of pathogen spores were taken from the atmosphere over important barley growing areas in Europe by means of an apparatus containing a Schwarzbach jet spore sampler. Single colony progenies were tested in the laboratory for their virulence against Mlra, Mlg, Mla6, Mla12, Mla7+Mlk, MlLa, Mla1, Mla9, Mla3, Mla13 and mlo. Virulence frequencies mostly showed regional differentiation and effectiveness of the resistance factors generally increased in the order given. Although according to previous results Va6 seemed to be fixed in the European mildew population, there was in 1986 a considerable decrease of the frequency of this virulence in areas of epidemiologic importance for Europe. In addition, the changes in frequencies of Va12, Va7, Va9 and Va13 point to possibilities for improved utilisation of host resistance. Frequency and level of resistance against triadimenol also showed significant regional differences. Against fenpropimorph the European population reacted similarly to standard isolates and there were few regional differences. The results are discussed with respect to current and previous conditions of cultivation, to the spread of the pathogen by wind, and towards strategies for more effective use of host resistance and of fungicides.

INTRODUCTION

A method for monitoring the occurrence and distribution of virulence and fungicide sensitivity characters in barley mildew was developed in Weihenstephan at the end of the 1970s with the support of the German Research Foundation. Since then the situation both in Germany (FRG) and in neighbouring countries has been monitored. The results on regional differentation and on the dynamics of the populations showed that vast parts of Europe have to be considered as an epidemiologic unit due to the spread of the mildew population by wind (Limpert et al., 1984b; Limpert, 1985; Limpert, this volume).

These results had two consequences. First, the Commission of the

---

*This work has been supported by the Commission of the European Community and by the German Ministry of Agriculture. Their help is gratefully acknowledged.

European Community (CEC) agreed to support investigations covering the countries of the European Community since 1985. Second, the CEC agreed to support this workshop to improve coordination between the groups already working in this area.

The aims of our supra-national investigation are

1) to enlarge and deepen the knowledge on the epidemiology of pathogens spread by wind (Limpert, in press, a) in order to improve strategies for more effective use of host resistance and of fungicides, and
2) to extend advice for the choice of varieties and of fungicides to those countries in which such investigations have not been made (Limpert, 1986a,b).

As the situation elsewhere is treated in other contributions, this paper reports mainly on the European countries investigated with additional information for the FRG. Recently, a similar approach was made to analyse the population of the wheat mildew pathogen in Europe (Felsenstein et al., this volume).

MATERIALS AND METHODS

Representative random samples of mildew spores were obtained from the air by means of an apparatus containing a jet spore sampler* (Schwarzbach, 1979), mounted on the roof of a car, while driving across regions of interest for epidemiology and for barley cultivation. Sampling distances were on average 100 km. They did not cross frontiers between countries nor frontiers between areas with different barley cultivation within a country (Fig. 1; Table 1). Trapped spores fall on to segments of primary leaves of the susceptible lines HJ or SM 4142, laid in a Petri dish of 12 cm diameter on 0.5% agar containing 15 ppm of benzimidazole. In order to obtain samples of the population at similar stages of crop development, trapping was started in the southernmost regions and extended northward, following the development of the vegetation.

For virulence analysis eleven differential varieties with the resistances indicated were used: Union (Mlg), Carina (Mla6+Mlg), Vada (MlLa), Ortolan (Mla7+Mlk), Sultan (Mla12), Sewa (Mla3), Gitte (Mla1), Welam (Mla9), Atem (mlo+MlLa) and Rupal (Mla13+MlRu3). Seed was tested prior to virulence analysis using two to three differential isolates of

*Product of Burkard Manufacturing Co. Ltd, Rickmansworth, Herts. WD3 1PJ, England.

mildew in order to ensure that it did not contain more than 3% aberrant seed.

Plants were grown for 10 d (19°C, 92 $\mu Em^{-2}s^{-1}$, 10,000 lux continuous light) under mildew-proof conditions. Test sets on agar generally contained one segment (1.2 cm) of each variety, cut from the middle part of primary leaves. After incubation of the sampled spores (8 d, 17°C, 5.5 $\mu Em^{-2}s^{-1}$, 960 lux, continuous light) each set was inoculated with the progeny of a single colony isolate by sucking the conidia of one colony into a pipette and blowing them into a miniature settling tower. Inoculum density was in the range of 250-400 conidia cm^{-2}.

After incubation, infection severity was scored for quality and quantity of attack, that means for reaction type (0-IV) and relative colony number (0-1.0). For estimation of virulence frequencies only high levels of attack were taken into account: type IV and a colony number of at least 50% (=0.5) compared with the most heavily attacked variety (=1.0). The epidemiologic reasons for this evaluation are as follows: a) type IV infections produce more than the double amount of conidia compared with the other types, thus outnumbering them in competition, b) if the mean of high quantitative levels of attack is 0.75 (the mean of 0.5-1.0) and that of the remainder is 0.25 (mean of 0-0.5) then within 10 generations a difference of 60,000:1 would have developed ($0.75^{10}/0.25^{10}$). Thus the term virulence is used in this context to designate the ability to create type IV infections, combined with a certain level of aggressiveness (sensu Vanderplank, 1968).

Fungicides under test were triadimenol and fenpropimorph, kindly provided by the respective manufacturers. The reaction of each isolate, tested on segments of primary leaves grown from treated seeds (Igri, Mlra), was compared with the reaction on leaf segments from untreated plants inoculated simultaneously. The highest dose of fungicide was determined that allowed at least 50% sporulation. Levels of resistance were determined by comparing infection at this dose with that obtained from four standard isolates, Al-1, Gi-1, Tr-2 and Ru-3, which were isolated from the field before introduction of these fungicides. The method is described in more detail elsewhere (Limpert et al., 1984a; Limpert, 1985); for designation of resistances and virulences see Jørgensen (this volume). The results from the regions investigated in Europe are given mainly in histograms. Different hachures accentuate small values in order to highlight those regions in which the resistance gene or fungicide is most effective.

RESULTS AND DISCUSSION

Distribution of virulence in Europe in 1985

The frequency of Va6 was close to 100% almost everywhere in Europe and the corresponding resistance was still of practical value only in the western region of France (Fig. 2). Virulence against Mlg was even more frequent (Table 1). As Va6 was selected mainly during the past in the FRG and the UK and as selection was nearly absent in Denmark, France and in the regions of Austria investigated, the high frequency of Va6 in the latter countries provides strong evidence for the spread of the fungus population by wind, which is treated in detail elsewhere (Limpert, this volume).

Following selection on successful varieties, Va12 has increased markedly during the last decade, especially on the continent. The lowest values in 1985 were found in the UK and in Denmark (Fig. 3), where Va7+Vk was most frequent (Fig. 4), mainly due to the popularity of the variety Triumph.

Mla9 still provides good disease control in most parts of Europe. Nevertheless the virulence increased, except in Denmark, during the preceding four years by approximately five-fold, leading to the values in Fig. 5.

Resistance from *Hordeum distichon laevigatum* (MlLa) has been used extensively over a long period, mainly in England and even more so in Denmark, where it covered up to 80% of the barley area at the end of the 1970s. Since the specific matching virulence nevertheless remained at a rather low level, the resistance is considered to be more durable than the race-specific resistances mentioned previously. Fig. 6, however, indicates that considerable specificity for MlLa-varieties does exist. Selection on varieties like Golf or Menuet led to unexpectedly high values of the matching virulence in France. This may threaten the large number of MlLa-varieties on the European market, because the mildew population in France is in a position favourable for spread by wind to other parts of Europe within the near future.

Virulence against Mla1, Mla3 or Mla13 was uncommon (Table 1), although the highest values for each were found in Denmark. The frequencies of Va1 generally ranged between 0-10%, but for southern Jutland and for Sealand/ Lolland 30% and 17%, respectively, were recorded. Va3 was found in England (0.2%), northern Germany (1.2%) and in Denmark (1.5%).

The frequency of Va13 generally was at or below the detection level (Table 1). The situation in three regions, however, is worth noting:
a) In Austria Va13 (7%) was found only in region A3 (Fig. 1), adjacent to Czechoslovakia, where corresponding varieties were already heavily attacked (according to Brückner, this volume). This points to immigration of Va13 from this neighbouring area by wind.
b) In Denmark Mla13-varieties were cultivated on increasing areas (0%, 0.5%, 3% and 14% from 1982 to 1985 respectively, Jørgensen, pers. comm.). The corresponding virulence frequency was 6% in 1985, whereas in 1983 and 1982 no virulent isolate had been found amongst, respectively, 108 and 382 isolates tested. Thus an early stage of a change in the composition of the pathogen population had been observed.
c) In Rheinland-Pfalz (south-western Germany) the frequency of Va13-virulence compared with that of the corresponding resistance was even lower. No virulent isolate was found amongst 165 isolates tested in 1985, although the variety Koral (Mla13) had been grown on 10 to 15% of the barley area from 1982 to 1985 (Table 3).

Distribution of virulence in Germany (FRG) in 1985

The distribution of host resistances in Germany (Table 3) was calculated from information given in the 'Beschreibende Sortenliste' and in the 'Besondere Ernteermittlungen'. The high frequency of Va12 in southern Germany is consistent with the popularity of Mla12-varieties in Rheinland-Pfalz and more so in Baden-Württemberg, where the frequency of Mla12 exceeded 30% during the recent past (Table 3). An additional influence of immigration of Va12-virulence from France probably occurred.

Concerning the frequency of Va7, Va9 and VLa, the situation was similar in that values were considerably lower in the south than in the north (Figs 4, 5 and 6). For Va9 and VLa this corresponded with the frequencies of the corresponding resistances (Table 3). In one region in Baden-Württemberg (D13) the frequencies of both Va7 (Fig. 4) and Va9 (Fig. 5) were notable as they were significantly higher than those in the adjacent regions. The assumption that this was due to a different choice of varieties in that particular region was confirmed by a local breeder (Frank, pers. comm.).

The frequency of Va7 has decreased considerably during the last seven years, most obviously in southern Germany and particularly in Rheinland-Pfalz where a decrease from 80% to 10% was observed. This indicates a possibility for recurrent use of race-specific resistance.

Distribution of fungicide resistance in Europe in 1985

Economically important systemic fungicides with specific action are the inhibitors of sterol biosynthesis. One representative was chosen for each of the two main groups, triadimenol for the group of inhibitors of demethylation, and fenpropimorph for the morpholines, which act differently. The variety Igri (Mlra) was used for fungicide tests because it is widely grown in Europe and only one out of 881 isolates was avirulent to it.

Regional differences in resistance to triadimenol were obvious (Table 2). The significance of different levels of resistance and of different compositions of mildew populations can be deduced from the decline of the concentration of seed-applied triadimenol in the growing plant and from the duration of effect of the treatment against the original, sensitive mildew population (Steffens et al., 1981). Whereas, for example, a resistance level of 4 should enable the pathogen to multiply on the new parts of the plant from tiller emergence, thus reducing the duration of effect of the fungicide by two to three weeks, a level of 40 or more must be considered to allow the growth of the pathogen in the field even on the first leaf of seedlings grown from seed treated with the recommended dose (Limpert, 1985).

Fig. 7 illustrates the distribution of frequencies of resistance >49. The frequency of these isolates was low in southern Germany with the exception again of region D 13, and even lower in Austria and in western France, whereas very resistant populations were found in southern England and northern Germany. A more detailed picture (Fig. 8) shows that low levels of resistance occurred only in the former regions. In general the levels and frequencies of resistance against triadimenol were correlated with the intensity of fungicide use.

Against the morpholine, fenpropimorph, regional differences were far less pronounced. The reaction of 420 isolates from different regions was similar to that of the standard isolates, and parallels between sensitivity of isolates and intensity of fungicide use were not established.

The mildew population in 1986 and main changes from 1985-1986

Sampling routes in 1986 (Fig. 10) were mainly the same as in 1985 (Fig. 1). The area investigated was extended, however, to Scotland, southern France and northern Italy.

Distribution of virulence: The population in southern France (described in more detail by Andrivon and de Vallavieille-Pope, this volume) and in northern Italy, differed markedly from that in remaining areas in north-western Europe (Table 4). The pathotypes from these regions had a smaller number of virulences against the resistance genes analysed. There are thus efficient means for mildew control with a number of varieties present on the European market. In northern Italy even the frequency of Vg was only around 50%, and also the low frequencies of the other virulences indicate that the Alps are an epidemiologic barrier for the exchange of mildew populations, at least in the direction of Italy.

The most striking difference was observed for a virulence which was regarded to be almost fixed in the population, Va6 (Fig. 2, Table 4). Its frequency decreased considerably in many regions. The greatest decline was noted in England, from where the population on the continent can be influenced significantly. There was a decrease of approximately 40 to 70% from northern England via the west to the south, and 30% in the east (Newcastle-Cambridge). On the other side of the Channel the decrease continued along both sides of the Belgian-French frontier, in Belgium and in northern and north-eastern France (minus 20%) and reached more than 30% reduction in northern Baden-Württemberg, region D 20, D 21 (Fig. 10). A decrease of 20 to 35% was observed on the Danish islands, and in the remaining area of north-western Europe the frequency remained rather constant.

Concerning the frequency of Va12 a reduction of 10 to 30% was observed in large areas (Fig. 3; Table 4), most obvious in the whole area of approximately 250,000 km^2, composed of northern France, Belgium, the southern Netherlands and south-west Germany, but also in Austria and Denmark. Only in England, northern Germany and in limited areas of southern Germany were the values for 1986 similar to those for 1985.

Conversely, but less pronounced and with slight changes in comparison with the preceding example, there was an increase, almost everywhere, of the frequency of Va7+Vk, most clearly expressed in northern England and in south-west Germany (plus 20-30%), and in Austria (plus 20%). Exceptions were the area of northern Germany (minus 10-30%) and the south-east part of England (minus 10%). In most regions there was also an increase of 10-20% in the frequency of VLa, which can be explained by the growing popularity of varieties with the corresponding resistance and by the spread of this virulence from France.

Compared with the former changes, changes in frequency of Va9 were regionally limited and less obvious. An increase of more than 20% was recognized in the eastern part of Lower Austria (A 4,5), and around 10% in the northern half of Baden-Württemberg (D 20, 21, 25; Fig. 10; Table 4) and in western UK (GB 2, 3, 4). Values from Reading in the south via Cheltenham to Birmingham, Preston and Carlisle up to Motherwell near Glasgow were between 14% and 23%, whereas values in the east, from Dover to Edinburgh, ranged between 0 and 10%. A pronounced decrease between 10 and 30% was observed in most parts of Denmark and of Schleswig-Holstein (D 1-3), whereas a slight decrease was observed in the area from southern Netherlands to Belgium and to the northern half of France.

The limited area of distribution of Va13 hardly increased, but there was a significant change in the frequency of this virulence. The increase from 6 to 14% in Denmark corresponds to increasing selection, as described above. The increase in Austria from 7% to nearly 30% cannot be explained by regional selection as the area of multiplication ranged rather constantly around 4% during the years 1982-1985 (Rossoll and Kweta, 1983-1986). The increase observed is better explained by immigration of the mildew population from neighbouring Czechoslovakia where Mla13 was no longer effective (Brückner, this volume). In Rheinland-Pfalz, again, no virulent isolate was found despite considerable use of cultivars with this resistance.

The virulence combination, Va13+Va9+Va7, is of significance in this respect, because it predominated in the Va13-subpopulation. Therefore three main changes in the composition of the mildew population in Lower Austria, causing an increase of a virulence frequency of approximately 20% each, can be explained by the change of frequency of this combination. Thus the efficiency of three main resistances is impaired, reducing considerably the diversity of varieties able to control mildew.

The results of 1986 as well as the development during the period 1985-1986, provided further indications that both the distribution and the development of the frequency of Va12 were contrary to those of Va7. The different development was described above. Concerning this distribution, the population investigated in Scotland can be regarded as an example. In the west (GB 5) the frequency of Va12, 19%, was rather low, whereas 81% for Va7 was the maximum in the UK. Conversely, the maximum for Va12 and the minimum for Va7 were in the east (GB 7). These are further indications for the unexpectedly low frequency of this virulence combination. This feature can be exploited agronomically, at least to a certain extent, by growing varieties with these resistances side by side, or in mixtures.

Additional support from the German Ministry of Agriculture facilitated trapping over a denser net of spore sampling routes in Germany and to obtain more reliable data from each region. In the northernmost part (Schleswig-Holstein) the frequency of Va6 and Va12 remained constant, whereas there was a decrease of approximately 20% of Va7 and Va9. In the north-western part of the country, not investigated in the previous year, the mildew population was better adapted to Mla9-varieties than elsewhere in the country. In that part, even Mla12 or Mla7 resistances were able to give better disease control.

In southern Germany the outstandingly high values of Va9 and Va7, first observed in 1985 only in one region, Baden-Württemberg (D 13; Fig. 1), were confirmed in 1986 (D 21; Fig. 10; Table 4). Their region of occurrence, however, had enlarged in the meantime. Va9 was now present with frequencies of 21 to 28% in three regions (D 20, D 21, D 25; Fig. 10). Va7 was also present in regions D 20 (62%) and D 24 (26%) besides the region of origin, where its frequency had increased from 48% to 75%. This increase was accompanied not only by a decrease of 30% in the frequency of Va12, as expected from the example described above, but also by a similar decrease in the frequency of Va6. In 1985, and probably in 1986, there was little selection by Mla12- and Mla6-varieties in Baden- Württemberg (Table 3), thus allowing for such a decrease in virulence frequency.

Distribution of fungicide resistance: For the investigation of the sensitivity of the mildew population against triadimenol a wide range of concentrations was used in 1985 (Table 2). Because of the levels of resistance encountered, the range was enlarged further in 1986, with a greater number of intervals (Table 5) in order to determine more exactly the composition of the mildew population.

The population in northern Italy which was investigated for the first time, differed markedly from that in the remaining areas also with respect to sensitivity against triadimenol. It was the most sensitive population observed and some isolates were more sensitive than the standard control isolates.

The results obtained in 1985 and the pattern described in Europe (Figs 7, 8; Table 2) were mainly confirmed in 1986 (Table 5). Thus the tendency towards increasing resistance observed in the beginning of the 1980s

(Limpert, 1985) was, in general, not continued. There was even an increase in sensitivity in the UK although the population remained considerably resistant. On the other hand a slight but significant increase of resistance was observed throughout Germany.

Outstandingly high levels of resistance were observed in a particular area of northern Germany (D 3a; Fig. 10; Table 5) in which fungicides are most intensively used; levels of resistance >512 were present there as well as in certain regions in Denmark, the UK and France. In France the highest values of the frequency of resistance >64 were observed in Champagne (F 12 St. Menehould-Reims). In this region, investigated for the first time in 1986, the values were similar to those in Schleswig-Holstein in northern Germany where the barley mildew population most resistant to triadimenol was found. A comparison of two random samples obtained in mid-June and mid-July, respectively, from the latter region revealed no substantial difference in the resistance of the population samples.

Investigations of the sensitivity of mildew against fenpropimorph were concentrated in 1986 in regions in which fungicides were intensively used for mildew control. Compared with results from the preceding year the same degrees of sensitivity were observed. The standard isolates again had a mean position. In some regions the distribution of the population was slightly but significantly displaced towards reduced sensitivities. These results should be treated with care, as they could indicate the first sign of a change in the composition of the pathogen population, but further investigations are needed because the results were not consistent in all the regions concerned. Even in regions such as Schleswig-Holstein, where fungicides are applied very intensively, the sensitivity of random samples was distributed symmetrically around that of the standard isolates and was similar to the sample from northern Italy, which represents an original, unselected population.

Perspectives

The decrease of Va7-frequency in Rheinland-Pfalz in southern Germany since the late 1970s suggests a possibility for recurrent use of host resistance. The same possibility may exist for Mla7-varieties in the UK (but see Wolfe et al., this volume), as well as for Mla9-varieties in Denmark and northern Germany, and for MlLa-varieties in France.

Detailed and sound results on combinations of virulences, obtained from a large European area, may give rise to fundamental improvements in the knowledge of the ability of the pathogen to combine diverse virulences. This knowledge may be used for the development of strategies for the improved utilization of host resistance. The data of Wolfe et al. (this volume) on negative associations between Va7 on the one hand, and Va12 and VLa on the other, were confirmed by our own data from northern and western regions (Figs 3, 4 and 6), but further evaluations have to be made to be able to draw the correct conclusions.

To be able to explain the structure of the pathogen population and its dynamics, there is often a crucial need for sound data of the barley varieties cultivated in a region and of their fungicide treatment. With respect to the distribution of varieties, such data exist for Denmark and in even more detail for the FRG and for France. From other countries, however, the data still need to be obtained together with data on the use of fungicides for the whole area.

Several results confirm the earlier view that large parts of Europe have to be regarded as a single epidemiologic unit. Depending on the geographic position, the pathogen population of a given region is more or less able to influence the development in another. Spread fom west to east has often been recognized whereas the distribution of Va13 and especially the frequency of Va13 in Rheinland-Pfalz below the detection level may be an additional indication for reduced dispersal of a virulence in the opposite direction. This confirms earlier data on the distribution of Va9 in Europe. The advantages for a coordinated use of host resistance and of fungicides in Europe, taking into account the dispersal of mildew mainly from west to east, are obvious. Even if the system depicted in Fig. 9 is not applicable as a whole it does provide a demonstration of the basic principles involved.

REFERENCES

Anonymous, 1986. Beschreibende Sortenliste Getreide-Mais-Ölfrüchte, Bundessortenamt, Strohte-Verlag, Hannover.

Anonymous, 1986. Besondere Ernteermittlungen bei Getreide und Kartoffeln, 1984-1986. Bundesministerium für Ernährung, Landwirtschaft und Forsten.

Jørgensen, J.H., 1983. Durability of barley powdery mildew resistance genes in Denmark 1963-1980. In: Durable resistance in crops, F. Lamberti, J.M. Waller and N.A. van der Graaff (eds.), 396-399.

Jørgensen, J.H and Torp, J., 1978. The distribution of spring barley varieties with different powdery mildew resistances in Denmark from 1960 to 1976. Kgl. Vet.-og Landbohøjsk. Årsskr. 1978, 27-44.

Limpert, E., 1985. Ursachen unterschiedlicher Zusammensetzung des Gerstenmehltaus, Erysiphe graminis DC. f.sp. hordei Marchal, und deren Bedeutung für Züchtung und Anbau von Gerste in Europa. Diss. München, 183 pp.

Limpert, E., 1986a. L'oidium de l'orge: quelles races en France? Une nouvelle approche pour aider aux choix des variétés et des fongicides. Perspectives Agricoles, 105, 35-39.

Limpert, E., 1986b. Gerstenmehltau: Pflanzenschutzstrategie überdenken. Pflanzenschutz-Praxis 2, 18-24.

Limpert, E. (in press, a). Spread of barley mildew by wind and its significance for phytopathology, aerobiology and for barley cultivation in Europe. Advances in Aerobiology, G. Boehm, R.M. Leuschner (eds.), Proc. 3rd Int. Conf. Aerobiology, Experientia, Birkhäuser, Basel.

Limpert, E. (in press, b). Frequencies of virulence and fungicide resistance in the European barley mildew population in 1985. J. Phytopath.

Limpert, E., Schwarzbach, E. and Fischbeck, G., 1984a. Influence of weather and climate on epidemics of barley mildew, Erysiphe graminis f.sp. hordei. In: Interactions between climate and biosphere, H. Lieth, R. Fantechi and H. Schnitzler (eds.), Progress in Biometeorology Vol. 3, Swets & Zeitlinger B.V., Lisse, 146-157.

Limpert, E., Schwarzbach, E. and Fischbeck, G., 1984b. Distribution of virulence in the European population of barley mildew, Erysiphe graminis f.sp. hordei, Proc. Eucarpia Cereal Sect. Meeting, Weihenstephan, Vorträge für Pflanzenzüchtung 6, 1-21.

Rossoll, E. and Kweta, G., 1982 (-1985). Feldanerkennungsflächen 1982 (-1985). Jahrbuch 1982 (-1985) der Bundesanstalt für Pflanzenbau, Wien. (1982, 231-238; 1983, 229-236; 1984, 209-217; 1985, in press).

Schwarzbach, E., 1979. A high throughput jet trap for collecting mildew spores on living leaves. Phytopath. Z. 94, 165-171.

Steffens, W., Führ, F., Kraus, G. and Scheinpflug, H., 1981. Aufnahme und Verlagerung von (^{14}C) Triadimenol in Sommergerste und Sommerweizen nach Saatgutbeizung. Mitt. Biol. Bundesanst. Land-Forstwirtsch. 203, 234-235.

Vanderplank, J.E., 1968. Disease Resistance in Plants. New York, Academic, 349 pp.

Wolfe, M.S., 1984. Trying to understand and control powdery mildew. Plant Pathology 33, 451-466.

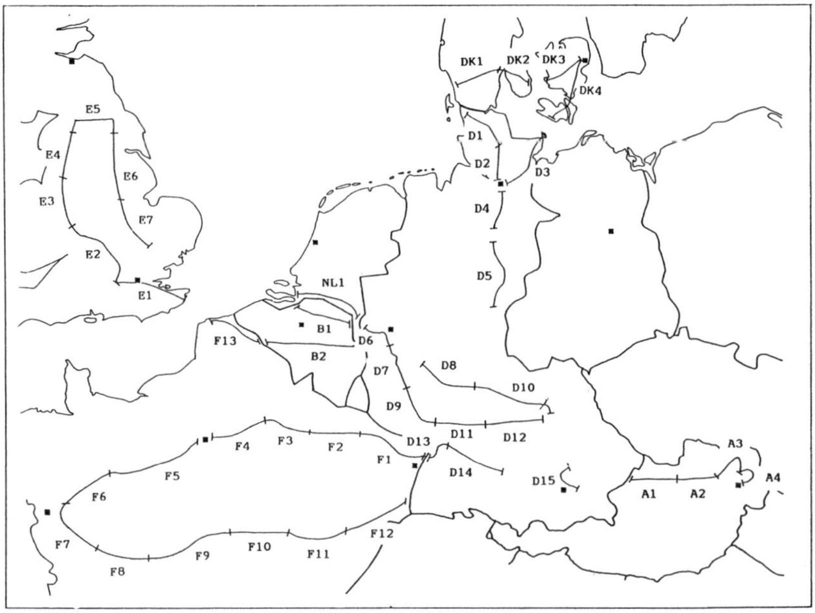

Fig.1: Network of spore sampling routes in Europe in 1985

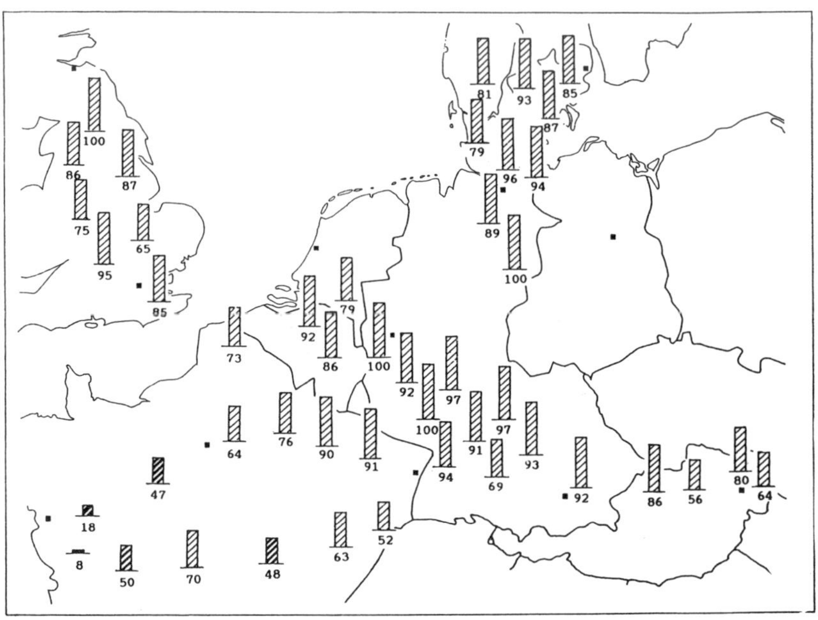

Fig.2: Distribution of <u>Va6+Vg</u> in Europe in 1985

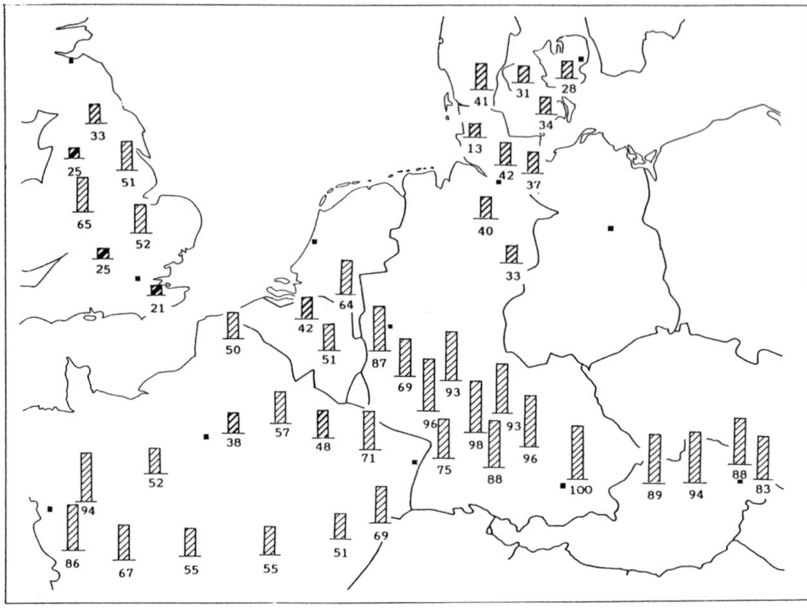

Fig.3: Distribution of Va12 in Europe in 1985

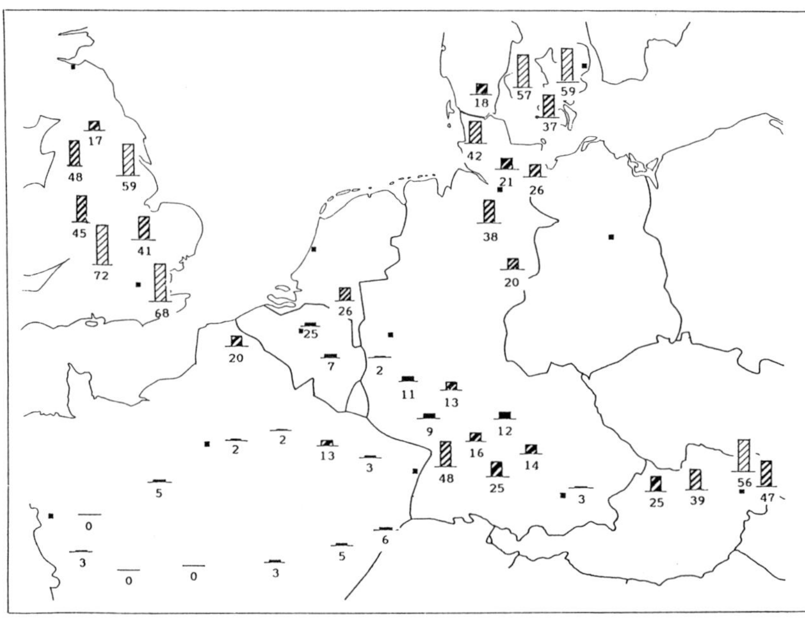

Fig.4: Distribution of Va7+Vk in Europe in 1985

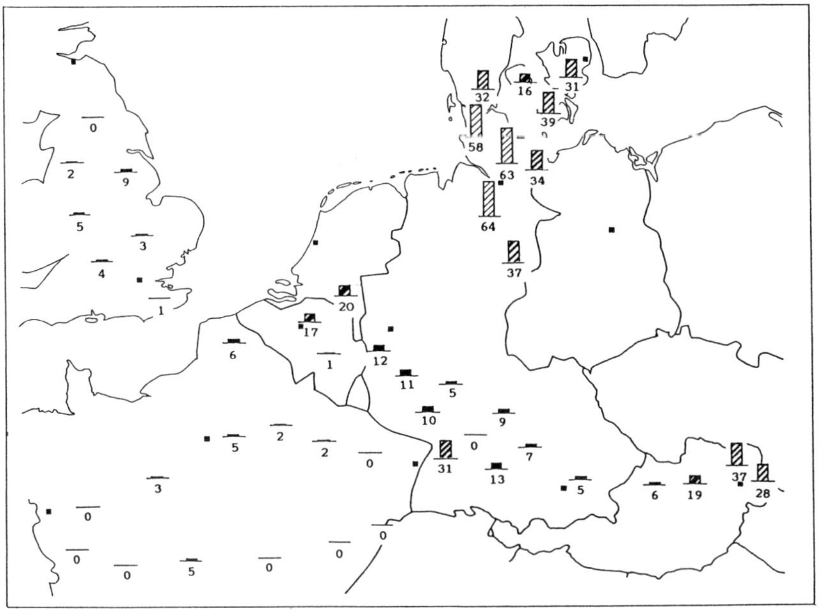

Fig.5: Distribution of Va9 in Europe in 1985

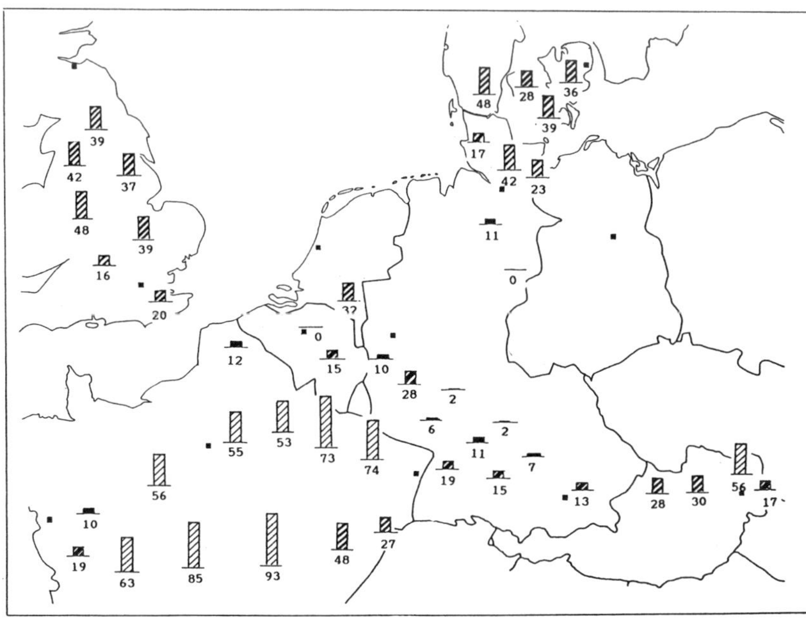

Fig.6: Distribution of Vv in Europe in 1985

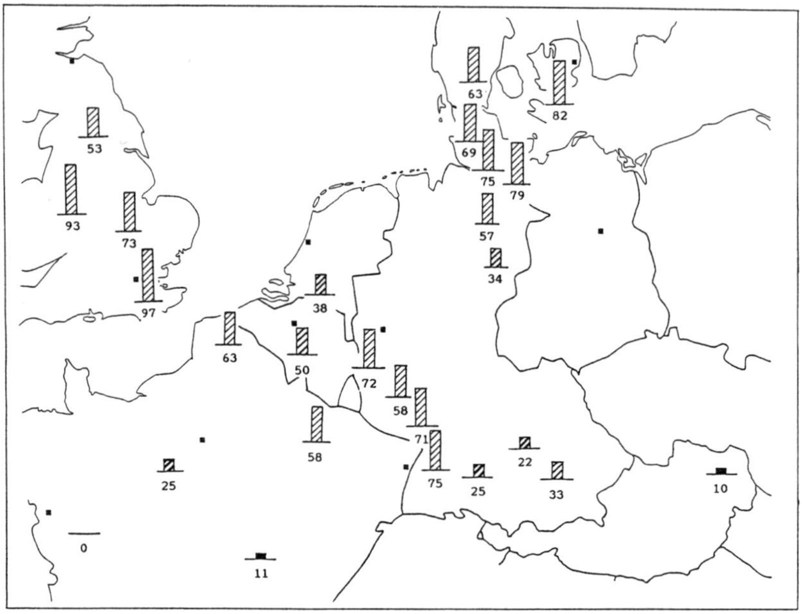

Fig.7: Survey on resistance against triadimenol in Europe in 1985. Indicated are frequencies of resistance factors ≥ 49.

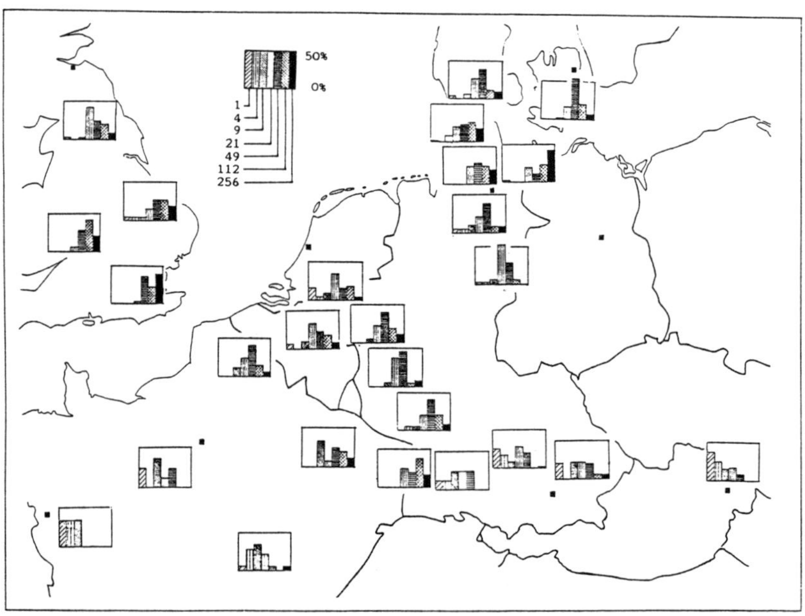

Fig.8: Resistance against triadimenol in Europe in 1985. Indicated are frequency distributions of resistance factors from 1 to 256.

TABLE 1: Barley mildew sampling in 1985 and results of virulence analysis.
n = number of isolates investigated individually

No.	Date	Distance	n	Number of virulent isolates							
				Vg	Va6 +Vg	VLa	Va7 +Vk	Va12	Va1	Va9	Va13
England											
E 1	0507	Dover-London	82	77	67/79*	16	56	17	2	1	0
E 2	0507	Watfd-Staffd (M1/6)	57	56	54	9	41	14	0	2	0
E 3	0507	Staffd-Preston (M6)	65	63	49	31	29	42	0/38	3	0/38
E 4	0507	Preston-Penrith (M6)	48	35/38	38/44	20	23	12	0	1	0
E 5	0607	Penrith-Newcastle	36	34	36	14	6	12	1	0	0
E 6	0607	Durham-Newark	93	90	81	34	55	46/90	0	8	0
E 7	0607	Newark-Cambridge	66	26/26	17/26	26	27	34	0	2	0
France											
F 1	2706	Haguenau-Metz	58	58	53	43	2	41	2	0	0
F 2	2706	Metz-St.Menehould	60	58	54	44	8	29	1	1	0
F 3	2706	St. Menehould-Reims	49	34	35/46	26	1	28	2	1	0
F 4	2706	Reims-Paris	56	47	36	31	1	21	5	3	0
F 5	2806	Paris-Le Mans	66	54	31	37	3	34	4	2	0
F 6	2806	Le Mans-Ancenis	116	115	21	12	0	111	0	0	0
F 7	2906	Ancenis-Partenay	36	32	3	7	1	31	1	0	0
F 8	2906	Partenay-Le Blanc	24	24	12	15	0	16	2	0	0
F 9	2906	Le Blanc-La Charite	20	17	14	17	0	11	0	1	0
F10	3006	La Charite-Beaune	29	20	14	27	1	16	0	0	0
F11	3006	Beaune-Besancon	60	54	38	29	3	31	1	0	0
F12	3006	Besancon-Mulhouse	48	48	25	13	3	33	4	0	0
F13	0707	Calais-front.Belg.	119	112	87	14	24	60	7/116	7/116	0
Belgium											
B 1	1906	Antwerp.-front.NL	12	11	11	0	3	5	0	2	0
B 2	0407	Mons - Liege	81	70	70	12	6	41	8	1	0
Netherlands											
NL1	1706	Heerlen-Zierikzee	47	37	37	15	12	30	0	9	0
Denmark											
DK1	1407	Romo-Ribe-Kolding	116	115	94	56	21	48	35	37	8
DK2	1407	Kolding-Knudshoved	100	99	93	28	57	31	-	16	-
DK3	1407	Halsskov-Riso	88	84	75	32	52	25	4/36	27	1/36
DK4	1507	Lyngby-Rodby	122	119	106	48	45	42	21	47	8
Germany											
D 1	1307	Rendsbg-Finsbg-Nib.	24	24	19	4	10	3	3	14	0
D 2	1307	Hamburg-Rendsburg	24	22	23	10	5	10	1	15	0
D 3	1607	Puttgarden-Hamburg	35	35	33	8	9	15	2	12	0
D 4	1607	Thieshope-Allertal	45	44/46	40	5	17	18	1	29	1
D 5	1607	Hannover-Hanmünden	30	29	30	0	6	10	7	11	0
D 6	1906	Meckenheim-Aachen	52	52	52	5	1	45	3	6	0
D 7	1706	Bingen-Meckenheim	83	75	76	23	9	57	14	9/80	0
D 8	1906	Limburg-Aschaffenbg	60	56	58	1	8	56	0	3	0
D 9	1706	Schifferstadt-Bingen	82	78	82	5	7	79	7	8	0
D10	1906	Aschaffenbg-Nürnberg	58	58	56	1	7	54	1	5	0/46
D11	1706	Schnelldf-Hockenheim	45	43	41	5	7	44	3	0	0
D12	1706	Nürnberg-Schnelldorf	72	67	67	5	10	69	0	5	0
D13	2706	Pforzheim-Rastatt	48	45	35/45	9	23	36	1	15	0
D14	2706	Ulm-Pforzheim	48	44	33	7	12	42	4	6	0
D15	1206	Freising-München	60	60	55	8	2	60	0	3	0
Austria											
A 1	1406	Wels-Braunau	36	36	31	10	9	32	0	2	0
A 2	1406	Tulln-Wels	70	67	39	21	27	66	3	13	0
A 3	1406	Mistelbach-Tulln	90	89	72	50	50	79	5	33	6
A 4	1406	Marchfeld	36	35	23	6	17	30	4	10	0

* 79 isolates investigated instead of 82

TABLE 2: Frequency and level of triadimenol-resistance of random samples of barley mildew from different parts of Europe in 1985

Origin of sample		n	Percentage of isolates per dose of triadimenol (mg a.c./kg seed, I) and resistance levels *, (II)						
		I	0	6	13,7	31,7	71,7	164	375
		II	<4,1	4,1	9,4	21,4	48,9	112	256
Standard-isolates		4	100	0	0	0	0	0	0
England									
E 1	Dover-London	36	0	0	0	3	36	22	39
E 3	Staffd-Preston	14	0	0	0	7	29	43	21
E 5	Penrith-Newcastle	36	3	0	3	42	25	19	8
E 7	Newark-Cambridge	26	4	4	4	15	27	27	19
Netherlands									
NL1	Heerlen-Zierikzee	32	16	3	9	34	16	19	3
Belgium									
B 2	Mons-Liege	48	6	0	10	33	23	19	8
France									
F 2	Metz-St.Menehould	24	0	0	33	8	25	21	13
F 5	Paris-Le Mans	8	25	0	37	13	25	0	0
F 7	Ancenis-Partenay	24	33	33	33	0	0	0	0
F10	La Charite-Beaune	18	6	28	33	22	6	0	6
F13	Calais-front.Belg.	48	0	0	13	25	42	15	6
Denmark									
DK1	Romo-Ribe-Kolding	56	4	0	7	27	39	13	11
DK4	Lyngby-Rodby	67	0	0	2	16	54	21	8
Germany									
D 1	Rendsbg.-Flnsbg.-Nib.	24	4	0	13	21	29	21	13
D 2	Hamburg-Rendsburg	24	0	0	0	25	29	25	21
D 3	Puttgarden-Hamburg	42	2	0	0	19	12	24	43
D 4	Thieshope-Allertal	44	5	5	11	23	39	9	9
D 5	Hannover-Hanmünden	32	3	3	6	53	28	6	0
D 6	Meckenheim-Aachen	47	0	0	4	23	40	19	13
D 7	Bingen-Meckenheim	24	0	0	4	38	46	4	8
D 9	Schifferstadt-Bingen	24	0	4	4	21	42	21	8
D12	Nürnberg-Schnelldorf	36	25	17	8	28	19	0	3
D13	Pforzheim-Rastatt	24	0	0	0	25	21	37	17
D14	Ulm-Pforzheim	24	13	13	25	25	25	0	0
D15	Freising-München	52	21	0	23	23	21	6	6
Austria									
A3/4	Mistelb.-Tulln-Marchf.	60	37	23	15	17	8	0	0

*) resistance levels: dose test isolate/dose standard;
dose standard isolates = 1,46 mg/kg, determined separately

TABLE 3: Distribution of mildew resistance (%) on the barley area in Germany (FRG) from 1983-1985.

		none	Ha	Mlg.	Mla6	MlLa	Mla7	Mla12	Mla9	Mla13	Gimpel (u)
Schl.-Holst.	83	59.5	2.8	11.7	19.0	2.1	0	2.2	7.2	0	-
	84	60.4	0	7.2	10.1	2.1	0	1.2	3.9	0	-
	85	46.4	9.5	15.9	12.1	8.0	0	0	7.9	0	-
Nieders.	83	42.9	3.0	28.9	16.1	1.6	0	10.9	9.9	0.4	-
	84	47.1	2.4	19.9	14.9	0.4	0.1	7.7	9.7	0.3	0.7
	85	37.0	3.4	27.1	15.8	4.0	0	5.2	13.4	0.9	0.2
Nordr.-Westf.*	83	34.3	8.2	-	36.7	-	-	-	-	-	-
	84	29.9	12.5	-	39.9	-	-	-	-	-	-
	85	27.5	8.8	-	41.1	-	-	-	-	-	-
Hessen	83	56.8	3.8	24.4	24.5	0	0.6	6.9	0	0	-
	84	53.7	1.3	21.7	31.6	0.2	0.5	6.3	0.2	0	0.2
	85	51.4	7.3	22.8	35.6	0	0.5	6.7	0.5	0.3	0.2
Rheinl.-Pfalz	83	27.9	2.2	47.5	28.4	0	0.8	16.3	0.8	10.3	-
	84	26.6	1.3	36.5	31.4	0	0	11.5	1.2	15.8	-
	85	12.9	1.4	38.3	36.2	0	0.8	10.2	2.5	14.6	0.8
Saarland	83	25.2	3.5	52.9	30.9	0	0	20.4	0	0	-
	84	32.1	0.8	47.8	37.0	0	9.7	17.1	0	0	-
	85	27.5	0.6	55.2	40.2	0	1.3	20.2	1.3	0	1.4
Baden-Wrtbg.	83	30.9	4.7	51.0	10.4	0.5	2.0	39.9	0	0.5	-
	84	31.3	4.1	40.2	14.8	0	3.2	31.1	2.7	0.4	6.0
	85	32.7	4.2	28.4	14.7	0.1	1.9	10.6	1.4	0.5	18.8
Bayern	83	41.6	4.9	43.2	34.5	0	1.2	6.6	0	0.3	-
	84	47.1	5.1	40.8	30.3	0	0.9	8.9	0	0	0.1
	85	42.9	5.8	39.5	31.1	0	1.1	8.3	0.8	0.1	0.3
FRG	83	41.3	4.7	31.5	27.1	0.6	0.7	11.0	3.2	1.0	-
	84	41.9	4.5	25.9	27.2	0.3	0.7	9.2	3.2	1.2	0.9
	85	36.7	5.6	27.6	28.5	1.7	0.7	6.3	4.6	1.5	2.6

*Data of winter barley only

Period	Zones			
	A	B	C	D
1	Ly	Ar	Sp	Ru, MC
2	Ar	Sp	Ru, MC	Ly
3	Sp, NR1	Ru, MC	Ly	Ar
4	Ru, MC	Ly	Ar, NR2	Sp, NR1
5 (≙ 1')	Ly	Ar, NR2	Sp, NR1	Ru, MC

Fig. 9: System for a Europe-wide use of race-specific resistance against barley mildew. The whole area of cultivation is divided into several zones (A-D) across the main wind direction (MWD), in each of which certain resistance genes will be used during a certain period. Every new period, cultivation moves for one stage against MWD or from A to D, respectively. Varieties with new resistance genes can be integrated at any position. During the fifth period positions of departure are reached again; diversity has increased, however, in the meantime by the addition of new resistance genes (NR1, NR2). Symbols of resistance genes: Ly = Lyallpur (Mla7+Mlk); Ar = 'arabische' (Mla12); Sp = Spontaneum (Mla6); Ru = Rupee (Mla13); MC = Monte Cristo (Mla9).

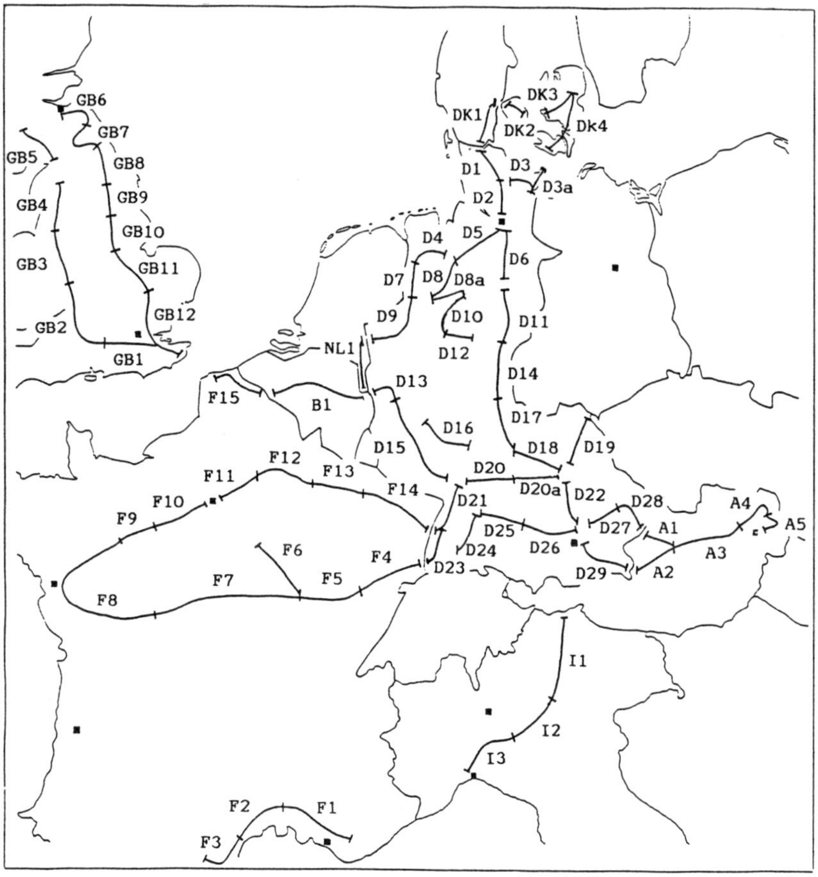

Fig. 10: Spore sampling routes in Europe in 1986.

TABLE 4: Barley mildew sampling in 1986 and results of virulence analysis.
n = number of isolates investigated individually; in certain cases a smaller number was scored (values in brackets)

No.	Date	Distance	n	Frequencies of virulent isolates (%) Vg	Va6 +Vg	VLa	Va7 +Vk	Va12	Va3	Va1	Va9	Va13
Great-Britain												
GB 1	0807	Dover-Reading	36	89	50	17	58	14	0	3	0	0
GB 2	0807	Reading-Birmingham	33	67	27	61	58	51	0	3	15	0
GB 3	0807	Birmingham-Preston	35	73(33)	26	60	66	49	0	9	17	0
GB 4	0807	Preston-Carlisle	35	80	37	63	80	31	0	0	14	0
GB 5	0807	Gretna Gr-Motherwell	26	58	31	54	81	19	0	0	23	0
GB 6	0907	Edinburgh-Chirnside	44	64	46	55	18	57	0	7	5	0
GB 7	0907	Chirns.-Kelso-Coldst.	46	80	57	57	7	73	0	5	7	0
GB 8	0907	Coldstream-Newcastle	47	81	60	62	34	49	0	0	6	0
GB 9	0907	Newcastle-Harrogate	48	85	50	52	60	38	0	0	10	0
GB10	0907	Harrogate-Newark	45	78	38	36	49	36	0	0	4	0
GB11	0907	Newark-Cambridge	47	83	38	55	34	62	0	2	6	0
GB12	1007	Cambridge-Dover	33	67	30	27	36	42	0	0	3	0

TABLE 4: ctd

No.	Date	Distance	n	Frequencies of virulent isolates (%)								
				Vg	Va6 +Vg	VLa	Va7 +Vk	Va12	Va3	Va1	Va9	Va13
France												
F 1	0305	Brignoles-Tarrascon	53	98	32	43	0	83	2	0	0	0
F 2	0305	Beaucaire-Montpell.	73	100	37	23	0	96	1	8	0	0
F 3	0305	Montpell.-Lezignan	9	100	22	44	0	67	11	11	0	0
F 4	2906	Mulhouse-Besancon	47	92	55	89	0	38	0	4	0	0
F 5	2906	Besancon-Beaune	42	83	57	91	7	31	0	2	0	0
F 6	0605	Beaune-Courtenay	18	100	56	78	6	22	0	6	0	0
F 7	2906	Beaune-Le Blanc	13	100	62	85	15	31	8	0	0	0
F 8	2906	Le Bl.-Ancenis-Le M.	15	73	40	73	0	73	7	7	0	0
F 9	3006	Le Mans-Illier	44	98	45	84	2	66	0	2	2	0
F 10	05-S	Grignon(stat.)	58	67	43	62	5	40	0	5	2	0
F 11	3006	Paris-Reims	23	96	78	70	17	30	0	4	0	0
F 12	3006	Reims-St. Menehould	36	86	69	83	6	36	0	6	0	0
F 13	3006	St. Menehould-Metz	96	81(95)	75	76	10	40	0	3	0	0
F 14	3006	Metz-Strasbourg	48	92	73	65	23	38	0	2	6	0
F 15	0707	Calais-fr.Mons	54	91	50	22	30	37	0	4	0	0
Belgium												
B 1	0707	Mons-fr.Aachen	69	87	77	17	28	33	0	6	7	0
Netherlands												
NL 1	1307	Venlo-fr.Aachen	30	100	87	50	33	27	0	0	13	0
Denmark												
DK 1	1407	Kolding-fr.Flensbg.	100	93	78	66	46	21	0	8	28	24
DK 2	1407	Kolding-Knudshoved	100	83	58	65	55	24	3	7	26	15
DK 3	1407	Halsskov-Riso	96	72	52	51	45	21	1	7	10	15
DK 4	1507	Riso-Rodby	92	77(89)	69	22	66	22	3	1	5	2
Germany (FRG)												
D 1	1407	Flensburg-Kiel	24	96	96	54	21	25	4	0	42	4
D 2	1407	Kiel-Hamburg	24	92	83	42	8	25	0	0	33	0
D 3	1607	Kiel-Neustadt	20	90	75	55	0	30	0	0	10	0
D 4	1307	Dörpen-Oldenbg.-Bre.	12	100(11)	83	50	17	8	0	8	42	0
D 5	1606	Hamburg-Bremen	47	100	98	83	9	40	2(41)	0	55	0
D 6	1506	Allertal-Thieshope	43	95(42)	84	26	23	35	0	2	33	0
D 7	1307	Rheine-Dörpen	18	100	89	22	50	22	0	0	78	0
D 8	1606	Bremen-Osnabrück	48	96	92	33	17	35	4	6	42	0
D 9	1307	Venlo-Rheine	54	96	94	22	19	57	0	6	22	0
D 10	1906	Minden-Soest	37	95	84	30	5	38	0	19	30	0
D 11	1506	Hannover-Hanmünden	47	98	96	15	13	81	0	2	19	0
D 12	1906	Soest-Warburg	46	100	100	54	15	50	4	10	39	0
D 13	1107	Aachen-Meckenheim	27	100	100	15	4	52	0	0	4	0
D 14	1506	Fulda-Kassel	93	94	88	18	1	76	0	5	5	0
D 15	1107	Meckenheim-Speyer	50	96	92	22	8	50	0	0	8	0
D 16	1307	Limburg-Aschaffenbg.	60	93	90	18	15	77	0	0	14	0
D 17	1506	Würzburg-Brückenau	44	95	95	18	11	84	0	9	14	0
D 18	1506	Würzburg-Nürnberg	48	98	98	2	4	90	0	2	2	0
D 19	0607	Nürnberg-Hof	32	100	94	48	22	91	0	3	9	0
D 20	0107	Kraichgau-Crailsheim	47	98	68	2	62	70	0	0	28	0
D 21	0107	Strasbourg-Bruchsal	48	92(47)	69	23	75	44	0	2	21	0
D 22	2006	Nürnberg-Freising	42	98	91	24	7	88	0	2	7	0
D 23	0605	Mulhouse-Renchtal	17	100	65	71	6	77	0	0	0	6
D 24	2806	Stuttgart-Donauesch.	51	100	96	24	28	77	0	0	2	0
D 25	2806	Stuttgart-Ulm	37	97	81	27	14	89	0	3	22	0
D 26	2806	Ulm-Freising	42	98	95	17	7	93	0	0	0	0
D 27	2306	Freising-Dingolfing	43	91	79	2	2	93	0	0	0	0
D 28	2306	Dingolf.-fr.Schärd.	42	93	81	12	12	91	0	0	5	0
D 29	2206	München-Salzburg	48	90	81	4	0	96	0	0	0	0
Italy												
I 1	0205	Brenner-Verona	8	63	0	13	25	13	0	13	0	0
I 2	0205	Verona-Piacenza	45	42	11	33	16	22	0	15	13	0
I 3	0205	Piacenza-Genova	9	78	22	44	0	33	0	0	0	0
Austria												
A 1	2306	Schärding-Linz	48	96	69	38	58	79	0	4	19	0
A 2	2206	Salzburg-Linz	45	93(43)	69	36	29	84	0	0	22	0
A 3	2306	Linz-St.Pölten	90	98(88)	68	47	34	82	1	0	13	0
A 4	2206	Krems-Mistelbach	95	96	58	44	71	64	0	8	60	23
A 5	2206	Marchfeld	44	100	59	32	75	52	2	5	50	34

TABLE 5: Frequency and level of triadimenol-resistance of random samples of barley mildew from different parts of Europe in 1986

No.	Date	Distance	n	Percentage of isolates per dose of triadimenol (mg a.c./kg seed, I) and resistance levels * (II), respectively										
				I 0.73 II 0.5	1.46 1	2.9 2	5.9 4	11.7 8	23.4 16	46.7 32	93.7 64	187.5 128	375 256	750 512
Standard-isolates			4	0	100	0	0	0	0	0	0	0	0	0
Great-Britain														
GB 1	0807	Dover-Reading	36	0	0	0	3	9	11	9	11	34	23	0
GB 3	0807	Birmingham-Preston	23	4	4	9	9	0	22	17	22	4	9	0
GB 7	0907	Chiras.-Kelso-Coldstr.	33	0	0	0	6	12	29	21	9	18	6	0
GB 8	0907	Coldstr.-Newcastle	36	0	3	0	3	6	6	47	25	11	0	0
GB11	0907	Newark-Cambridge	22	0	0	0	0	9	14	23	23	5	18	9
France														
F 2	0305	Beaucaire-Montpellier	24	0	33	50	13	0	4	0	0	0	0	0
F 4	2906	Mulhouse-Besancon	31	0	0	10	23	17	0	17	30	3	0	0
F 5	2906	Besancon-Beaune	37	0	0	0	3	14	25	19	14	22	3	0
F 7	2906	Beaune-Le Blanc	13	0	0	15	0	0	8	47	15	15	0	0
F 8	2906	Le Bl.-Ancenis-Le Mans	13	0	0	8	8	23	46	15	0	0	0	0
F 9	3006	Le Mans-Illier	20	0	15	10	5	45	20	5	0	0	0	0
F 11	3006	Paris-Reims	23	0	0	4	0	0	17	62	17	0	0	0
F 12	3006	Reims-St.Menehould	35	0	0	0	0	2	3	9	83	3	0	0
F 13	3006	St.Menehould-Metz	48	0	2	0	0	6	11	29	23	29	0	0
F 14	3006	Metz-Strasbourg	24	0	4	4	4	5	8	8	50	17	0	0
F 15	0707	Calais-fr.Mons	46	0	0	0	9	28	11	13	13	17	6	2
Belgium														
B 1	0707	Mons-fr.Aachen	25	0	4	8	8	12	16	36	12	0	4	0
Denmark														
DK 1	1407	Kolding-fr.Flensburg	48	0	10	8	2	2	10	15	8	10	22	10
DK 3	1407	Halsskov-Riso	48	0	4	12	7	2	17	12	12	15	17	2
Germany (FRG)														
D 1	1407	Flensburg-Kiel	22	0	0	0	0	0	9	14	59	14	4	0
D 2	1407	Kiel-Hamburg	16	0	0	0	0	0	12	13	25	50	0	0
D 3a	1507	Puttgarden-Neustadt	32	0	0	0	0	6	3	9	9	22	31	19
D 4	1307	Dörpen-Oldenbg.-Bre.	12	0	0	0	0	8	0	42	42	8	0	0
D 8a	1706	Osnabrück-Minden	24	0	0	0	0	8	12	25	42	8	5	0
D 9	1307	Venlo-Rheine	24	0	0	0	4	13	13	37	25	8	0	0
D 13	1107	Aachen-Meckenheim	27	0	0	0	0	0	0	33	30	15	22	0
D 15	1107	Meckenheim-Speyer	12	0	0	0	0	0	0	50	50	0	0	0
D 16	1307	Limburg-Aschaffenbg.	22	0	0	0	0	0	18	18	41	18	5	0
D 20a	0107	Crailsheim-Nürnberg	35	0	0	3	3	3	23	40	20	3	5	0
D 21	0107	Bruchsal-fr.Strasburg	36	0	0	0	0	3	6	36	33	14	8	0
D 24	2806	Stuttgart-Donauesch.	36	0	8	22	0	3	11	33	6	11	6	0
D 27	2306	Freising-Dingolfing	37	0	5	0	8	8	3	49	13	11	3	0
D 29	2206	München-Salzburg	36	0	6	3	0	6	14	42	14	14	2	0
Italy														
I 2	0205	Verona-Piacenza	27	4	48	41	4	4	0	0	0	0	0	0
Austria														
A 4	2206	Krems-Mistelbach	54	0	50	31	9	0	4	6	0	0	0	0
A 5	2206	Marchfeld	21	0	24	43	14	0	5	9	5	0	0	0

* resistance levels: dose test isolate/dose standard; dose standard isolates = 1.46 mg/kg

BARLEY MILDEW IN EUROPE: EVIDENCE OF WIND-DISPERSAL OF THE PATHOGEN AND ITS IMPLICATIONS FOR IMPROVED USE OF HOST RESISTANCE AND OF FUNGICIDES FOR MILDEW CONTROL*

E. Limpert

Technische Universität München, Lehrstuhl für Pflanzenbau und Pflanzenzüchtung, 8050 Freising-Weihenstephan, Deutschland (BRD)

ABSTRACT

Some basic features of wind-dispersal of the barley mildew pathogen *Erysiphe graminis* f.sp. *hordei* are briefly outlined. The considerable proportion of unnecessary virulence reflects in a quantitative way the importance of the spread of barley mildew throughout Europe. This is true even in regions where winter barley is grown and where mildew is regarded as endemic. The dispersal is enhanced in the main wind direction, i.e. in general from west to east, and reduced against it. The survival of spores after drift depends mainly on the virulences they possess on the one hand and on the selection due to the host resistances present in the region of immigration on the other. If these conditions are favourable, entire populations rather than single spores migrate and cover, in the case observed, approximately 110 km per year. Therefore, successful cultivars with valuable mildew resistance should not be introduced from west to east, but against the main wind direction, from east to west; the same is true for fungicides with specific action. The data are able to outline in which way large areas of Europe have to be regarded as an epidemiologic unit.

The barley mildew population was monitored in large parts of Europe from 1979 to 1986, using the method described by Limpert (1985) and Limpert and Fischbeck (this volume). The patterns of virulence frequencies revealed, in general, highest values in the regions of selection and lower ones in the others. In the latter, the virulence was unnecessary, pointing to immigration from the former by wind. Thus spores labelled with unnecessary virulence provide a valuable tool for investigations of the dispersal of the barley mildew pathogen.

High frequencies of the unnecessary virulence Va6 were present in France, Denmark and Austria (Limpert, in press; Limpert and Fischbeck, this volume) and further evidence of the influence of drift is available from other virulences, both for barley (Gacek, this volume; Limpert and Fischbeck, this volume) and for wheat mildew (Karjalainen, this volume; Felsenstein et al., this volume).

---

*This work has been supported by the German Research Foundation (DFG) and by the Commission of the European Communities. Their help is gratefully acknowledged.

The distribution observed in Europe of Va1 and Va9 in 1981 (Limpert, 1985) and of Va13 in 1986 (Limpert and Fischbeck, this volume), indicate that the spread of spores is enhanced in and reduced against the main wind direction.

Several examples (Limpert, 1985) provide evidence that in regions of immigration the unnecessary virulence is able to persist in the population if combined with necessary virulence, whereas in regions of selection the latter may occur uncombined in considerable frequencies. Therefore both the virulence combinations of the immigrating mildew and the composition of the host population causing selection in the region of immigration are the two main factors influencing changes in the composition of the population already present.

Two further observations indicate in a more detailed way the quantitative influence of wind-dispersal of the mildew population. They are both from southern Germany, which can be roughly described as a square with Bavaria in the east and Rheinland-Pfalz and Baden-Württemberg in the north-west and in the south-west, respectively.

In 1979 a steady decrease of the frequency of Va7+Vk was noted, from approximately 80% in the north-west to 20% in the south-east (Limpert, unpublished), pointing to Rheinland-Pfalz being the region of origin and to wind dispersal from the west. Although the virulence was unnecessary in Bavaria, its frequency, observed in the region of Weihenstephan about 330 km away from Rheinland-Pfalz, followed a maximum-function during the following four years. The evolution was rather similar to that partly observed, partly estimated in the region of origin, indicating that not just single spores, but rather the entire mildew population from Rheinland-Pfalz followed the drift and passed Weihenstephan. The interval of three years between these developments corresponds to a displacement of the population of approximately 110 km on an average per year (Limpert, 1985).

There were even short-term influences of long distance dispersal of spores apparent in the composition of the population in a region. In southern Germany Va12 was abundant, due to selection, only in Baden-Württemberg and was mostly unnecessary in Bavaria. The increase of the frequency of Va12 near Weihenstephan from approximately 10 to 30% from 1980 to 1982 exhibited a pronounced annual rhythm, with a significant increase between August and October and a slight decrease up to spring and early summer in the following year (Limpert, 1985). Having in mind the geographical situation and the distribution of Va12 in southern Germany, this rhythm was consistent with that of changes in the main wind direction (data from the German meteorological service, DWD, unpublished), which is from west to south-west during the period of increase of Va12, but from northwesterly directions in spring and early summer. These data suggest that even after primary inoculation, when the mildew population in a field is supposed to be established (Wolfe and Schwarzbach, 1978), its composition might still change considerably due to immigrating spores sedimenting on to the crop.

The increase of approximately 30% in the frequency of Va7+VAb and of Vk, as well as a similar decrease in the frequency of VLa and of Va12, all observed from March to July 1985 in Cambridge (Wolfe et al., 1986), could possibly be explained in a similar way. In regions E1 and E2 (Limpert and Fischbeck, this volume, Figs 1, 3, 4, 6, Table 1) in the south and west of Cambridge, the mildew population was so composed that it could have created these changes in the region of Cambridge after drift on the prevailing winds coming into the area concerned from the south-west.

The implications of these examples are as follows: widely used fungicides with specific action as well as successful cultivars with valuable race-specific resistance should not be introduced from west to east, but rather against the main wind direction from east to west. A system for the Europe-wide use of host resistances and of fungicides is depicted in Fig. 9 (Limpert and Fischbeck, this volume) to demonstrate the basic principles involved.

The results presented specify the way in which large parts of Europe need to be regarded as a single epidemiologic unit, and the way in which the decisions taken in one part may influence the success of mildew control in another. It also stresses the need for coordinated efforts in Europe as well as for the development of strategies to make the best use of host resistance and of fungicide for mildew control.

REFERENCES

Limpert, E. 1985. Ursachen unterschiedlicher Zusammensetzung des Gerstenmehltaus, _Erysiphe graminis_ DC. f.sp. _hordei_ Marchal, und deren Bedeutung für Züchtung und Anbau von Gerste in Europa. Diss. Weihenstephan, 183 pp.

Limpert, E., in press. Frequencies of virulence and fungicide resistance in the European barley mildew population in 1985. J. Phytopath.

Wolfe, M.S. and Schwarzbach, E., 1978. Patterns of race changes in powdery mildews. Ann. Rev. Phytopathol. _16_, 159-180.

Wolfe, M.S., Slater, S.E. and Minchin, P.N., 1986. Mildew of barley. U.K. Cereal Pathogen Virulence Survey, 1985 Annual Report, 27-34.

HOST RESISTANCE AND PATHOGEN VIRULENCE IN POWDERY MILDEW OF SPRING BARLEY IN AUSTRIA

B. Zwatz

Federal Institute of Plant Protection, Vienna, Austria

ABSTRACT

Spring barley is one of the most important cereal crops in Austria, occupying 31% of the cereal-growing area. The most damaging disease is powdery mildew (*Erysiphe graminis* f. sp. *hordei*). For the integrated control of mildew, both resistant varieties and fungicides are used. Officially, there is a preference for the use of resistant varieties (biological control of mildew) and so, in 1984, the Federal Institute of Plant Protection began an analysis of resistance genes in varieties and of virulence genes in mildew populations. A survey of these analyses is given.

INTRODUCTION AND METHODS

Spring barley occupies 31% (266,000 ha) of the cereal area in Austria compared with 9% (73,000) for winter barley. With this background, the epidemiological conditions for powdery mildew (*Erysiphe graminis* f. sp. *hordei*) are optimal. Indeed, mildew is the most serious and damaging disease of spring barley. For this reason, control of mildew is an important problem; it is attempted mostly by breeding and growing resistant varieties and by using fungicides.

Plant breeding in Austria is carried out only by private plant breeders. Growing resistant varieties is the most important part of integrated plant protection. The Federal Institute of Plant Protection is responsible for testing the varieties to determine their susceptibility to different diseases.

In 1984, we began to determine resistance genes in barley varieties and virulence genes in powdery mildew populations (the introduction of this work was kindly promoted by Doz. Dr. E. Schwarzbach). Mildew spores are collected by using the Schwarzbach jet spore sampler fitted to a car roof. For determination of resistance and virulence we use segments of primary seedling leaves on water agar. The tests are carried out in a controlled environment chamber (17-18°C, permanent light).

RESULTS AND DISCUSSION

Table 1 provides a survey of the frequency distribution of virulence genes in the pathogen population for 1984 and 1986. Except for a large decrease in virulence for Ml-(La) and an increase in virulence for Ml-g, the values for each year are similar.

Table 2 provides a survey of resistance genes and of their field susceptibility together with the proportion of the barley area of spring varieties approved in Austria. The latter data are based on seed production figures from 1985.

TABLE 1 Virulence frequencies in powdery mildew of spring barley in Austria, 1984 and 1986

Test varieties	Resistance genes and symbols		Virulence % 1984	1986
Igri	Ml-ra	Ra	100	100
Union, Adele	Ml-g	We	55	81
Carina	Ml-a6, Ml-g	Sp,We	30	36
Optima	Ml-a6	Sp	31	-
Vada, Varunda	Ml-(La)	La	54	19
Aramir	Ml-a12, Ml-g	Ar,We	37	46
Adora	Ml-a7, Ml-g	Ly,We	22	19
Welam	Ml-a9	MC	15	12
Atem, Apex	ml-o	Mlo	1	4
Serva	Ml-a13	Ru	9	10
Gitte	Ml-a	Al	8	3

In Table 3, the data from Table 2 are re-arranged to give the total area of each resistance gene, for comparison with the virulence frequencies in Table 1.

In comparing Table 3 with Table 1, it is notable that the frequencies of virulence for Ml-(La) and ml-o are lower than might be expected from the seed production figures. In all other comparisons, the virulence frequency is greater than that of the seed production area. These differences point to the usefulness and durability of Ml-(La) and ml-o. Although virulence for ml-o is uncommon, there was a slight increase from 1984 to 1986. Further analysis of this change is the subject of a joint project (see Joint Proposal 5, this volume).

TABLE 2 Approved varieties of spring barley in Austria with their resistance gene symbols, susceptibility and area occupied, based on seed production data from 1985

Variety	Resistance symbols	Susceptiblity	% area of seed production
Adele	We	7	2.9
Adora	Ly,We	7	4.3
Anita	Ar,s*	6	0.9
Apex	Mlo,We	2	6.1
Aramir	Ar,We	7	11.6
Atem	Ml,La	2	16.8
Berta	La,We,s	6	6.4
Carina	Sp,We	7	11.6
Carmen	Ar,We,ul	5	-
Ebra	Ly,ul	6	1.9
Eva	none	7	<0.1
Florina	La,MC	5	<0.1
Grand Prix	u2	5	3.7
Havila	La,We	6	7.7
Irma	Ar	7	0.9
Jutta	Ar,We	6	1.1
Keti	Ru,We	2	1.3
Kym	La,We	6	8.5
Lea	We	6	1.0
Patty	Ar,s	6	0.5
Quantum Plus	ul	6	1.3
Robin	Ly,ul	5	-
Roland	MC	7	2.1
Ronda	u2	5	2.4
Serva	Ru,We	6	4.3
Valenta	We,ul	6	1.9
Welam	MC	7	0.8
total area			5636 ha

1 = immune, 2-5 = resistant, 6-9 = susceptible
s = slow-mildewing, ul = moderately effective resistant gene,
u2 = highly effective resistance gene

TABLE 3 Areas of seed production classified by resistance gene

Resistance gene	Area of seed production (%)
Ml-g	66.4
Ml-a6	11.6
Ml-(La)	39.4
Ml-a12	14.9
Ml-a7	6.2
Ml-a9	2.9
ml-o	22.9
Ml-a13	5.6
Other, unknown	17.1

EXPERIENCES WITH BREEDING OF SPRING BARLEY FOR RESISTANCE TO POWDERY MILDEW (ERYSIPHE GRAMINIS F.SP. HORDEI) IN CZECHOSLOVAKIA

Dr Frantisek Brückner

Cereal Research and Breeding Institute, Kromeriz, Czechoslovakia

ABSTRACT

Older cultivars with Ml-g, Ml-a6, or Ml-a7, were quickly overcome by mildew in Czechoslovakia. Cultivars derived directly from the original donor of resistance, such as Spartan (Ml-a9) and Koral (Ml-a13 plus other genes), were more durable. However, the most durable resistance so far appears to be ml-o.

DISCUSSION

Up to 1964 nearly all Czechoslovak cultivars of spring barley were derived from local spring barleys of the Haná district, in which there were two genes for powdery mildew resistance, Ml-a8 and Ml-h. These cultivars, however, were highly susceptible to powdery mildew under field conditions, and so intensive breeding for resistance to this disease was begun in the years after World War II.

Weihenstephan strains and the first cultivars with the Ml-g gene bred from these strains were used as donors of resistance. The first Czechoslovak cultivar with the Ml-g gene was Merkur (1964), but its contribution to improved resistance in comparison with old cultivars was very small.

This was explained by the first analysis of the pathogen population of powdery mildew in Czechoslovakia in 1960-61. The majority of identified pathotypes (races) fell into groups A and D, i.e. without virulence to Ml-g on the basis of Nover's differential collection; in the western part of Czechoslovakia, however, the most frequent race was C_3, virulent to Ml-g. This race was not found in the eastern part of the country (in Slovakia).

Further cultivars with the Ml-g gene, Jantar (1966), Denár (1969), Dukat (1971), Topas (1971) and Favorit (1973), were highly susceptible to powdery mildew at the time of their release. Introduction of Ml-a6 in the cultivar Ametyst (1972) was only a little more successful, since this cultivar could be considered susceptible after only 3 years of cultivation. Cultivar Rapid (1976) with Ml-g and Ml-a6 genes combined did not contribute to resistance at all. A similar situation occurred with introduction of cultivars with resistance conditioned by the Ml-a7 gene:

Atlas (1976), Diabas (1977), and with the Ml-al2 gene, Zeffr (1981).

Introduction of the cultivar Spartan (1977) with the Ml-a9 gene was more successful. This cultivar maintained its high resistance and high productivity until 1980. In 1981, when it reached the largest distribution (31% of the whole spring barley area), there occurred a complete breakdown of resistance, correlated with a large reduction in yield. In contrast with the cultivars mentioned above, the breeders of Spartan used the original donor of the Ml-a9 gene, cultivar Monte Cristo. The same donor was also used at approximately the same time in Sweden. In comparison with Swedish cultivars with this resistance, it took more time to breed Spartan, because this cultivar has high malting quality.

Another cultivar deriving its resistance to powdery mildew directly from the donor is Koral (1978). The donor is the six-row local cultivar I 25 (Hor 805). Stability of high resistance (infection type 0) in two two-row strains and in crosses with I 25, I 265 and P 1647 was tested in the Expanded European Barley Disease Nursery. Other Czechoslovak cultivars carrying this resistance are Krystal (1981), Karat (1981), Bonus (1983) and Zenit (1985).

Czechoslovak lines with this resistance formed the basis for breeding cultivars in the German Democratic Republic - Tamina, Defra, Femina, Lenka, and in Scotland - Casino and Merchant.

The first pathotype to overcome the resistance of these cultivars was isolated in Czechoslovakia in 1980, and was followed by further pathotypes in subsequent years. These pathotypes had one common trait, represented by their virulence even to cultivar Rupee, which is a donor of the Ml-al3 gene and of another gene with infection type ln.

Cultivar Koral seemed to Schwarzbach and Fischbeck to be moderately resistant after inoculation with a pathotype virulent to Rupee. They thought therefore that this cultivar had in addition the resistance from Hordeum laevigatum. Czechoslovak pathotypes, however, lack virulence for Ml-(La) but they are able to attack cultivar CI 7672 with infection type 2-3n, i.e. similar to gene Ml(La). Our results confirm the results of Danish research workers, that the strains I 265 and P 1647 contain three closely linked genes with Ml-al3 epistatic to the others.

Currently, cultivars with resistance derived from I 25 take about 65% of the whole spring barley area in Czechoslovakia. This, of course, evokes considerable selection for pathotypes virulent to these cultivars. We must now consider that this resistance is overcome. Large yield reductions, however, have not yet occurred because in Czechoslovakia conditions have not been suitable for severe epidemic development of powdery mildew since 1981.

Other Czechoslovak cultivars, Rubin (1982) with Ml-a and Ml-a7 (two-line cultivar), Mars (1983) with the Ml(La) gene, Opal (1980) and Jaspis (1986) with Ml-a6 and Ml-(La) occupy only small areas. Pathotypes able to attack these cultivars are not commonly distributed, so that the degree of attack differs according to locality. No doubt the resistance of these cultivars would lose its importance entirely if the cultivars were to occupy larger areas.

The following observations result from long-term experience with breeding spring barley for mildew resistance in Czechoslovakia.

1. Breeding based on commercial cultivars grown in neighbouring states is completely ineffective.
2. Relatively good results can be achieved only by means of breeding based on original donors of resistance from geographically distant localities. Durability of the resistance of these cultivars however, is to a large extent dependent on their distribution. If they are distributed quickly on to large areas, the resistance will not be durable, even if they are conditioned by more genes for specific resistance.
3. Breeding for horizontal resistance based on polygenes is very exacting with respect to method and time, and its results are to a great extent influenced by environmental conditions. For example, the resistance of the cultivar Asse with horizontal resistance is unsuitable in most cases in Czechoslovak conditions.
4. Breeding for non-specific resistance conditioned by the ml-o gene sems to be the most promising with respect to durability of resistance (but see Schwarzbach, this volume). We bred for this resistance simultaneously with specific resistance, but up to now, there are no practical results. We quickly stopped using mutants with the induced ml-o gene, because the strains had low yield and low 1000 kernel weight, even with a high degree of resistance. The use of Ethiopian barleys with the spontaneously occurring ml-o gene was more successful. It is true, however, that although some strains (e.g. G 1028 in E.E.B.D.N.) had normal 1000 kernel weights, they still did not give the same yields as strains and cultivars with effective specific resistance.

DISTRIBUTION OF POWDERY MILDEW RESISTANCE AND VIRULENCE IN DENMARK

J. Helms Jørgensen[1)] and Mogens S. Houmøller[2)]

1) Risø National Laboratory, DK-4000 Roskilde, Denmark
2) Institute of Plant Pathology, DK-2800 Lyngby, Denmark

ABSTRACT

A Danish national fungicide resistance survey was initiated in 1983 and a virulence survey in 1985 because farmers, advisory service, and plant breeders need the information. The paper briefly summarises the results, lists the distribution of powdery mildew resistances in wheat and barley, and gives references to more detailed information. It is emphasised that the Danish surveys will benefit from a European coordination and rapid information exchange because the same fungicides and resistances are used extensively in most of Europe.

WHEAT

The wheat area in 1985/86 comprised about 400,000 ha of winter wheat and 10,000 ha of spring wheat. The winter wheat area has increased from about 100,000 ha in 1979.

The frequency of virulence in wheat powdery mildew (Table 1) was determined at four locations in 1985 and at five locations in 1986 (cf. Spring Barley and footnote 2 below).

TABLE 1. Distribution of powdery mildew resistance in wheat and of virulence, Denmark 1985 and 1986.

Variety	Per cent area (sown) 1984/85	1985/86	Resistance genes(s)	Virulence frequency[2] 1985[3]	1986
Kraka[1]	79	73	Ml-i	86	116
Anja[1]	7	9	?	70	121
Longbow[1]	4	11	Pm2	24	61
Vuka[1]	2	1	-	159	197
Kanzler	1	1	-	136	-
Disponent[1]	-	-	Pm8	80	104
Bert	-	-	Pm6+Ml-i	18	27
Brigand	-	-	Pm2+6	12	11
Sappo	-	-	Pm2+4	3	11
Kosack[1]	-	1	?	8	16

[1] On Danish 'List of Varieties' 1986
[2] No. colonies/leaf as per cent of those on 'Hobbit'
[3] Houmøller (1986a)

WINTER BARLEY

The winter barley area increased from nil in 1978 to 204,000 ha in 1984. Due to severe winter killing in 1984/85 only about 50,000 ha survived for harvesting in 1985. In the autumn of 1985 60,000 ha were sown, but only one third survived the winter of 1985/86. The frequency of virulence genes in the powdery mildew population (Table 2) was determined by colony count at seven locations (cf. Spring Barley).

TABLE 2. Distribution of powdery mildew resistance in winter barley and of virulence, Denmark 1985 and 1986.

Variety	Per cent area (sown) 1985	1986	Resistance code[1]	gene(s)[2]	Virulence frequency[3] 1985	1986
Igri[1]	75	57	We	Ml-ra[4]	98	174
Gerbel[1]	15	9	We	Ml-ra	-	-
Hasso[1]	3	12	We+Sp	Ml-ra+h+a6+a14	-	-
Mammut[1]	2	4	We+Sp	Ml-ra+a6+a14	94	97
Tapir[1]	-	4	We	-	-	-
Differential variety				Ml-h	76	103
Differential variety				Ml-a6+a14	85	94
Differential variety				Ml-ra	92	118

[1] On Danish 'List of Varieties' 1986
[2] Jensen and Jørgensen (1981)
[3] No. colonies/leaf as per cent of those on 'Pallas'
[4] Former gene designation Ml-(41/145) (Doll and Jensen, 1986)

SPRING BARLEY

Distribution of resistance

The spring barley area has decreased from about 1.6 M ha in 1979 to 1.0 M ha in 1986 when it constituted 36 per cent of the agricultural area. The importance of spring barley is also reflected in the large number, about 60, of varieties on the Danish 'List of Varieties'. The varieties originate from breeders in DK, S, D, DDR, NL, B and GB. All the varieties have one or more resistance genes (Table 3).

The decline in La resistance is due mainly to its decreased effectiveness in 1978-1979. The increase in Ly resistance is due to the popularity of the variety 'Triumph'. Recent varieties each have, on average, nearly 1.5 resistances. Variety mixtures cover approximately eight per cent of the barley area.

The grouping of varieties in Table 3 is based on the Danish 'List of

TABLE 3. Distribution of powdery mildew resistance in spring barley (per cent of the spring barley area) in Denmark 1979-1986.

Resistance	Code[1]	1979	1980	1981	1982	1983	1984	1985	1986
Laevigatum	La	71	50	36	39	38	40	39	37
Lyallpur	Ly	3	1	1	5	14	26	37	43
Monte Cristo	MC	14	27	38	26	32	23	10	3
Arabische	Ar	9	16	18	17	6	4	3	11
Algerian	Al	2	6	8	18	16	11	6	3
Rupee	Ru	1	0	0	0	0	3	14	17
Ricardo	Ri	0	0	0	0	0	0	0.01	1
Turkish	Ty	0	0	0	0	0	0	1	1
Mlo	Mlo	0	0	0	0	0	0	0.1	0.1
Weihenstephan	We	15	18	14	27	37	40	36	23
Total		115	118	115	132	143	147	146	139

[1] According to Danish 'List of Varieties' 1986

Varieties', which gives the source of resistance for each variety. A detailed account for resistance genes in spring barley varieties grown in Denmark 1960-1978 is given by Torp, Jensen and Jørgensen (1978). The distribution of resistance in former years is accounted for by Jørgensen and Torp (1978).

Distribution of virulence

The frequency of virulence in the airborne powdery mildew population at Risø has been monitored each year since 1978 (Munk, 1984).

In 1985 a Danish national virulence survey was initiated and financed by the Ministry of Agriculture. The purpose is to monitor virulence frequencies over time and space, and to provide qualitative data on powdery mildew for disease forecasting. The main results for wheat and winter barley are summarized in Tables 1 and 2 above, and those for spring barley are briefly described in the following.

The methods comprise estimation of virulence frequencies from colony counts on seedlings in mobile nurseries exposed at seven selected locations, two to four times a year, and determination of virulence gene frequencies by genotype-testing of single colony isolates. The differential varieties used are near-isogenic lines in a 'Pallas' background (Kølster *et al.*, 1986) and selected commercial varieties. The main results from 1985 and 1986 are given in Table 4.

The figures indicate regional differences, e.g., for Ly, Ar and Al virulence. Further, the "old" resistances We and Sp are virtually

TABLE 4. Virulence frequency in barley powdery mildew in two regions of Denmark 1985 and 1986.

Resistance	Code	Gene(s)	Virulence frequency[1] Sealand 1985[2]	1986	Jylland 1985[2]	1986
Laevigatum	La	Ml-(La)	59	38	70	75
Lyallpur	Ly	Ml-a7+Ml-k	37	67	16	34
Monte Cristo	MC	Ml-a9+Ml-k	32	26	28	32
Arabische	Ar	Ml-a12	25	30	43	51
Algerian	Al	Ml-a	8	1	33	6
Rupee	Ru	Ml-a13+Ml-(Ru3)	0.6	2	0.2	7
Ricardo	Ri	Ml-a3	0.7	8	0.8	6
Turkish	Ty	Ml-a3+Ml-(Ty2)	1	0	0.1	0
Mlo	Mlo	ml-o	1	1	2	2
Weihenstephan	We	Ml-g+Ml-(CP)	75	76	108	90
Spontaneum	Sp	Ml-a6+Ml-a14	84	101	87	86
Kwan	-	Ml-k	45	87	47	68

[1] No. colonies/leaf as per cent of those on 'Pallas'
[2] Houmøller (1986b)

ineffective, most of the more recent resistances, e.g., La, Ly and MC are partly effective, and the "new" resistances Ri, Ty and Mlo, are very effective. The results also show that Ri and Ru virulences have increased about ten times from 1985 to 1986. Al virulence decreased substantially during the same period.

Fungicide resistance

During 1986/87 the virulence survey project will be closely coordinated with the research project on fungicide resistance in powdery mildew initiated in 1983 at the Institute of Pesticides.

Until now the 'test-tube'-method has been used. Subpopulations of mildew collected in the field are tested in glass tubes on seedlings of cv. 'Golden Promise'. In the future this method will be supplemented by testing single colony isolates on leaf segments. These isolates are also analysed for virulence patterns.

Results obtained with barley powdery mildew in 1984 and 1985 have shown mean ED-50 values from 4 to 8 for propiconazole, but there was a wide variation between and within fields (Christensen and Nielsen, pers. comm.). Preliminary results from 1986 indicate the same tendency.

The frequency of powdery mildew isolates able to grow on seedlings grown from seeds treated with the recommended dose of Baytan IM of 1.5 gr/kg seed varies from less than 1 per cent in some fields to 30-35 per cent in others (Junker, pers. comm.).

CONCLUDING REMARKS

The Danish national virulence and fungicide resistance surveys were initiated from a need from breeding, advisory service and farming in Denmark. The resistance genes and fungicides used in Denmark are also used in other European countries, and the pathogen population(s) move freely by wind-borne spores all over northern and central Europe. The Danish surveys are therefore only a part of a common European activity related to a common European agricultural problem. An improved coordination and rapid exchange of data between the national programmes will strengthen the value of the Danish national efforts.

REFERENCES

Doll, H., and Jensen, H.P. 1986. Localization of powdery mildew resistance gene Ml-ra on barley chromosome 5. Hereditas 105, 61-65.

Houmøller, M.S. 1986a. Undersøgelser af hvedemeldug ved virulensanalyser i Danmark 1985. Tidsskr. Statens Planteavlsforsøg, Meddelelse no. 1866.

Houmøller, M.S. 1986b. Undersøgelser af bygmeldug ved virulensanalyser i Danmark 1985. Tidsskr. Statens Planteavlsforsøg, Meddelelse no. 1865.

Jensen, H.P., and Jørgensen, J.H. 1981. Powdery mildew resistance genes in North-west European winter barley varieties. Tidsskr. Planteavl 85, 303-319.

Jørgensen, J.H., and Torp, J. 1978. The distribution of spring barley varieties with different powdery mildew resistances in 1960 to 1976. Kgl. Vetr. og Landbohøjsk. Årsskr. 1978: 27-44.

Kølster, P., Munk, L., Stølen, O., and Løhde, J. 1986. Near-isogenic barley lines with genes for resistance to powdery mildew. Crop Science 26, 903-907.

Munk, L. 1984. A survey of barley powdery mildew in Denmark 1978-1983. Chapter 4 In Heterogene Bygsorters Indflydelse på Meldugpopulationens Sammensætning. Ph.D. Thesis, Royal Vet. Agric. Univ., Copenhagen, pp 93.

Torp, J., Jensen, H.P., and Jørgensen, J.H. 1978. Powdery mildew resistance genes in 106 Northwest European spring barley varieties. Kgl. Vetr. og Landbohøjsk. Årsskr. 1978: 75-102.

WHEAT MILDEW IN EUROPE: VIRULENCE ANALYSIS FOR IMPROVEMENT OF DISEASE CONTROL

F.G. Felsenstein, E. Limpert and G. Fischbeck

Technische Universität München, Lehrstuhl für Pflanzenbau und Pflanzenzüchtung, 8050 Freising-Weihenstephan, Deutschland (BRD)

ABSTRACT

Representative random samples of Erysiphe graminis f.sp. tritici from several European regions were analysed for their virulence on hosts carrying the mildew resistance genes Pm2, Pm4b, Pm8, Mli and the combinations Pm2+Pm6, Pm4b+Pm8, Pm2+Pm4b+Pm8. The regional patterns of virulence differed markedly and were in general in good agreement with the cultivars grown regionally. Furthermore, some signs of migration were noticed, indicating that large parts of Europe should be considered as an epidemiologic unit.

INTRODUCTION

As mildew is a wind dispersed disease, it is possible to obtain representative random samples of regional populations from the atmosphere, in order to monitor characters such as virulence. For this purpose, the method successfully used for several years for barley mildew (Limpert and Schwarzbach, 1981; Limpert et al., 1984a; Limpert and Fischbeck, this volume) was developed to survey the wheat mildew pathogen. The first results, obtained in 1986, are presented for this pathogen which is important in most European countries.

MATERIALS AND METHODS

The method is very close to that used for barley mildew (Limpert, 1985; Limpert and Fischbeck, this volume; Limpert et al., 1984b). Random samples of the regional mildew population were obtained along transects through regions of importance using an apparatus containing a Schwarzbach jet spore sampler fitted to the roof of a car, and progenies of single colony isolates were subsequently analysed in the laboratory.

For the investigation of wheat mildew the incubation time was extended to 10 days; benzimidazole concentrations in agar media were 60 ppm for trapping and 30 ppm for virulence analysis; the leaf segments cut out of the middle part of the primary leaf were 3 cm long. The cultivars used were Kanzler for trapping, and the differential set included Avalon (Pm2), Heiduck (Pm2+Pm6), Armada (Pm4b), Disponent (Pm8), Kronjuwel (Pm4b+Pm8),

Kormoran (Mli) and Apollo (Pm2+Pm4b+Pm8), with Kanzler as the susceptible control. For analysis of UK populations, cultivars Brimstone and Aquila replaced Heiduck and Kormoran, respectively. Virulence frequencies were determined as the proportion of isolates showing a level of attack of at least 50% in comparison with that on Kanzler.

RESULTS

Three kinds of distribution of virulence were observed in Europe. In the first, frequencies decreased from west to east. For virulence against Pm2 (Fig. 1) the highest values were found in England and intermediate levels in Denmark and northern parts of Germany. The values of the combined virulence against Pm2+Pm6 (Fig. 2) were generally lower. A similar distribution was observed for virulence against Pm4b (Fig. 3). In England the frequencies ranged around 50%, while in the other investigated regions proportions of 0 to 30% were found, with the exception of 64% in Bavaria.

For virulence against Pm8 (Fig. 4) a quite different pattern was observed. The highest proportions of virulent isolates of approximately up to 90% were found in the Federal Republic of Germany and Austria, and the lowest, between 0 and 35%, in England and neighbouring regions on the continent.

A more homogeneous distribution was found for virulence against Mli (Table 1), and for the combined virulence against Pm4b+Pm8 (Fig. 5) and Pm2+Pm4b+Pm8 (Fig. 6). Due to the contrasting distributions of the single virulences, combined virulence against Pm4b+Pm8 and against Pm2+Pm4b+Pm8 were relatively rare. The outstandingly high frequencies of these combinations in Bavaria (48% and 18%, respectively), were in agreement with the values for each virulence separately, indicating that there, Pm2 was the most effective gene. In contrast, relatively high proportions of isolates virulent against resistance Mli were found throughout the investigated regions. An exception was northern Italy, where no isolates virulent against any of the investigated host resistances were found. In southern France virulences against only Pm2 and Mli were observed.

DISCUSSION

The results show that the proportion of virulent isolates corresponded mostly with the current and previous distribution of host resistances used in the respective regions. In the UK, Pm2-varieties like Avalon, Fenman,

Norman, Galahad and Longbow are commonly cultivated, thus selecting the matching virulence. Even the outstandingly high value of virulence against Pm4b in southern Bavaria seems to be in agreement with varietal distribution. The winter variety Kronjuwel (Pm4b+Pm8) is grown on 10-12% of the wheat area, and nearly the same area is occupied by spring varieties all of which are supposed to have Pm4b (Straß, Zimmermann, pers. comm.). So the occurrence of Pm4b both in spring and in winter varieties is a disadvantageous complementation in favour of selection for the matching virulence.

In some cases, however, the intensity of selection cannot explain the frequency of the virulence concerned, for example, the relatively high values of virulence against Pm2 in northern France, the Netherlands, Denmark and some regions of northern Germany. These high values may be due to spores dispersed by wind from Britain, where a high frequency of virulence is present. The high values of virulence against Pm8 in the regions from Denmark via Germany to Austria seem also to be influenced by wind drift, but from eastern European countries where varieties with Pm8 have been used commonly for some decades. The data from Finland from recent years are also consistent with selection by Pm8-varieties in eastern countries. Kurtto (cited by Karjalainen, this volume) observed virulence against Pm8 (differential Salzmunde) predominating in the Finnish wheat mildew population. These data give the first indications of interactions between populations of wheat mildew within the Europe-wide agroecosystem. More indications for such relationships between pathogen populations have been demonstrated for barley mildew by Limpert et al. (1984a), Limpert (1985) and Limpert (this volume).

All host resistances investigated, with the exception of Mli, seem to be very effective in the southernmost regions. The pathotypes there were very simple compared with those from the other regions. On average they had only one virulence, compared with two in Austria, FRG, Denmark and northern France, and four in the UK. A similar south-north gradient was monitored in barley mildew (Limpert and Fischbeck, this volume; Andrivon and de Vallavieille-Pope, this volume).

Despite the high values of the virulence against Mli, cultivars with this resistance still showed intermediate mildew control in the field (Anon, 1986). This difference between high attack on seedlings and lower attack in the field might be due to adult plant resistance.

The data presented here are in good agreement with the investigations at a national level by Wolfe and van Kints (this volume) in England, and by Jørgensen and Houmøller (this volume) in Denmark. The results can be explained by selection caused by variety utilisation and by spread of mildew populations by wind. Thus it is obvious that the precision of the interpretation depends crucially on knowledge of the utilisation of varieties and of their resistance genes. Data on resistance genes in varieties are available for the UK (Summers and van Kints, 1986) and for the FRG (Heun and Fischbeck, this volume). Reliable data on variety distribution are also available from Denmark, France and FRG. However, data are missing on resistance genes in French varieties and are scarce on variety distribution in the UK. They should be made generally available because the wheat mildew population in Europe overall may be influenced considerably by those in the UK and in France.

These data show that the method used is able to provide a reliable European mildew survey. It can be expected that more extensive investigations would improve our knowledge of the epidemiology of wheat mildew in Europe and thus improve selection of strategies for the effective use of host resistance.

ACKNOWLEDGEMENTS

We wish to express our thanks to Dr. Straß and Dr. Zimmermann, Bayerische Landesanstalt für Bodenkultur und Pflanzenbau, Weihenstephan, and to Prof. Wolfe, Plant Breeding Institute, Cambridge, for provision of seeds and for valuable discussions.

REFERENCES

Anon. Beschreibende Sortenliste für Getreide-Mais-Ölfrüchte, 1986. Bundessortenamt, Strohte-Verlag, Hannover.

Limpert, E. and Schwarzbach, E. 1981. Virulence analysis of powdery mildew of barley in different European regions in 1979 and 1980. Barley Genetics IV, Proc. 4th Int. Barley Genet. Symp., Edinburgh, 458-465.

Limpert, E., Schwarzbach, E. and Fischbeck, G. 1984a. Distribution of virulence in the European population of barley mildew, *Erysiphe graminis* f.sp. *hordei*. Proc. Eucarpia Cereal Sect. Meeting, Weihenstephan, Vorträge für Pflanzenzüchtung 6, 1-21.

Limpert, E., Schwarzbach, E. and Fischbeck, G. 1984b. Influence of weather and climate on epidemics of barley mildew, *Erysiphe graminis* f.sp. *hordei*. In: Interactions between climate and biosphere, H. Lieth, R. Fantechi and H. Schnitzler (eds.). Progress in Biometeorology Vol. 3, Swets and Zeitlinger B.V., Lisse, 146-157.

Schwarzbach, E. 1979. A high throughput jet trap for collecting mildew spores on living leaves. Phytopath. Z. 94, 165-171.

Summers, R.W. and van Kints, T. 1986. Mildew of wheat. UK Cereal Pathogen Virulence Survey. 1985 Annual Report, 7-12.

Vanderplank, J.E. 1968. Disease Resistance in Plants. New York Academic, 216 pp.

Wolfe, M.S. 1984. Trying to understand and control powdery mildew. Plant Pathology 33, 451-466.

Wolfe, M.S., Slater, S.E. and Minchin, P.N. 1986. Mildew of barley. U.K. Cereal Pathogen Virulence Survey. 1985 Annual Report, 27-34.

TABLE 1: Wheat mildew sampling in 1986 and results of virulence analysis
n: number of isolates investigated individually
in certain cases a smaller number was scored (values in brackets)

Date	Distance	n	Frequencies of Virulence (%) against						
			Pm2	Pm2+Pm6	Pm4b	Pm8	Pm4b+Pm8	Mli	Pm2+Pm4b+Pm8
England									
08.07.	Dover-Reading	21	67	57	45(20)	0	0	55(20)	0
09.07.	Newark-Cambridge	20	90	65	55	35	10	100	5
France									
03.05.	Beaucaire-Montpellier	11	22	0	0	0	0	64(09)	0
07.07.	Mons(front)-Calais	38	68	29	26	11	0	66	0
Netherlands									
13.07.	Venlo-fr. Aachen	13	54	39	31	23	0	77	0
Denmark									
14.07.	Kolding-front Flensbg.	27	23(22)	4	15	59	0	89	0
15.07.	Riso-Rodby	11	46	9	9	46	0	100	0
Germany N--S									
14.07.	Kiel-Flensburg	20	29(17)	32(19)	5	80	0	75	0
16.07.	Neustadt-Kiel	13	33(12)	8	0	85	0	77	0
15.07.	Puttgarden-Neustadt	47	15(44)	6	2	87	6	61	0
16.06.	Neustadt-Hamburg	48	18(43)	6	29	63	17	75	0
16.06.	Hamburg-Bremen	7	14	0	29	71	14	100	0
17.06.	Osnabrück-Minden	32	21(29)	16	13	58	19	80(31)	0
17.06.	Minden-Braunschweig	53	35(52)	10	17	70	12	82	4
13.07.	Venlo-Rheine	16	50	0	13	25	0	75	0
19.06.	Soest-Warburg	28	14	7	11	54	11	71	4
15.06.	Hannover-Hanmünden	18	0(10)	6	0	56	0	78	0
15.06.	Hanmünden-Fulda	24	5(20)	9(23)	8	39	4	79	0
15.06.	Würzburg-Nürnberg	10	20	10	20	60	0	90	0
01.07.	Heilbronn-Crailsheim	9	25(08)	0	0	56	0	89	0
15.06.	Nürnberg-Freising	23	15(20)	0	64(22)	83(22)	48	83	18(22)
Italy									
02.05.	Verona-Piacenza	6	0	0	0	0	0	0	0
Austria									
22.06.	Linz-Krems	31	21(27)	20	29(30)	74	10	97	0
22.06.	Marchfeld	7	0	17	0	86	0	100	0

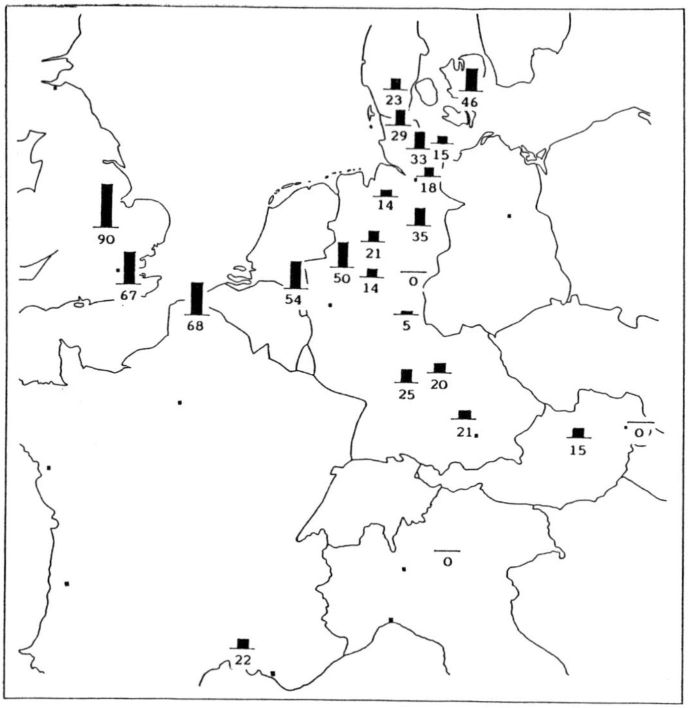

Fig. 1:
Distribution of virulence against Pm2 in 1986.

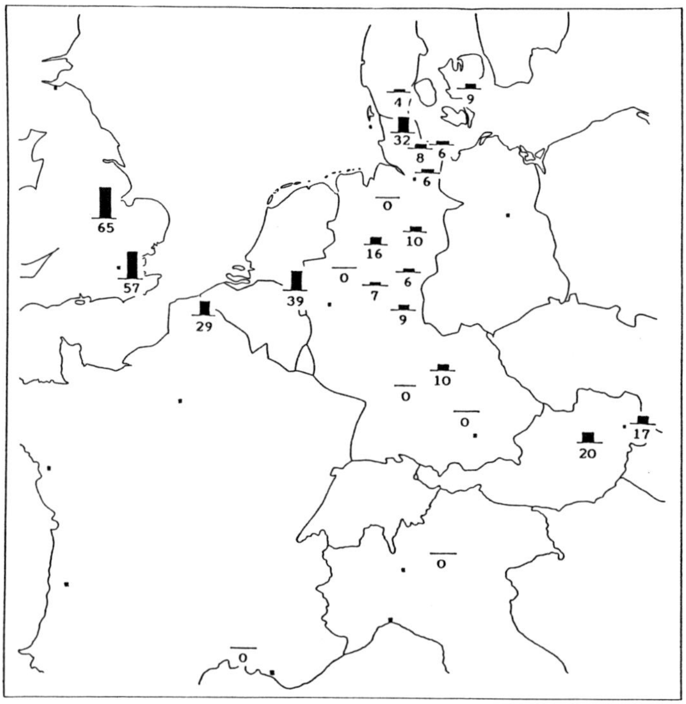

Fig. 2:
Distribution of virulence against Pm2+Pm6 in 1986.

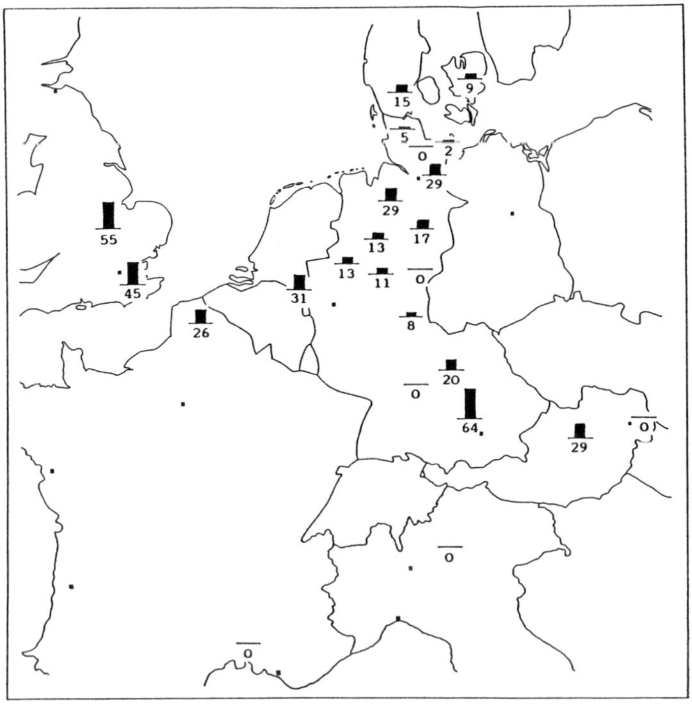

Fig. 3:
Distribution of virulence against Pm4b in 1986.

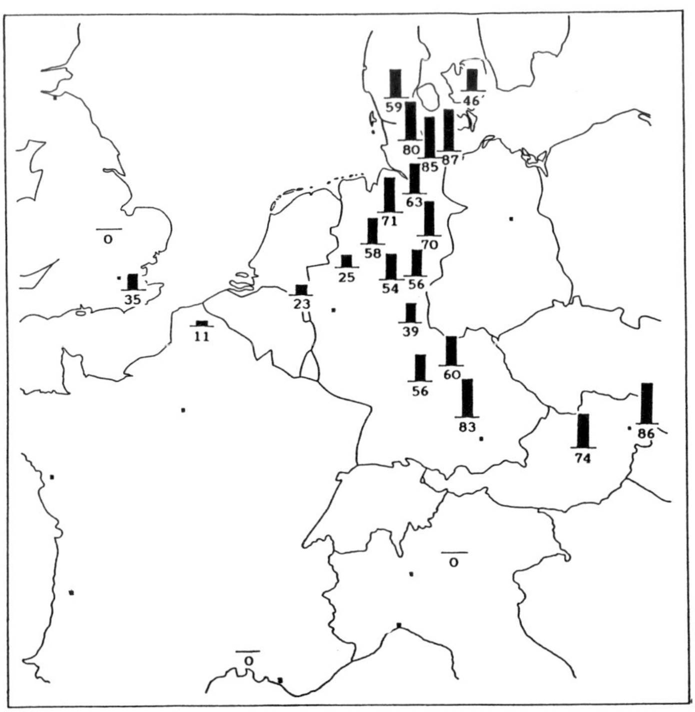

Fig. 4:
Distribution of virulence against Pm8 in 1986.

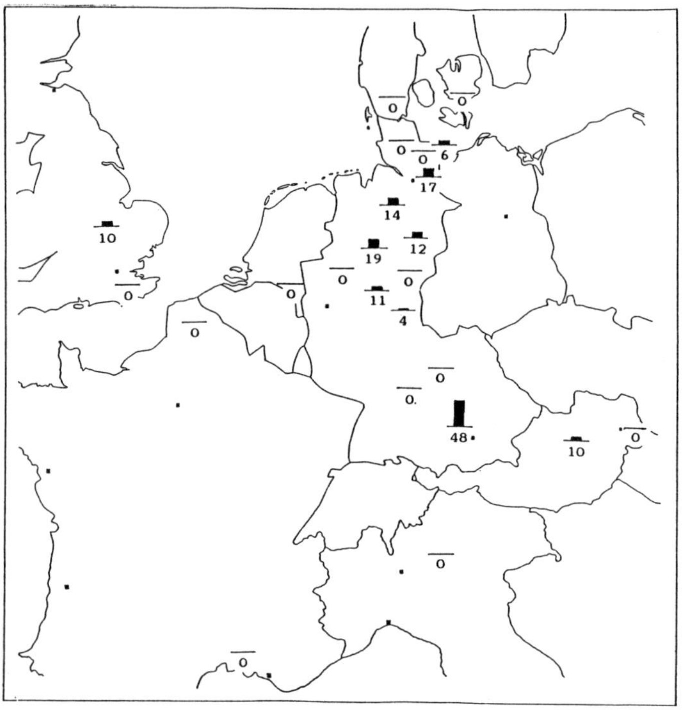

Fig. 5:

Distribution of virulence against Pm4b+Pm8 in 1986.

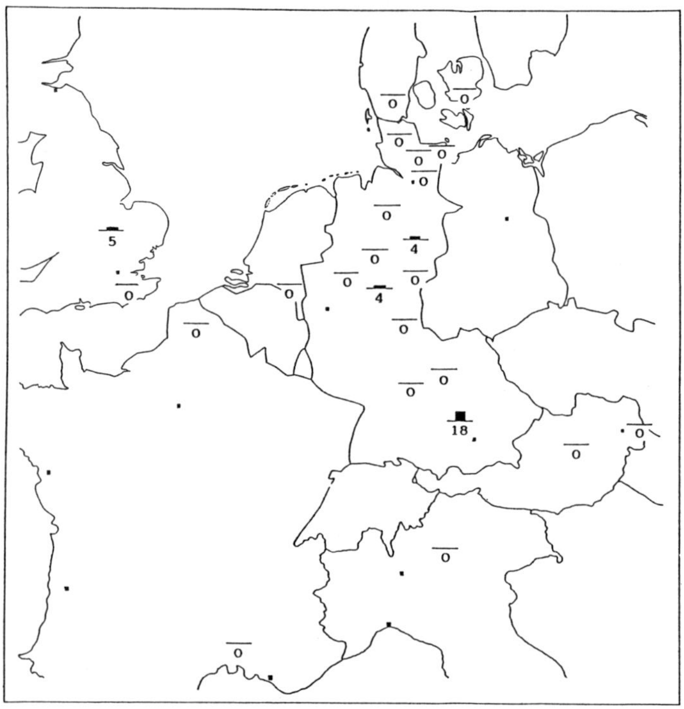

Fig. 6:

Distribution of virulence against Pm2+Pm4b+Pm8 in 1986.

IDENTIFICATION OF RESISTANCE GENES: WHEAT POWDERY MILDEW

M. Heun and G. Fischbeck
Technische Universität München, Lehrstuhl für Pflanzenbau und Pflanzenzüchtung, 8050 Freising-Weihenstephan, FRG

ABSTRACT

Using the gene-for-gene concept and 11 tester isolates of the wheat mildew pathogen, the mildew resistances of 59 winter wheat cultivars were analysed, including all of those currently registered in the FRG.

INTRODUCTION

To determine which resistance genes are responsible for differences in mildew attack, the gene-for-gene concept (Flor, 1955) and pedigree relationships (Fischbeck, 1975) may be useful. The more sophisticated alternative, a genetic analysis of crosses between a great number of cultivars and lines possessing defined resistance genes, is hardly practicable. An attempt has been made, therefore, to develop a tester set of mildew cultivars to differentiate between known genes for mildew resistance and to determine the genes incorporated in the present set of winter wheat cultivars registered in the FRG. Powdery mildew resistances in barley have been identified in a similar way since 1980 and published each year in "Beschreibende Sortenliste".

MATERIALS AND METHODS

Seed of registered winter wheat cultivars was provided by the "Bundessortenamt", Hannover, FRG from standard seed samples obtained from plant breeders for registration purposes. A set of isogenic lines with known resistance genes was obtained from J.G. Moseman, Beltsville, USA. The powdery mildew isolates nos. 1, 3, 4, 6, 7, 8, 9 and 10 were chosen out of approx. 2000 isolates obtained in 1984-86. They were purified, multiplied and tested repeatedly for differentiation between hosts with different resistance genes. These isolates, together with isolate W72/27 (provided by R.W. Summers, Cambridge, U.K), 85072 and 85135 (provided by P.M. Fried, Zürich, Switzerland), allow the identification of all the genes for mildew resistance included in these experiments. Furthermore, all combinations of these resistance genes can be determined together with additional information about pedigree relationships.

For inoculation, 2 x 3 primary leaves (each 3 cm long) of each of the host genotypes were placed on agar containing 50 ppm benzimidazole. Spores formed on infected leaves of cv Kanzler were dispersed using a settling tower aiming at inoculation densities between 200-400 spores/cm^2. The leaves were kept under spore free conditions at 17°C and low light intensity and infection grade and infection type were assessed ten days after inoculation (for isolates 1 and 7 fourteen days were necessary). Each experiment was repeated twice.

Based upon these observations, three groups of host-pathogen interactions were defined:

resistant (r)
intermediate (i)
susceptible (s)

RESULTS

The host-pathogen interactions of the six near isogenic and the seven wheat cultivars carrying known genes and gene combinations for mildew resistance with 11 powdery mildew isolates are shown in Table 1 (from Heun and Fischbeck, in press).

TABLE 1 Reactions of wheat lines with known powdery mildew resistance genes after inoculation with eleven different mildew isolates (r = resistant, s = susceptible, i = intermediate host reaction and - = not tested). The resistance patterns of three lines with 'Blaukorn' resistance are also shown.

Cultivar	major resistance gene/s	powdery mildew isolate no. 1	3	4	6	7	8	9	10	W72/27	85072	85135
Axminster/8CCa	Pm1	s	s	i	s	r	s	i	r	s	i	s
Ulka/8CC	Pm2	r	r	r	s	s	s	s	r	s	r	s
Asosan/8CC	Pm3a	s	r	s	i,r	r	r	s	r	r	i	r
Chul/8CC	Pm3b	r	r	s	i,r	r	r	i,r	r	r	i,s	r
Sonora/8CC	Pm3c	s	s	s	s	s	s	s	s	s	s	r
Khapli/8CC	Pm4a	s	s	s	s	s	s	s	i	r	r	s
Armada	Pm4b	s	s	r	s	s	s	s	r	r	r	s
Hope	Pm5	-	i	-	i	-	s	s	-	i	-	i
Timgalen	Pm6	i	s	s	s	i	i	s	i	s,i	s	s
Disponent	Pm8	s	s	s	r,i	r,i	s	s	s	r,i	s	r
Normandie	Pm9, Pm1, Pm2	r	r	r	r	r	s	r	r	s	r	s
M. Huntsman	Pm2, Pm6	r	r	r	s	r,i	r,i	s	r	i	r	s
Kormoran	Ml-i	s	i	s	i,s	i	s	s	s	i	i	i
Kronjuwel	Pm4b, Pm8	s	s	r	r	r	s	s	r	r	r	r
623/65	PM4b, Pm8	s	s	r	r	r	s	s	r	r	r	r
641/65	Pm4b	s	s	r	s	s	s	s	r	r	r	s

a = eight times backcrossed with cv 'Chancellor'

Mildew reactions against the 11 cultures in Table 1 were determined for 59 winter wheat cultivars. Of these, 27 showed susceptible reactions to all isolates and therefore do not carry any of the identified major genes for mildew resistance (Ares, Artus, Basalt, Caribo, Carstacht, Castell, Diplomat, Dozent, Farmer, Florida, Frühgold, Futur, Gambrinus, General, Granit, Heinricht, Ignaz, Jaguar, Jubilar, Kanzler, Kobold, Monopol, Nimbus, Oberst, Okapi, Ural, Vuka). The reactions of the remaining 32 cultivars are listed in Table 2 (from Heun and Fischbeck, in press). Based upon differences in reaction patterns, they form 10 different groups involving 6 of the 12 major genes analysed.

TABLE 2 Reaction of the winter wheat cultivars registered in the FRG to eleven different powdery mildew isolates. The resistance genes were identified by comparing the reaction patterns with those in Table 1.

Cultivar	identified resistance gene/s	powdery mildew isolate no. 1	3	4	6	7	8	9	10	W72/27	85072	85135
Aquila	Ml-i	s	i	s	i,s	i	s	s	s	i	i	i
Carimulti	Ml-i	s	i	s	i	i	s	s	s	i	i	i,s
Cariplus	Ml-i	s	i	s	i	i	s	s	s	i	i	i
Dolomit	Ml-i	s	i	s	i	i	s	s	s	i	i	i
Falke	Ml-i	s	i	s	s	i	s	s	s	i	i	i
Kormoran	Ml-i	s	i	s	i	i	s	s	s	i	i	i
Kraka	Ml-i	s	i	s	i,s	i	s	s	s	i	i	i,s
Markant	Ml-i	s	i	s	i	i	s	s	s	i	i	i
Milan	Ml-i	s	i	s	i	i	s	s	s	i	i	i
Reiher	Ml-i	s	i	s	i,s	i	s	s	s	i,s	i	i,s
Rektor	Ml-i	s	i	s	i	i	s	s	s	i	i	i
Severin	Ml-i	s	i	s	i	i	s	s	s	r,i	i	i,r
Sperber	Ml-i	s	i	s	i,s	i	s	s	s	i	i	i,s
Tukan	Ml-i	s	i	s	s	i	s	s	s	i	i	i
Urban	Ml-i	s	i	s	i	i	s	s	s	r,i	i	r
Wattines	Ml-i	s	i	s	i	i	s	s	s	i	i	i
Disponent	Pm8	s	s	s	i	i	s	s	s	r	s	r
Götz	Pm8	s	s	s	r,i	i,r	s	s	s	r	s	r
Merkur	Pm8	s	s	s	r	r	s	s	s	r	s	r
Niklas	Pm8	s	s	s	r+s	r+s	s	s	s	r+s	s	r+s
Odilo	Pm8	s	s	s	r,i	r	s	s	s	r	s	r
Boxer	Ml-i, Pm4b	s	i	r	i	i	s	s	r	r	r	i
Mission	Ml-i, Pm4b	s	i	r	i	i	i,s	s	r	r	r	i
Bert	Ml-i, Pm6	i	i	s	i	i	i	s	i	i	i	i
Kristall	Ml-i, Pm8	s	i	s	r	r	s	s	s	r	i	r
Granada	Pm8, Pm5	s	i	s	r	r	s	s	s	r	i	r
Sensor	Pm8, Pm5	s	i	s	r	r	s	s	s	r	i	r
Kronjuwel	Pm8, Pm4b	s	s	r	r	r	s	s	r	r	r	r
Heiduck	Pm2, Pm6	r	r	r	s	i	i	s	r	i,s	r	s
M. Huntsman	Pm2, Pm6	r	r	r	s	i	i	s	r	i	r	s
Sorbas	Pm4b, Pm6	i	s	r	s	i	i	s	r	r	r	s
Apollo	Pm2, Pm4b, Pm8	r	r	r	r	r	s	s	r	r	r	r

REFERENCES

Fischbeck, G. 1975. Genetics of disease resistance in European winter wheat breeding. Proc. 2nd Intern. Winter Wheat Conf., Zagreb, 449-460.

Flor, H.H. 1955. Host-parasite interaction in flaxrust - its genetics and other implications. Phytopathology 45, 680-685.

Heun, M. and Fischbeck, G. Identification of wheat powdery mildew resistance genes by analysing host-pathogen interactions. Plant Breeding, in press.

THE POWDERY MILDEW SITUATION ON BARLEY AND WHEAT IN FINLAND

R. Karjalainen

Department of Plant Pathology and Plant Breeding
University of Helsinki
SF-00710 Helsinki, Finland

ABSTRACT

Cereal production in Finland operates under northern conditions. Spring barley and spring and winter wheat are grown on 0.6 and 0.2 million hectares, respectively. Cereal mildews occur in wheat and barley fields every year, but they induce a serious yield reduction only every fourth or fifth year. About 20% of the total barley area is grown with two-rowed barley cultivars which carry resistance genes derived from Impala, Monte Cristo, and Tellus. Virulence gene frequencies of barley mildew populations occurring in Finland are not known, but they are apparently determined by the corresponding resistance genes in barley varieties grown abroad because winter barleys are not grown in Finland. Race surveys of wheat mildew carried out in the 1970s showed that virulence against resistance genes Pm1, Pm3b, and Pm5 (Axminster, Hope, and Chul) was quite frequent, while virulence against test cultivars Halle 13471 and Weihenstephan was less common. Up to now no problem of fungicide resistance has emerged because of limited use of chemical mildew control. However, the Tilt-treated area is increasing in Finland, which may affect the structure of the mildew population.

INTRODUCTION

In relation to the European perspective, cereal production in Finland operates under extreme northern conditions. Barley, for instance, is grown from latitude 60°C up to 67°C. Numerous barley cultivars are grown mainly because of the variable environmental conditions. Domestic barley breeding aims at improving grain yield, lodging resistance, earliness, sprouting resistance, grain quality, and resistance to diseases. Resistance to powdery mildew was the most important goal in the 1970s, but in the 1980s necrotrophic pathogens, particularly *Rhynchosporium secalis* and *Drechslera teres*, have received more attention than mildew in barley breeding programmes in Finland.

Wheat production in Finland is concentrated on the best soils in the southern parts of the country, and about five cultivars account for most of the spring wheat area. Resistance breeding to control wheat mildew was started in the 1960s, and it is still important, although more effort now goes into improving resistance to *Septoria nodorum*, which has become the most serious wheat pathogen in Finland in the past few years.

Distribution of resistance genes to mildew in wheat and barley cultivars grown in Finland

The barley growing area in Finland is approximately 0.6 million hectares. The majority of cultivars are multi-rowed and of domestic origin. About a third of the barley area is cultivated with Swedish varieties. Up to now, there have been no known major genes for mildew resistance in the multi-rowed cultivars currently grown in Finland, and most of them are very susceptible (Table 1). Kustaa, a Swedish variety (Svalof AB), which is one of the most popular cultivars, carries resistance genes derived from Impala (Ml-g + Ml-a6?). Cultivar trials indicate that Kustaa is still considerably resistant to mildew despite its wide cultivation in southern Finland during the past five years. Ida, another

TABLE 1 Distribution of mildew resistance genes in barley and wheat cultivars grown in Finland in 1985.

	% area	Resistance genes	Mildew severity 0-100
Spring wheat cultivars (0.2 million ha)			
Ruso	28.2	Reward x Pika (Pm1?)	40
Tapio	28.2	Hja c3929 x Kolibri (Pm?)	25
Luja	12.5	No res. genes	34
Kadett	11.3	Kolibri x WWV 439-66 (Pm? + Pm1)	4
Drabant	9.6	CI 12633 x Ring (Pm2 + Pm?)	20
Taava	4.5	No res. genes	38
Tahti	1.0	Karn x Aurora x Pika (Pm1?)	33
Apu	0.3	Garnet x Pika (Pm1?)	31
Spring barley cultivars (0.6 million ha)			
Pokko	13.5	No res. genes	42
Hankkija 673	12.5	"	35
Kustaa	11.7	Mari x MC x Impala/3 (Ml-g + Ml-a6?)	5
Arra	11.7	No res. genes	37
Agneta	9.0	"	46
Hankkija's Potra	8.6	"	30
Otra	6.1	"	34
Pomo	5.4	"	29
Kajsa	4.6	"	40
Hankkija's Eero	2.6	"	35
Ida	1.9	(Monte Cristo x Arla^{6}) x Tellus (Ml-a9 + Ml-g?)	7
Kilta	1.5	No res. genes	28

Swedish two-rowed cultivar (Weibull Ab), carries resistance genes derived from Monte Cristo and Tellus (Ml-a9 + Ml-g?). Ida is a new cultivar in Finland and its area is increasing. At present it is highly resistant to the mildew populations occurring in Finland.

Wheat is grown in Finland on about 0.2 million hectares, 85% of which is occupied by spring wheat. The majority of spring wheat cultivars are of domestic origin, and about 20-25% are from Sweden and Norway. Detailed data on mildew resistance genes in spring wheat cultivars (see e.g. Torp _et al._, 1978) are not available. However, it is evident that from the 1950s to mid-1970s, cultivars grown in Finland did not carry any specific resistance genes, or at most, the Pml gene derived from Pika. Today these cultivars, such as Ruso, Apu and Tahti, are very susceptible to mildew. In the late 1970s and early 1980s, late and high yielding cultivars with mildew resistance genes rapidly became popular in southern Finland. Hankkija's Tapio, a Finnish cultivar, carries resistance genes derived from Kolibri. At the beginning it was fairly resistant to mildew, but today it is as severely diseased as the susceptible cultivar Ruso. Drabant, a Swedish cultivar (Weibull Ab), carries mildew resistance genes derived from CI 12633 (Pm2,6?). Similar to Tapio, it is no longer resistant to mildew, but it is somewhat more resistant than Tapio and considerably more resistant in many cases than the most susceptible Finnish cultivars. Kadett, another Swedish cultivar, which is very popular in southern-most Finland, carries resistance genes derived from Kolibri and WWV 439-66 (Pompe, Pml). Kadett still appears to possess a considerable resistance to mildew (Table 1) despite its wide cultivation during the past five years. Very recently (summer 1986), there were some signs of outbreaks in single fields. A small area (0.05-0.07 million ha) of winter wheat is cultivated mainly with domestic varieties, which possess no known mildew resistance genes.

Virulence of mildew populations in Finland

The population structure of cereal mildews in Finland is very poorly known. Not a single study has been made on barley powdery mildew, and the few studies of wheat mildew date back to the 1970s.

Powdery mildew of barley occurs in Finland every year, but it usually attacks barley at later stages of crop growth and thus induces only a small yield reduction. About every fourth or fifth year mildew comes early and induces a serious yield reduction. Winter barleys are not grown in

Finland, and thus the primary inoculum originates from barley fields in the neighbouring countries. Facing Sweden, the west coast of Finland probably receives mildew conidia from Sweden and perhaps from Denmark, while the southern and eastern parts of Finland apparently receive them from Estonia, Germany and Poland, though this is only speculation. It is, however, obvious that the virulence gene frequencies of mildew populations found in Finland are determined by the corresponding resistance genes in barley varieties grown in the "source" countries. Currently only two cultivars grown in Finland, Ida and Kustaa, carry resistance genes, derived from Monte Cristo, Tellus, and Impala. In official trials there are some new breeding lines, particularly of Swedish origin, which carry genes for mildew resistance, and it is very likely that some of these lines will be grown in Finalnd in the near future.

Two fungicides, Tilt and Bayleton, are available for the chemical control of cereal leaf diseases, but so far they have been used very little, and no problem of fungicide resistance has yet emerged. However, the Tilt-treated area is currently increasing in Finland, and this may have some consequences for the mildew population structure.

TABLE 2 Race surveys of wheat powdery mildew carried out in Finland in 1974-1979 (Kurtto, 1984).

Race number	Nisu contr. (susc.)	Salzmunde	Ulka	Axminster	Halle 13471	Weihen-stephan M1	Hope (sel)	Chul	Number of isolates
15	S	R	R	S	R	R	S	R	1
16	S	R	R	S	R	R	S	S	13
19	S	R	R	R	R	R	S	S	4
27	S	S	R	S	R	R	S	R	10
35	S	S	R	S	R	R	S	S	239
43	S	S	S	S	S	R	S	R	1
44	S	S	R	R	R	R	S	S	11
46	S	S	S	S	R	R	S	R	1
47	S	S	S	S	S	R	S	S	8
49	S	R	S	S	S	S	S	R	1
54	S	S	S	S	R	R	S	S	3
58	S	S	S	S	R	R	R	S	1
59	S	S	R	R	R	R	S	R	1
62	S	S	R	S	R	R	S	S	1
66	S	S	S	S	R	S	S	R	1
69	S	R	S	S	S	R	S	S	3
?	S	S	R	S	R	S	S	S	3
								Total	302

Powdery mildew of wheat occurs both in winter and spring wheat in Finland every year, but it attacks wheat crops severely approximately every third or fourth year causing significant yield losses. The population structure of wheat mildew has been analysed on a small scale only by Mr Kurtto at the Agricultural Research Centre in 1974-1979. This race survey was based on ideas prevailing in the 1960s and used test varieties to differentiate physiological races (e.g. Leijerstam, 1962). Table 2 summarises the results of the collection of 302 isolates from southern Finland. A noticeable feature in this data is that race 35 is so dominant. In general, this data and further studies by Nissinen (1973) suggest that virulence against the resistance genes based on Axminster, Hope and Chul is quite frequent, while virulence against resistance from Halle 13471 and Weihenstephan is much less frequent. Race surveys carried out in the GDR during the same period show the same general pattern in the virulence of wheat mildew (Haller et al., 1980). However, it is important to note that the surveys in Finland were carried out at a time when the current varieties carrying mildew resistance genes (Kadett, Drabant and Tapio) were not yet in cultivation. In the past five years cultivars have changed radically in favour of resistant ones, and there is now an urgent need to examine the present situation of virulence genes occurring in Finnish wheat fields.

In Finland, we are going to start a three-year-study on factors affecting the population structure of wheat and barley powdery mildew using methods currently employed in other European countries. In addition, some analyses of mildew populations using molecular markers are also an aim of this project.

REFERENCES

Haller, G., Klein, M., Meyer, H. and Wallstabe, E. 1980. Das Rassenspektrum von Weizen- und Gerstenmehltau in den Jahren 1975 bis 1977 in der DDR. Arch. Phytopathol. u. Pflanzenschutz.

Kurtto, J. 1984. Vijjaharma. kasvitauti- ja tuhoelainpaiva. Kasvinsuojeluseura r.y.

Leijerstam, B. 1962. Studies in powdery mildew on wheat in Sweden. I. Physiological races in Scandinavia in 1960-1961. Stat. vaxtskyddsanst. 94, 279-293.

Nissinen, O. 1973. Effect of powdery mildew (*Erysiphe graminis* f.sp. *tritici*) on spring wheat in Finland. J. Scient. Agric. Soc. Finl. 45, 461-468.

Torp, J., Jensen, H.P. and Jorgensen, J. Helms. 1978. Powdery mildew resistance genes in 106 northwest European spring barley varieties. Kgl. Vet.- og Landbohojsk. Årsskr. 1978: 75-102.

BARLEY POWDERY MILDEW IN FRANCE

D. Andrivon and Claude de Vallavieille-Pope

INRA, Laboratoire d'Epidémiologie, INA PG,
F-78850 Thiverval-Grignon, France

ABSTRACT

Cereal powdery mildews were regarded until recently as secondary diseases in France, mainly due to widespread fungicide treatment. First results on virulence and fungicide resistance in barley mildew in France were provided by Limpert for 1985; a national programme began this year, in close cooperation with the Weihenstephan team. Results show a situation somewhat unusual in Europe, characterised by the overwhelming importance of V-(La) and the scarce incidence of V-a7 + V-k, due to locally grown cultivars. V-g and V-Ra were very common, while the frequencies of V-a6 and V-a12 varied from 30 to 80% depending on the regions investigated. High levels of triadimenol resistance, causing losses of efficiency in eastern France, were detected in 1985 and confirmed in 1986. With respect to fenpropimorph, a slight deviation in sensitivity of populations from eastern France was observed for the first time in 1986.

INTRODUCTION

It is difficult to be precise about the problem of cereal mildews in France, due to the little amount of French work done on these diseases. To assess their importance, the best information is provided by ONIC* enquiries about cereal crops (varieties used, areas, cultural and pesticide treatments) in the major growing areas, but also by observations of regional field trials made by ITCF* or GEVES*. Other useful sources are reports from breeders, plant pathologists and plant protection specialists, including foreign research (mainly German, British and Danish) concerning virulence distribution and cultivar genotypes.

1. Estimation of the agricultural importance of cereal mildews in France

Fifteen years ago, cereal mildews were an important problem for growers; Rapilly *et al.* (1973) reported that losses could reach 6 to 14% in

---

*

ONIC : Office National Interprofessionnel des Céréales (National Interprofessional Office for Cereals)

ITCF : Institut Technique des Céréales et Fourrages (Technical Institute for Cereals and Fodder Crops)

GEVES : Groupe d'Etude des Variétés et Semences (Group for Varieties and Seed Studies)

wheat, 20% in barley and 50% in oats. Since that time, the main revolution in cereal cropping has been the extension of fungicide use during the vegetative period of the crop. According to the ONIC enquiry for 1985 (Anon., 1986), the number of pesticide treatments (fungicide + insecticide) on wheat ranged from an average of less than one (0.7) in extensive areas (southern and western parts of the country) to 2 to 4 in the intensive farming regions of central, northern and eastern France. On barley, the average varied between 0.3 in north-eastern France (Lorraine) to 2 in Beauce (around Paris) and 3 in Champagne. The most widely used fungicides were triadimenol, triadimefon, propiconazole, prochloraz and fenpropimorph, which are effective against the major foliar diseases. Until recently, mildew seemed to be almost completely controlled over the majority of the 7 million ha (4 million ha of wheat and 1.5 of barley) of small grain cereal crops grown in France, and appeared to many advisers, growers and breeders to be no longer an important disease.

Nevertheless, two main considerations justify further epidemiological studies on mildew in France. First of all, the French cultivar pattern, as revealed by ONIC enquiries, is very homogeneous throughout large regions, both in wheat and in barley. For instance in barley, the few resistance genes used (mainly Ml-g, Ml-Ra, Ml-(La) and Ml-a12) occupy large contiguous areas, while others (namely Ml-a7, Ml-k, Ml-a9, Ml-a1 and ml-o) cover limited areas and some (Ml-a3, Ml-a13) are not used. Furthermore, when monitored over some three or four years, the proportion of each gene in each growing area is quite stable, and the few changes that do occur often tend to accentuate the prevalence of a single resistance (Table 1). In other words, there is diversification neither in space nor in time, contrary to the conditions considered by Robinson (1971) as important for the lasting effectiveness of vertical resistance. The French situation is thus likely to exert selection in each region for homogeneous populations of the pathogen, matching the prevailing resistance genes. The current practice of relying only on fungicide protection to control such populations could lead, as soon as fungicide efficiency "breaks down", to wide epidemics resulting in possible important crop damage.

The second reason is economic; for three years now, the EEC guaranteed cereal prices have been constantly falling. To maintain profitable crops, growers must thus restrict expenditure, mostly by reducing pesticide inputs which remain quite expensive.

To propose alternative ways for mildew control, it is necessary to monitor the patterns of virulence and fungicide resistance in the main, if not all, French barley growing regions. Our investigations, coupled with those of Limpert (this volume), began this year in Grignon and were continued in Weihenstephan. The results, added to those of Limpert (1986), provide a first impression of the current structure of the French barley mildew population.

2. The barley mildew population in France

2.1 Methods

The methods used, including spore-trapping with a Schwarzbach jet spore sampler (Schwarzbach, 1979) and analysis of virulence and fungicide response of single colony isolates on detached leaves kept on benzimidazole agar, are almost the same as those used by Limpert (Limpert and Fischbeck, this volume). They are now used successfully in Grignon.

2.2 Virulence distribution

Limpert's 1985 survey in the northern half of France provided a first indication of the distribution of virulence (Limpert, 1986). This has now been extended by the 1986 tours, and by our own results at Grignon.

The population structure is highly differentiated, and may be summarised as follows. In 1985, while V-g was present at high levels in all regions investigated, V-a12 was prevalent in the western part of the country, where Aramir (Ml-a12 + Ml-g) remained popular. V-(La) was mostly important in central and eastern France, regions where Menuet (Ml-(La)) was widely grown. V-a7 + V-k was of little importance, in spite of two maxima in the northern and eastern areas, where the cultivation of Triumph (Ml-a7 + Ml-(Ab)) was beginning. V-a6 + V-g was most common in central and eastern France, where it was probably imported from Great Britain and/or Germany, since in the last ten years no cultivar carrying "Spontaneum" resistance has been grown widely in the country. V-Ra was present in all regions at very high levels (around 100%). All other virulences occurred at negligible frequencies.

There were few changes in this pattern in 1986 (Fig. 1). The outstanding features in this year were the increase in frequency of V-(La) in all north-eastern regions, reaching 70 to 90% in Beauce, Bourgogne, Lorraine, Champagne and Alsace. The frequencies of V-Ra, V-g and V-a6 + V-g remained roughly constant, while V-a12 appeared to decrease slightly. V-a7 + V-k continued to increase, particularly in northern and eastern France, but remained scarce elsewhere. In southern regions, which were investigated for the first time this year, V-a12 was very common (more than 80%). This part of the country showed somewhat simpler races than the rest of France (average around 3 virulences per race in the south, around 5 in the north).

2.3 Distribution of fungicide resistance

Two ergosterol biosynthesis inhibitors were tested (triadimenol and fenpropimorph) which belong to two chemical families differing in their site of action.

Although less dramatic than in other European regions such as northern Germany or England, the situation with respect to triadimenol is of concern, especially when considering the outstanding role currently played by triazoles and other C-14 demethylation inhibitors (DMIs) in mildew control in France. The 1985 data showed the emergence of problems in northern and eastern France (Limpert, 1986). These problems were even more acute this year. Despite the generally poor conditions for mildew epidemic development, failure of disease control by DMIs, already noticed in 1985 (Leroux, pers. comm.), extended in Champagne and Lorraine in barley. Important parts of the mildew populations of these regions, as well as of northern France, showed a high degree of resistance to triadimenol (Andrivon et al., in press). Much attention should be paid to the further development of this phenomenon, particularly in large regions of intensive agriculture like Champagne, where many fungicide treatments are applied each year. Fig. 2 shows the proportion of the populations sampled in different regions with a resistance factor (R.F.) greater than 16 or 64. The latter, measured in our test conditions, is regarded as the limit of control capability of the dose of triadimenol recommended in France (30 g a.i. per 100 kg seed).

In 1985, Limpert's investigation of mildew populations from several regions of France showed no noticeable differences in sensitivity to fenpropimorph (Limpert, 1986). This year, the investigations focused on three parts of the country (Table 2), and showed a slight but significant deviation in populations from Champagne, compared with those from northern and southern France (Andrivon et al., in press). This could be related to a probable greater use of fenpropimorph in Champagne, where triazoles and related compounds are now largely ineffective. It is still too early to state whether these results are already the first sign of emerging resistance, since the resistance factors observed were low (2 to 4 on average).

TABLE 2. Sensitivity of barley mildew isolates to fenpropimorph from three regions in France. n = number of isolates analysed individually

origin of samples	n	Percentage of isolates per dose of fenpropimorph (mg a.c./kg seed, I) and resistance factors (II)					
		I	4.69	9.37	18.75	37.5	75
		II	0.25	0.5	1	2	4
standard isolates	4		0	0	100	0	0
Beaucaire-Montpellier	13		15.4	23.1	50	11.5	0
Mons (front.) Lille-Calais	36		1.4	23.6	58.3	9.7	6.9
Reims-Ste Menehould	36		0	1.4	51.4	40.3	6.9

3. Prospects

Further studies are necessary in France to develop a system of integrated protection against mildew adapted to our conditions. The problems emerging, noticed during the last two years, should stimulate growers to become more interested in such a system. The risks due to the disequilibrium of the cultivar pattern and to the over-important part devoted to chemical protection should now be taken into account and pathologists as well as breeders should consider other ways of coping with mildew. Furthermore, the French geographical position, between Great Britain and the FRG, provides a good opportunity for studying population movements.

The main directions of our research programme for the coming few years could then be described as:

* monitoring virulence and fungicide resistance patterns in major French barley cropping areas;

* studying the new cultivars released in France, as well as some older ones, with respect to their resistance genes;

* experimenting with cultural ways for limiting mildew, such as variety mixtures,

with the aim of guiding farmers in the choice of the cultivars they grow and of the fungicides they use, and to integrate this choice in a coherent system adapted both to agricultural conditions and to mildew population dynamics.

REFERENCES

Andrivon, D., Limpert, E., Felsenstein, F.G. 1987. Sensibilité au triadiménol et au fenpropimorphe dans des populations françaises d'*Erysiphe graminis* DC. f.sp. *hordei* Marchal. Agronomie, in press.

Anon. 1986. Enquete céréalière par sondage 1985: techniques. ONIC, Paris, 39 p.

Limpert, E. 1986. L'oidium de l'orge: quelles races en France? Perspectives Agricoles, 105, 35-39.

Rapilly, F., Lemaire, J.M., Cassini, R. 1973. Les principales maladies d'origine cryptogamique des céréales en France. INRA - ITCF, 192 p.

Robinson, R.A. 1971. Vertical resistance. Review of Plant Pathology, 50 (50), 233-239.

Schwarzbach, E. 1979. A high throughput jet spore sampler for collecting mildew spores on living leaves. Phytopathologische Zeitschrift, 94, 165-171.

Fig. 1 : Distribution of virulences in several regions of France in 1985 and 1986, expressed by the frequency of isolates able to overcome Ml-g (1a), Ml-a12 (1b), Ml-(La) (1c), Ml-a7 + Ml-k (1d) or Ml-a6 + Ml-g (1e).

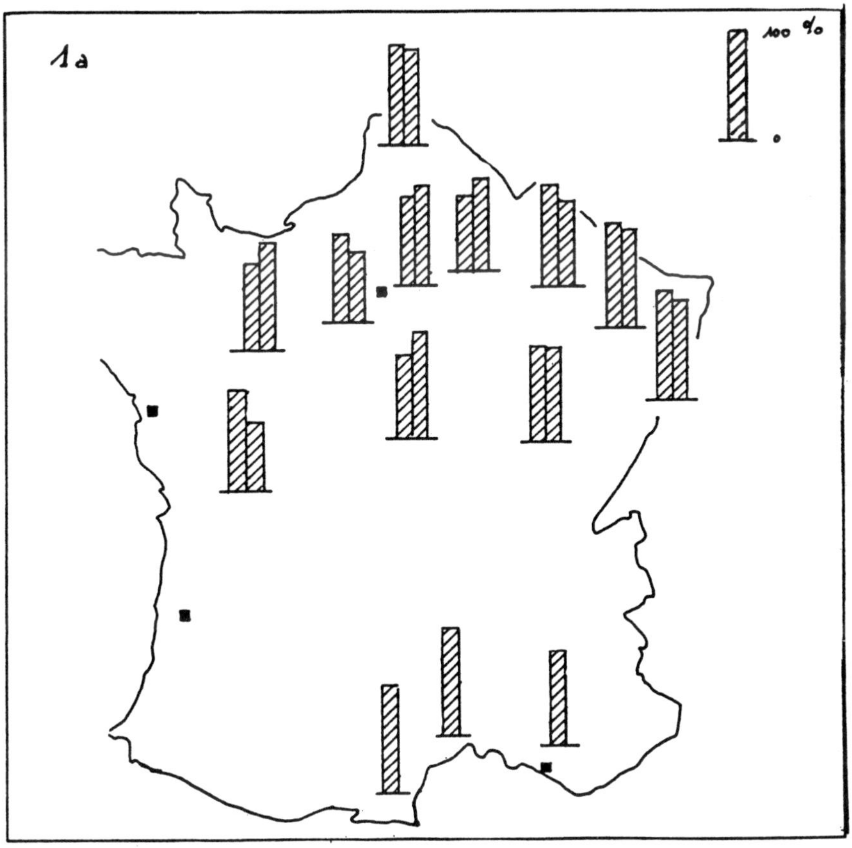

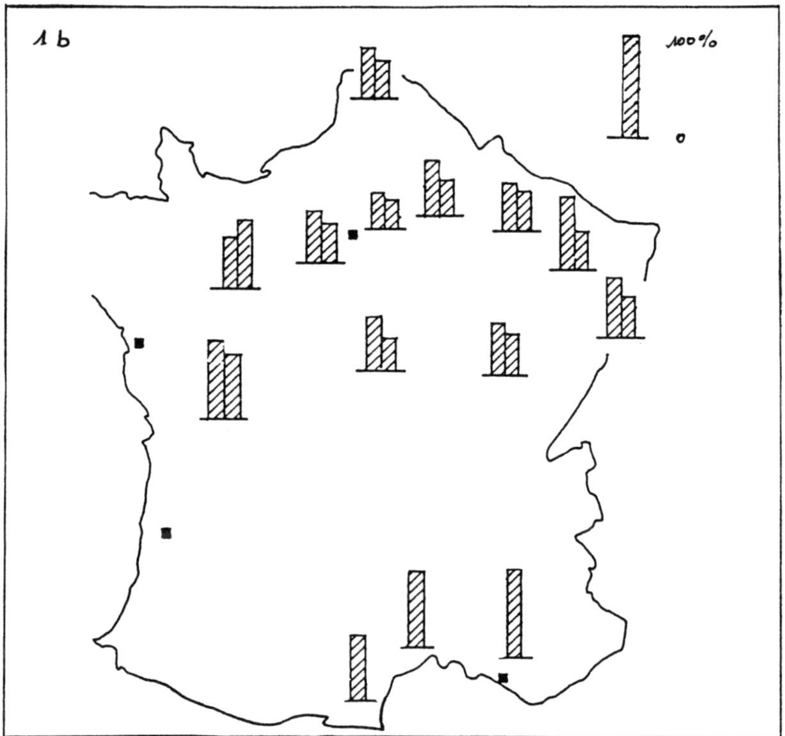
1 b
100%
0

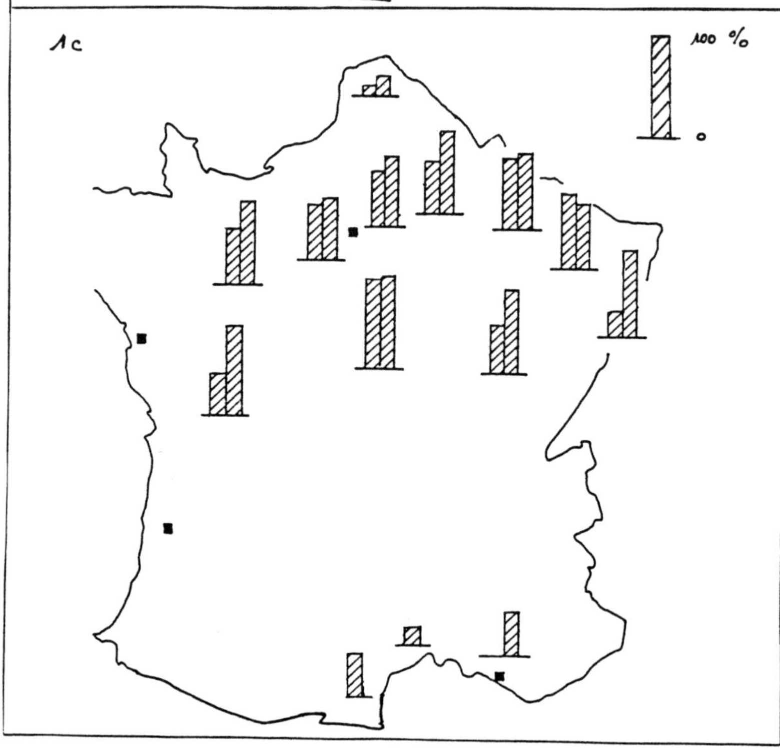
1 c
100 %
0

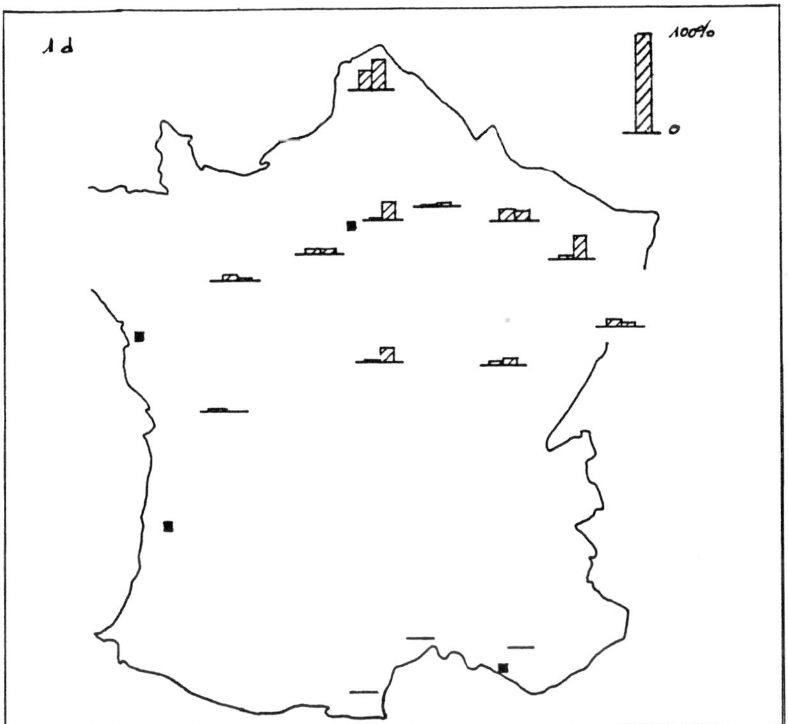
1d
100%
0

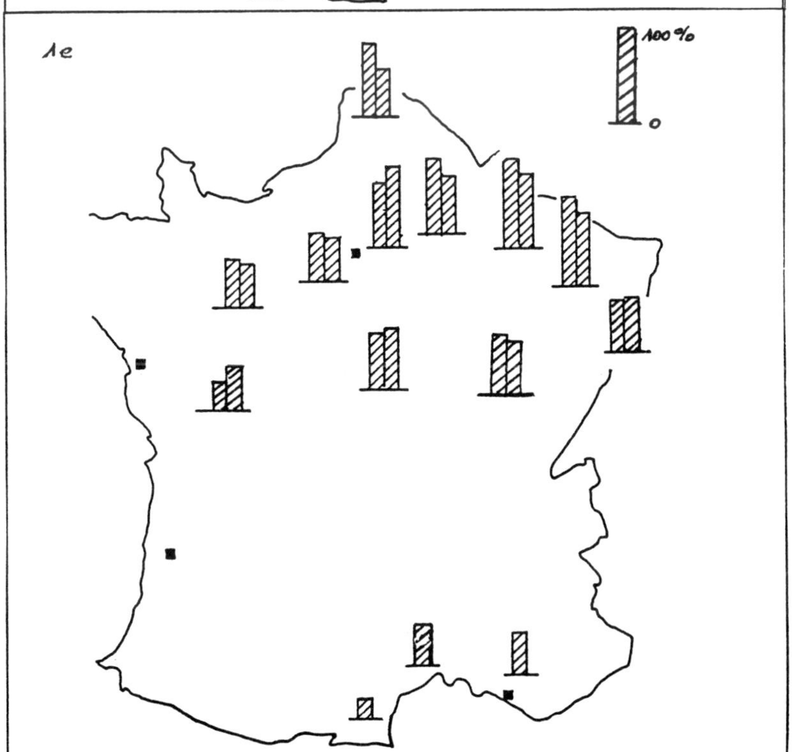
1e
100%
0

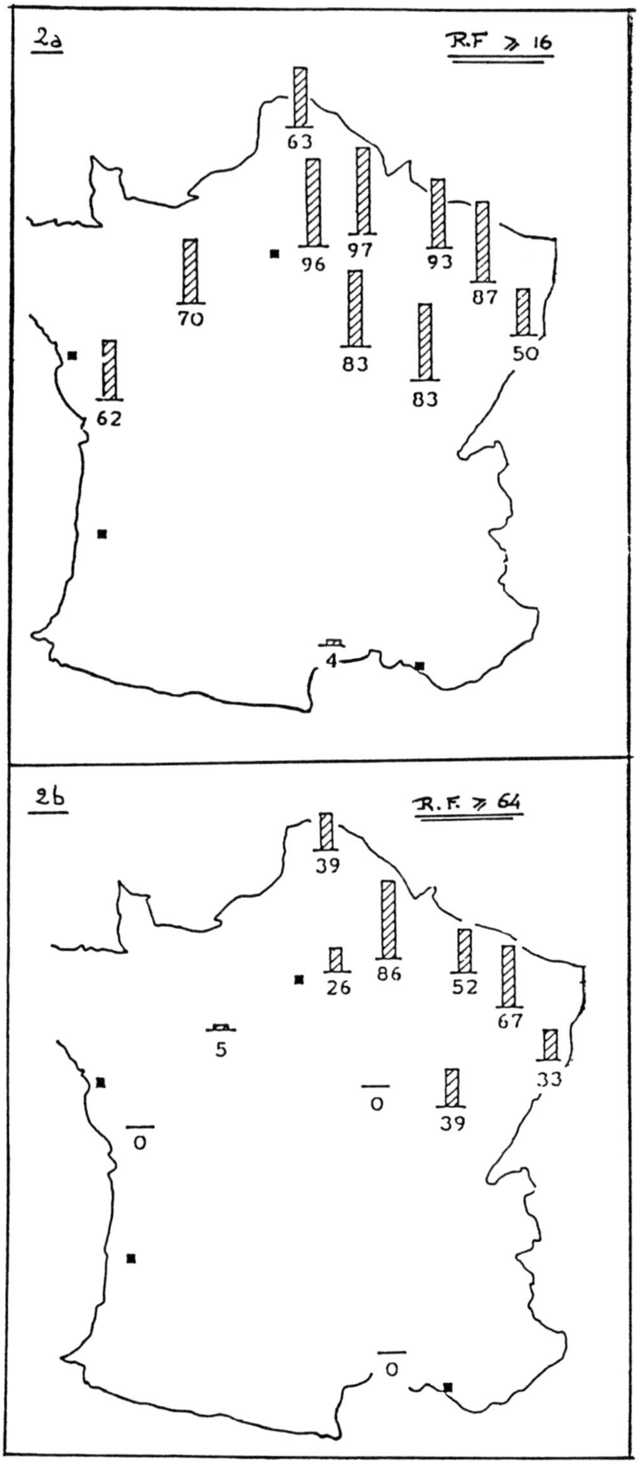

Fig. 2 : Sensitivity of barley mildew isolates from several regions of France, expressed by the proportion of the population showing a resistance factor greater than 16 (fig 2a) or 64 (fig 2b).

		Ml 41/145	Ml 37/136	Mlg	Mlv	Mla12	Mla7 + Mlk	Others	Number of R- geres
North (59)	'83	29.5	4.1	18.1	0	0	0	0	3
	'84	26.7	0	24.2	0	0	0	0	2
	'85	17.7	11.0	31.6	0	0	7.6	0	5
Lorraine (54)	'83	54.2	3.3	9.0	22.3	9.0	2.7	0	7
	'84	40.4	13.0	4.2	21.6	4.2	8.3	0	6
	'85	37.0	21.0	0	26.0	0	16.0	4.0	5
West (44)	'83	51.6	0	47.1	3.0	9.1	0	0	4
	'84	35.2	2.4	8.3	7.9	8.3	0	3.6	6
	'85	25.5	0	32.3	5.0	23.0	0	7.1	5
Beauce (45)	'83	29.2	0	10.3	53.0	5.2	0	5.1	5
	'84	29.5	0	7.8	57.6	0	3.4	6.6	5
	'85	20.0	3.9	2.7	64.4	0	2.7	0	6
Champagne(51)	'83	64.1	45.5	6.5	3.3	6.5	0	0	5
	'84	68.7	40.5	4.6	4.5	4.6	0	0	5
	'85	59.5	59.5	8.5	10.0	8.5	0	0	5
Centre (18)	'83	60.8	4.3	5.6	18.9	5.6	0	0	5
	'84	52.5	0	6.0	17.2	6.0	0	0	4
	'85	37.8	0	3.8	24.8	3.8	0	0	4
Southeast(04)	'83	70.1	0	93.6	2.1	3.9	0	0	4
	'84	69.5	0	94.5	1.5	9.3	0	0	4
	'85	61.2	0	55.7	5.6	15.0	0	0	4
Southwest(32)	'83	38.1	0	71.5	0	17.8	0	6.3	4
	'84	26.0	0	56.0	0	5.9	0	0	3
	'85	29.0	0	57.0	10.0	28.0	0	0	4

Tab. 1 : Percentage of the barley area occupied by the main resistance genes.
Numbers in brackets refer to the French department chosen as reference.

FUNGICIDE RESISTANCE IN CEREAL MILDEWS IN IRELAND

Brendan Dunne

Agricultural Institute
Oak Park Research Centre, Carlow, Ireland

ABSTRACT

Evaluation of fungicides for disease control has been carried out in Ireland for a number of years. Seed dressings and foliar spray treatments have been used and disease levels and yields have been recorded. From this work information has been extracted suggesting shifts in sensitivity to different fungicides in the pathogen population.

INTRODUCTION

Since the early 1970's a programme for the evaluation of fungicides has been carried out in Ireland at the Agricultural Institute, Carlow. While specific work on fungicide resistance only began in 1986 it is possible, by extracting information from fungicide trials, to indicate that the mildew population may have shifted over the years in response to fungicide use.

RESULTS AND DISCUSSION

Ethirimol (Milstem) was widely used in Ireland as a seed dressing and as a foliar spray to control mildew on spring barley from 1972. In the early years of use of this fungicide, disease control was excellent and so were the yield responses (Table 1). However, by 1978 the effectiveness of ethirimol diminished and it no longer gave an economic response.

TABLE 1 Response to ethirimol seed dressing on spring barley (cultivar Emma).

Year	% yield increase
1973	9.6
1975	5.1
1976	5.9
1978	0.0
1979	1.5

Evidence that the reduced effectiveness of ethirimol was due to a shift in the mildew population towards insensitivity to the compound was provided by the introduction of triazole fungicides in the 1979 and 1980 seasons. Triadimenol (Baytan) was then used as a seed dressing and its analogue triadimefon as a foliar spray. The yield response to triadimenol as a seed dressing was much superior to that obtained from ethirimol (Table 2).

TABLE 2 Response to ethirimol and triadimenol seed dressing on spring barley (cultivar Emma).

Year	% yield response Ethirimol	Triadimenol
1979	1.5	3.4
1980	4.6	16.0

These results indicate that the mildew population in 1980 had a large measure of insensitivity to ethirimol but was sensitive to the triazoles.

The triazole fungicides were widely used from 1979 onwards both as seed dressings and as foliar sprays. The first evidence that the mildew population was becoming insensitive towards them came in 1981. One of the indications of insensitivity towards fungicides is a reduced period of effectiveness of the fungicide towards the target organism. In 1979 and 1980 seed treatments outyielded foliar spray applications when triazole fungicides were used. However, from 1981 onwards the foliar spray applications of triazole fungicides outyielded the seed treatments indicating a reduced period of effectiveness for these fungicides (Table 3).

TABLE 3 Comparison of yield responses from seed dressing and sprays on spring barley (triazole fungicides; yield in t/ha).

Year	Seed dressing		Foliar sprays		Control
1980	6.27	114	5.91	109	5.48
1981	6.00	104	6.15	107	5.76
1982	5.42	96	6.02	106	5.67
1983	5.48	108	5.66	112	5.07
1984	5.34	110	5.63	116	4.84
1985	6.30	105	6.70	111	6.01

Further evidence that the reduced effectiveness of the triazole fungicides was caused by a change in the mildew population was provided by the use of morpholines in these same trials. In 1980 there was less disease on the triazole treatments than on the morpholine treatments. However, by 1981 the situation changed and since that year the morpholines have given greater control of the disease (Table 4).

TABLE 4 Performance of triazole and morpholine fungicides (foliar applied) in spring barley.

Year	% mildew Triazole	Morpholine
1980	22.7	24.3
1981	15.2	10.8
1982	37.6	34.0
1983	29.0	22.2
1984	47.6	36.0

The morpholine fungicides do not appear to have lost any of their effectiveness over this period.

In 1984 and 1985 a new seed treatment, Ferrax, combining ethirimol and flutriafol was introduced. In two years of trials the fungicide mixture gave improved mildew control and increased yield over the triazole alone when both chemicals were applied as seed dressings (Table 5).

TABLE 5 Comparison of seed dressings containing fungicides with different modes of action (spring barley).

	% mildew		Yield (t/ha)	
	1984	1985	1984	1985
Triadimenol	54.7	10.5	5.0	5.2
Flutriafol + ethirimol	34.1	4.0	5.6	5.4

Presumably (Table 5), the pathogen population with reduced sensitivity to triazoles was controlled more by the combination of fungicides with different modes of action.

In 1986 specific work on mildew sensitivity to triazoles began using the method developed by Ciba-Geigy. Wheat (cultivar Zenith) and barley (cultivar Golden Promise) are sown in test tubes placed in a growth chamber. After one week of growth, propiconazole in the concentrations 0.3, 1, 3, 10 and 30 ppm approx. is added to the test tubes. Plants are inoculated with the mildew fungus and disease assessment is carried out 14 days later. Preliminary work in 1986 with wheat gave an average ED_{50} of between 1 and 3 ppm propiconazole.

WHEAT AND BARLEY POWDERY MILDEW IN ITALY

L. Corazza, M.T. Ialongo

Istituto Sperimentale per la Patologia Vegetale, Rome, Italy

ABSTRACT

Due to the extreme variability of environment in Italy and the introduction of recently released varieties, it is necessary to record regularly the mildew reaction of cultivars, of both wheat and barley, in different growing areas. Effectiveness of known resistance genes is monitored as well as the existence of factors of virulence which make the field behaviour of varieties unpredictable. Sources of resistance to be used in breeding programmes are analysed.

INTRODUCTION

The varietal composition of *durum* and bread wheat in Italy is changing as a consequence of breeding and introduction of new varieties that are high yielding and well-adapted to marginal areas, and of the attempts to grow *durum* wheat in central and northern regions. The area of barley has increased in recent years and new varieties have also been introduced. Up to now, barley has been mostly winter sown and there is a trend to substitute old "local" varieties with new highly yielding varieties, most of foreign origin.

Particular attention is given to varietal behaviour to powdery mildew, since this pathogen is frequent both on wheat and barley.

Intense nitrogen fertilization, use of dwarf and semi-dwarf varieties, increased monoculture and irrigation in some arid areas in the south, are all factors which play a role in the incidence of infection by powdery mildew.

WHEAT

Due to the extreme variability of environment in Italy it is necessary to evaluate the field reaction of cultivars to powdery mildew in different growing areas and to identify the virulence factors present in natural populations of *Erysiphe graminis* (1) f. sp. *tritici*.

Field trials are organised yearly in several localities to compare *durum* and bread wheat cultivars, foreign varieties, new selections, related

---

(1) *Blumeria graminis* (DC.) Speer *comb. nov.* has been recently proposed for this fungus (Speer, 1975).

species, and isogenic and near-isogenic lines. Severity is recorded as percentage of powdery mildew infection on the plants according to the modified Cobb's scale (0-100%). This research aims at:

- monitoring the field behaviour of varieties and lines with respect to powdery mildew;
- obtaining indications of the effectiveness of known resistance factors and to lay emphasis on the presence of unexpected virulence factors in the natural populations of the pathogen;
- detecting the sources of resistance to be used in wheat improvement;
- supplying useful indications both to breeders and farmers.

During the past season, which was not particularly favourable for powdery mildew epidemics, attacks from the pathogen were more severe on _durum_ wheat than on bread wheat (Pasquini _et al._, 1986). Only two _durum_ wheat varieties of the Val group (Valforte and Valnova) maintained their resistance (Table 1).

TABLE 1 Field reaction to powdery mildew of several bread and _durum_ wheat cultivars.

Bread wheat		_Durum_ wheat	
Aniene	MS(*)	Aldura	MS
Aquileia	MS	Appio	MS
Argelato	MS	Appulo	S
Autonomia B	MS	Berillo	S
Avalon	R	Capeiti 8	S
Chiarano	MS	Cappelli	MS
Concordia	MS	Creso	MS
Dorado	MS	Grifoni B. 52	MS
Farnese	S	Isa I	MS
Gallo	S	Karel	S
Gemini	MR	Kid	MS
Leopardo	MS	Latino	MR
Manital	R	Maristella	MS
Mara	MS	Procace	S
Marzotto	MS	Produra	S
Mec	MS	Tito	S
Orso	MS	Trinakria	S
S. Pastore	S	Valforte	R
Salmone	S	Valnova	R
Valle d'oro	MS	Vezio	S

(*) R = Resistant
MR = Moderately resistant
MS = Moderately susceptible
S = Susceptible

Reactions in several localities of varieties and lines carrying known resistance genes has demonstrated the resistance of Khapli, Vernal and Wisconsin sel C.I. 12632, with T. timopheevi derived resistance (Table 2).

TABLE 2 Field reaction of several varieties and lines carrying known genes for resistance to powdery mildew of wheat.

Varieties and lines	Genes for resistance	YEARS 1982	1983	1984	1985	1986
Axminster x Cc^8	Pm1	23(a)	23	12	27	19
C.I. 13836 x Cc^8	Pm1	33	17	20	27	20
Norka		27	17	18	0	22
Norka x Cc^8	Pm1	27	23	22	0	25
Ulka x Cc^8	Pm2	27	10	16	30	12
Winsconsin sel.C.I.12632	Pm2,Pm6	0	0	0	0	0
Winsconsin sel.12632 x Cc^8	Pm2	10				
Asosan		20	0	0	7	0
Asosan x Cc^8	Pm3a	10	0	4	23	10
Chul		20	27	18	27	10
Chul x Cc^8	Pm3b	20	13	12	0	0
Sonora		3	7	0	10	18
Sonora x Cc^8	Pm3c	0	13	12	13	0
Khapli		5	0	0	13	0
Khapli x Cc^8	Pm4a	5	0	0	13	0
Yuma		10	7	4	7	27
Yuma x Cc^8	Pm4a	13	0	6	13	0
Hope x Cc^8	Pm5	0	35	10	17	20
M 1		17	13	8	43	10
Normandie	Pm1,Pm2,Pm9	27	10	15	23	18
Redcoat		0	7	10	0	0
Vernal		13	0	0	0	-
Arthur		-	-	-	-	16
Clement	Pm8	-	-	-	-	15

(a) Mean of severity recorded as a percentage according to the modified Cobb's scale.

However, the presence of virulence matching Yuma, Khapli, Valgerardo and Valnova has been demonstrated experimentally. This fact is important, since Khapli and some of its derivatives, like the Val group varieties all carrying gene Pm4, are generally considered to be highly resistant (Ceoloni et al., 1978; Pasquini, 1986). Recently, also in the field, a slight infection on Khapli and isogenic lines carrying Pm4 was recorded (Zitelli et al., 1985).

Two durum wheat varieties, 'Belfuggito' (with Agropyron intermedium derived resistance) and 'Belvedere' (with Triticum timopheevi derived resistance), not extensively grown, were resistant in the field (Zitelli,

1982) and to ET10, an isolate virulent on most of the known genes for resistance as well as on many wheat cultivars (Pasquini, 1926).

The possibility of introducing resistance factors from related and wild species has been exploited in Italy. Several Graminaceae species may be used as sources of resistance genes, for example, Aegilops spp. and Dasypyrum villosum (L.) P. Gandergy (= Haynaldia villosa (L) Schur) (Pasquini et al., 1978; Zitelli and Pasquini, 1984). Different strains of T. dicoccon Schrank and T. monococcum L. have been analyzed, both in the field and under controlled conditions and their utilisation in wheat genetic improvement programmes appears promising (Corazza et al., 1986).

BARLEY

Erysiphe graminis f. sp. hordei is widespread in Italy, even if its annual frequency is strictly related to climatic conditions. The field behaviour of both two-rowed and six-rowed barley varieties is recorded regularly (Corazza, 1984, 1985; Corazza and Chilosi, 1986) (Table 3).

TABLE 3 Field reaction of several barley varieties to powdery mildew.

Two-rowed cultivars		Six-rowed cultivars	
Alpha	MS(*)	Arma	MS
Aramir	MR	Barberousse	MS
Arda	MS	Etrusco	MR
Cornel	MS	Gerbel	MR
Havila	MS	Mirco	MR
Georgie	MR	Plaisant	MS
Igri	MS	Robur	S
Porthos	MS	Selvaggio	S
Tipper	MS	Thibaut	MS

(*) R = Resistant
MR = Moderately resistant
MS = Moderately susceptible
S = Susceptible

It is difficult to find complete resistance, probably due to the existence of virulence factors able to match many of the available genes for resistance to barley powdery mildew in Italian populations of the pathogen (Ceoloni and Vallega, 1979).

CONCLUSIONS

Instability of resistance and continuous emergence of new and sometimes more virulent genotypes within populations of powdery mildew show the need for appropriate approaches to control this pathogen.

Some of the control measures that are being developed in Italy are:

- evaluation of field behaviour of cultivars, of both wheat and barley, in order to recommend use of partially resistant varieties and to reduce the effects of infection to an acceptable level;
- introduction of genes for resistance derived from wild or related species, since many of them have been recognized to be useful;
- trials with different fungicides, since yield increase due to treatment is significant in susceptible varieties (Pancaldi, 1986);
- investigation of the influence of cultural practices on level of infection by powdery mildew.

REFERENCES

Ceoloni, C. and Vallega, J. 1978. Presenza di geni di virulenza in popolazioni di Erysiphe graminis tritici rispetto al frumento Khapli (Triticum dicoccum). Atti Giornate Fitopatologiche, 1, 355-362.

Ceoloni, C. and Vallega, J. 1979. Reaction of barley varieties and lines to virulence genes carried by two Italian mildew races. Genetica Agraria, 33, 53-74.

Corazza, L. 1984. Varietà di orzo a confronto per la resistenza alle principali crittogame parassite. L'Informatore Agrario, 33, 63-66.

Corazza, L. 1985. Il comportamento in campo di varietà di orzo nei confronti dei parassiti fungini nel 1984-85. L'Informatoire Agrario, 34, 69-71.

Corazza, L. and Chilosi, G. 1986. Comportamento in campo di varietà di orzo nei confronti dei principali parassiti fungini nel 1985-86. L'Informatore Agrario, 34, 51-55.

Corazza, L., Pasquini, M. and Perrino, P. 1986. Resistance to rusts and powdery mildew in some strains of Triticum monococcum L. and Triticum dicoccum Shubler cultivated in Italy. Genetica Agraria, 32, 359-361.

Pancaldi, D. 1986. Interventi anticrittogamici contro le malattie fogliari del frumento tenero: confronto tra varietà ed epoche di applicazione. Atti Giornate Fitopatologiche, 2, 253-264.

Pasquini, M. 1986. Nuovi fattori di virulenza isolati in Italia entro le popolazioni di Puccinia recondita tritici, P. graminis tritici ed Erysiphe graminis tritici: reazioni di varietà e linee di frumento duro e tenero. Informatore Fitopatologico, 4, 33-37.

Pasquini, M., Gras, M.A. and Vallega, J. 1978. Haynaldia villosa (L.) Schur come fonte di resistenza alle ruggini e all'oidio, da incorporare nelle specie di frumento coltivate. Atti Giornate Fitopatologiche, 349-351.

Pasquini, M., Biancolatte, E., Cecchi, V., Valli, M., Calcagno, F., Iudicello, P., Lendini, M., Corazza, L., Romani, M. and Arduini, F. 1986. Prove epidemiologiche 1985-86: comportamento in campo rispetto alle malattie di frumenti duri e teneri. L'Informatore Agrario, 39, 77-84.

Speer, E.O. 1975. Untersuchungen zur Morphologie und Systematik der Erysiphagen I. Die Gattung Blumeria Golovin und ihre Typusart Erysiphe graminis DC. Sidowia, 27, 1-6.

Zitelli, G. 1982. Fattori limitanti la produzione del grano duro in Italia: strategie da adottare per il controllo delle fitopatie. Monografie di Genetica Agraria, 5, 201-216.

Zitelli, G. and Pasquini, M. 1984. Transfer of disease resistance genes in cultivated durum wheat from related species. Annali Istituto Speriqmentale per la Cerealicoltura, 15, 107-119.

Zitelli, G., Pasquini, M., Biancolatte, E., Cecchi, V. and Valli, M. 1985. Risultati delle prove eseguite nel 1984-85 sul comportamento in camp di frumenti duri e teneri nei confronti delle malattie fungine. L'Informatore Agrario, 37, 69-78.

MILDEW RESISTANCE AND VIRULENCE IN THE NETHERLANDS

A.G. Balkema-Boomstra
Foundation for Agricultural Plant Breeding SVP
P.O. Box 117, Wageningen, Netherlands

ABSTRACT

Details are given for the screening of mildew resistances both for breeding programmes and for the official List of Varieties.

INTRODUCTION

Powdery mildew is the most important disease of barley in the Netherlands. It occurs every year, in particular on sandy soils. Recent estimates of yield losses caused by powdery mildew are not available. Most of the barley grown in the Netherlands is sown in the spring (Table 1).

TABLE 1 1985 acreage of barley

Winter barley	: 6,800 ha ;	Hasso	: 65%
		Masto	: 11%
		Flamenco	: 11%
Spring barley	: 32,100 ha ;	Trumpf	: 37%
		Havila	: 23%

Barley has traditionally been grown rather extensively, i.e. with little nitrogen fertilizer and little or no fungicide treatment. Thus mildew resistance is an important criterion of a variety for admittance to the Descriptive List of Varieties. This may be even more so if barley is grown more intensively to increase yield and, consequently, diseases may become more pronounced.

At the SVP, research on mildew resistance has been carried out since 1966. A large number of primitive varieties has been tested and resistance from many different sources has been introduced into breeding material. This material has been released to the Dutch barley breeders at regular intervals.

The Government Institute for Research on Varieties of Cultivated Plants (RIVRO) carries out the research necessary for the compilation of the List of Varieties. When a barley variety reaches its last year of trials, it is tested at the SVP to determine its mildew resistance genes

TABLE 2 Spring barley varieties on 61st Descriptive List of Varieties (1986)

Variety	Mildew resistance symbol*	Mildew resistance genes
Apex	L	ml-o
Efron	w,MC	Ml-g + Ml-CP, Ml-k + Ml-a9
Grit	e	Ml-a12
Robin	w,e,E"	Ml-g + Ml-CP, Ml-a12, Ml-a6
Menuet	m	Ml-v
Trumpf	l,E"	Ml-a7 + Ml(Ab)
Bellona	w,e,m	Ml-g + Ml-CP, Ml-a12, Ml-v
Golf	w,m	Ml-g + Ml-CP, Ml-v
Havila	w,m	Ml-g + Ml-CP, Ml-v

* explanation of symbols :

L = L100
w = Weihensteplaner CP 127 422
MC = Monte Cristo
l = Lyallpur 3645

e = Emir
E" = Ep 72
m = Minerva

TABLE 3 Virulence of barley mildew isolates maintained at the SVP

isolate codes	Differentiating varieties																
	W.CP	W.37/136	W.41/145	Vdd.	M.511	M.501	Hor 1063	Hor 1657	Hor 1036	Rabat	Atlas	Marco	Amsel	Emir	Vada	Trumpf	Welam
	1	2	3	4	5	6	7	8	9	10	11	12	13	14	15	16	17
V 4003			+							+	+				+		
AV 4012			+				+				+		+		+		
E 4023			+					+		+	+			+			
AV 6530		+	+	+		+	+	+					+		+		
A 7012	+	+	+				+				+		+				
A 7014	+	+	+				+					+	+				
EW 7020	+	+	+					+						+			+
E 7021	+	+	+					+		+				+			
E 7240	+	+	+		+				+					+			
AE 7330	+	+	+	+	+		+	+					+	+			
AEVT 7432	+	+	+			+	+	+			+		+	+	+	+	
E 7500	+	+	+	+		+								+			
7502	+	+	+	+		+					+						
AT 7510	+	+	+	+		+	+						+			+	
AW 7510	+	+	+	+		+	+						+				+
E 7520	+	+	+	+		+		+						+			
AV 7530	+	+	+	+		+	+	+					+		+		
AEV 7532	+	+	+	+		+	+	+			+		+	+	+		
7550	+	+	+	+		+	+		+								
AEVT 7630	+	+	+		+	+	+	+					+	+	+	+	
7704	+	+	+	+	+	+						+					

(Table 2). For this, and our own research, the SVP maintains a series of mildew isolates with as many different combinations of virulence genes as possible (Table 3). For their identification, we use a binary code completed with a letter code (Table 3, 4). Resistance genes are determined on seedlings, grown in flats filled with soil in the greenhouse under more or less standard conditions (temperature, light).

In 1986 the mildew virulence distribution was estimated from mildew collected at the SVP winter barley trial field at Wageningen. On 12 May, infected leaves were collected in the field and shaken on to cv 'Golden Promise' seedlings growing in flats in the greenhouse. Conidia from these plants were used to estimate the virulence distribution. For this purpose ten nurseries, each consisting of one flat filled with soil in which eleven differentiating varieties were sown, were prepared. The nurseries were inoculated with conidia from the 'Golden Promise' seedlings on 22 May, and were evaluated on 30 May. The size of each leaf was measured and the pustules on each leaf were counted. The number of pustules per cm^2 leaf area was calculated and expressed as the percentage of the number of pustules per cm^2 on cv 'Golden Promise' (Table 5).

From this table the high degree of infection of Voldagsen appears to be the most striking result. The fact that the trial field was situated adjacent to a farmers field of cv 'Hasso' may explain this high proportion of Ml-a6 virulence.

Secondly, a relatively high proportion of Ml-a9 virulence was detected which was also found in earlier years. This is somewhat curious, since varieties with this resistance have not or are hardly being grown in the Netherlands. It is felt that knowledge of the virulence distribution in surrounding countries may help to explain these results.

TABLE 4 Set of differentiating varieties used for the identification of mildew isolates

Variety name	Resistance gene	Value for isolate coding
1. Weih.CP 127.422	Ml-g + Ml-CP	1
2. Weih.st 37/136	Ml-h	2
3. Weih.st 41/145		4
4. Voldagsen st 8141/44	Ml-a6	1
5. Gat.mutant 511		2
6. Gat.mutant 501		4
7. Hor. 1063	Ml-k	1
8. Hor. 1657	Ml-c	2
9. Hor. 1036	Ml-a3	4
10. Rabat	Ml-a	1
11. Atlas	Ml-at	2
12. Marco	Ml-h + Ml-marco	4
13. Amsel	Ml-g + Ml-a7	A
14. Emir	Ml-a12	E
15. Vada	Ml-v	V
16. Trumpf	Ml-a7 + Ml-ab	T
17. Welam	Ml-k + Ml-a9	W

TABLE 5 Virulence distribution of powdery mildew collected in a winter barley trial field in 1986

Variety	Resistance genes	Mean number of pustules per leaf*	Mean leaf area cm^2	Mean number of pustules per cm^2	Relative number of pustules per cm^2
Weih.CP	Ml-g + Ml-CP	25.3	6.9	3.7	28
Vold.8141/44	Ml-a6	33.2	5.5	6.0	57
Hor.1036	Ml-a3	1.1	7.4	0.15	1
Gunhild	Ml-g + Ml-a	4.9	7.3	0.7	7
Aramir	Ml-g + Ml-a12	7.3	6.7	1.1	10
Heine 4808	Ml-k + M-a7	13.3	6.4	2.1	20
Trumpf	Ml-a7 + Ml-Ab	14.8**	7.0	2.1	20
Welam	Ml-k + Ml-a9	20.1	6.8	3.0	29
Grand Prix 4	Ml-a13	0	5.5	0	0
Apex	ml-o	0.5	7.1	0.1	1
Golden Promise	-	46.3	4.4	10.5	100

* = mean of 10 varieties

** = virulent and avirulent pustules were difficult to distinguish

DISTRIBUTION OF BARLEY POWDERY MILDEW RESISTANCE AND VIRULENCE IN POLAND 1984-1986

E. Gacek

Experimental Breeding Station of the IHAR
46-233 Bakõw near Kluczbork, Poland

ABSTRACT

Data are provided on the distribution of mildew resistances in Poland and on the frequencies of their corresponding virulences. Virulences corresponding to the Weihenstephan, Spontaneum, Lyallpur and Arabische resistances are common. Virulences against Ml-a9 and Ml-a13 are more common than expected from the area occupied by these resistances. This anomaly may be due to wind transport of spores from neighbouring countries.

INTRODUCTION

Powdery mildew (caused by *Erysiphe graminis* DC. f.sp. *hordei*) is the principal foliar disease of barley in Poland and occurs annually on spring and winter barley fields. Its importance will probably increase in the future due to the increasing cultivation of winter barley. Breeding and simultaneous cultivation of many varieties with different resistances, as well as growing cultivar mixtures, seems to be the best way of controlling this disease in Poland.

The barley varieties grown in Poland originate from breeders in Czechoslovakia, the GDR, the Netherlands, Poland and the UK. All the varieties possess one or more resistance genes.

Barley mildew populations in Poland have been studied qualitatively since the mid-1960s (Ralski and Mikolajewicz, 1969; Czembor, 1981; Czembor *et al*., 1979). In 1981 at Exper. Stn. Bakõw, quantitative methods (sensu Wolfe and Schwarzbach, 1975) were introduced for the first time.

MATERIALS AND METHODS

Using appropriate test isolates of *Erysiphe graminis* f.sp. *hordei*, the resistance characters (phenotypes) of commercial barley varieties and the most advanced breeding lines were identified. Each year, the results are sent to the Research Centre for Varieties of Agricultural Crops (COBORU), where "Recommended Lists" of varieties are prepared.

The relative frequencies of specific pathogenicity characters in Polish barley mildew populations were investigated according to the concept of "virulence analysis" (Wolfe and Schwarzbach, 1975). Direct and indirect quantitative methods of mildew population analysis have been applied at Bakőw. The direct method included mobile seedling nurseries. Healthy seedlings of differential sets were exposed on many occasions at different places in the Bakőw area (south-west Poland), in both commercial crops representing different resistances and in breeders trials. Colony numbers occurring on test seedlings were expressed as percentages of the colony numbers produced on a susceptible control. In the indirect method, conventional leaf samples received from specified barley cultivars around Poland were tested to a differential set (see Table 2) in the glasshouse.

In both cases bulk isolates were used so that many genotypes probably contribute to the estimates of mean pathogenicity of the samples. Mean pathogenicity is equivalent to virulence measured as the number of colonies produced by an isolate on a given cultivar, expressed as a percentage of the number of colonies produced on a susceptible control. Each year, a summary of the data is distributed to provide breeders and barley growers with information on the pathogen population structure and dynamics.

RESULTS AND DISCUSSION

The most important results are reported here; further details are in the original papers (in press).

Distribution of powdery midlew resistance in barley cultivars grown in Poland

The spring barley area in Poland is about 1.1 million hectares, and is increasing. The following spring barley cultivars are grown: Diva, Grit (Ml-al2); Aramir, Ars (Ml-al2 + Ml-g); Havila, Koru (Ml-(La) + Ml-g); Bielik (Ml-(La)); Polon (Ml-g); Roland (Ml-a9); Kristal (Ml-al3) and Apex (ml-o). Winter barley is grown on about 0.2 million hectares. Its area has increased rapidly in recent years and is still increasing. This has occurred largely through the availability of high yielding winter cultivars with good winter-hardiness. The following cultivars are grown in Poland: Goplanski (Ml-a6); Maron, Popiel (Ml-h); Borwina and Erfa (not determined); and a newly introduced Polish variety Drop (Ml-ra + ?).

From 1978 to 1982, about a third of the barley area was cultivated with Ml-al2 varieties. The severity of mildew on these cultivars was very low at their introduction but increased quickly when the varieties became widely distributed. In 1983-1986 the high yielding cultivars (Bielik, Havila, Koru) with Laevigatum resistance rapidly became popular in Poland. At

TABLE 1 Distribution of powdery mildew resistance in barley, Poland 1984-1986.

Resistance	Code	Gene	Per cent area sown 1984	1985	1986
Spring barley					
Laevigatum	La	Ml-(La)	45	59	62
Arabische	Ar	Ml-a12	52	34	29
Rupee (J 25)	Ru	Ml-a13	0	0.1	3
Monte Cristo	MC	Ml-a9	0.01	0.1	1
Ml-0 (L 100)	Mlo	ml-o	0	0	0.01
Weihenstephan	We	Ml-g	63	62	58
Winter barley					
Spontaneum	Sp	Ml-a6	18	5	1
Weihenstephan 1	Ha	Ml-h	53	61	54
Weihenstephan 2	Ra	Ml-ra	0	0	1
Not determined	-	-	13	20	27

present they occupy about 60 per cent of the spring barley area. During the recent past, the varieties Roland (Ml-a9) and Kristal (Ml-a13) were introduced into agriculture. Cultivar Apex (ml-o) was approved in 1986.

Despite the limited distribution of these cultivars (Table 1), the disease severity on them is relatively high. Pathogenicity for them is much higher than could be explained from the extent of their cultivation. My explanation is that the corresponding virulences originate from barley fields in nieghbouring countries. The northern part of Poland probably receives mildew conidia from Denmark and perhaps from Sweden, while the western and southern parts of the country may receive them from GDR, FRG, Austria and Czechoslovakia. However, this is only speculation and remains open to confirmation.

The currently grown winter varieties possess only a small range of rather ineffective mildew resistance genes. Pathogenicity characters for these resistances are common in mildew populations.

Improvement of the durability of powdery mildew resistance

The major emphasis in our mildew resistance breeding in recent years has been on the incorporation of many major resistance genes into varieties and on the strategies of their efficient use in time and space. In order to improve durability of the resistance of commercial cultivars we are trying to obtain as wide a range of host diversity as possible. We intend to provide crop heterogeneity by encouraging the use of field-to-field diversification (Priestley, 1981), growing cultivar mixtures (Wolfe, 1978), and integrated control (genetical resistance + fungicides).

The results obtained indicate that spring and winter barley cultivar mixtures can be exploited to provide a simple way of controlling diseases and sustaining high yields. Some of the mixtures yielded up to 1 tonne per hectare more than the mean of their components in pure stands.

Distribution of virulence

The population structure of barley mildew in Poland was analysed by Prof. H.J. Czembor at Exper. Stn. Bakőw from 1971-1983. This survey was based on the race analysis concept then prevailing and the interactions between powdery mildew and barley were studied qualitatively. The analyses showed that races virulent to resistances widely used at that time were dominant, while those virulent to newly introduced R-genes were infrequent. The most common mildew races belonged to group C and were highly dangerous to commercial barley cultivars (Czembor, 1981; Czembor and Kudla, 1983). Race surveys carried out during the same period in the German Democratic Republic (Frauenstein _et al._, 1979) and in Czechoslovakia (Brückner, 1975; 1982) showed almost the same general pattern in the virulence of barley mildew.

The frequency of virulence in the powdery mildew populations has been monitored each year since 1981 and the varieties currently in use as differentials are shown in Table 2.

TABLE 2 Characteristics of differential cultivars.

Cultivar	Resistance	Gene	Corresponding virulence
Union	Weihenstephan	Ml-g	V-g
Carina	Spontaneum	Ml-a6	V-a6
Ortolan	Lyallpur	Ml-a4/7	V-a4/7
Emir	Arabische	Ml-a12	V-a12
Welam	Monte Cristo	Ml-a9	V-a9
Lofa Abed	Laevigatum	Ml-(La)	V-(La)
Tyra	Algerian	Ml-a	V-a
Karat	Rupee (J 25)	Ml-a13	V-a13
Atem	Mlo	ml-o	V-mlo
Skrzeszowicki	-	none	all

The first information on barley midlew virulence distribution in our country came from Schwarzbach's consultancy to Poland in 1981. He collected representative samples of airborne spores of barley mildew by means of a jet spore trap and described the pathogen population in terms of frequencies of important virulences. It was found that mildew populations in Poland had much more complex virulence than in other European countries (Schwarzbach, 1981).

The more recent situation in frequency of the major virulences is shown in Table 3 and 4. The figures indicate that virulences corresponding with resistances Weihenstephan, Spontaneum, Lyallpur and Arabische, are very common in mildew populations occurring in Poland. During recent years, there was a rapid increase of virulences against Ml-a9 and Ml-a13 resistances. Their frequencies are much higher than could be explained from the extent of cultivation of varieties with corresponding resistances. The possible explanation for this is the movement of corresponding virulences from neighbouring countries by the prevailing westerly and southerly winds (Gacek, 1983).

TABLE 3 Virulence frequency in barley powdery mildew at Bakow region, 1984-1986.

Virulence	Frequency[1]		
	1984	1985	1986
V-g	93	87	97
V-a6	96	101	87
V-a4/7	71	88	61
V-a12	90	77	86
V-a9	8	19	47
V-(La)	46	60	58
V-a	2	7	11
V-a13	3	11	29
V-mlo	0	1	1

[1] No. colonies as per cent of those on "Skrzeszowicki" in mobile nurseries.

TABLE 4 Mean pathogenicity of bulk isolates from leaf samples received in 1986.

Isolate source	Barley mildew virulences							
	V-g	V-a6	V-a4/7	V-a12	V-(La)	V-a9	V-a13	V-a1
Weihenstephan	92	83	67	81	61	31	18	11
Arabische	84	90	45	94	43	23	21	14
Laevigatum	74	44	50	36	87	33	10	9
Monte Cristo	62	67	38	76	18	68	30	7
Rupee/J 25	86	53	41	60	8	20	76	4

Pathogenicity for Ml-a and especially for ml-o occurs at low frequency, but pathogenicity for ml-o has been found in the field. Barley breeding has to consider the observed virulence pattern. Many of our breeding lines carry the Ml-a and ml-o genes.

REFERENCES

Brückner, F. 1975. Resistance of some European spring barley varieties to seven races of powdery mildew. Sbor. UVTIZ, Ochr. Rostl. 11, 253-259.

Brückner, F. 1982. Nalez rasy padli na jecmeni (Erysiphe graminis DC. var. hordei Marchal) prekonavajici geny resistance Ml-a9 a Ml-a14. Sbor. UVTIZ Ochr. Rostl. 18, 101-106.

Czembor, H.J. 1981. Rasy fizjologiczne maczniaka jeczmienia (E. graminis f.sp. hordei) wystepujace w Polsce w latach 1975-1979. Hod. Roslin Aklim. Nasienn. 25, 215-226.

Czembor, H.J. and Kudla, M.M. 1983. Rasy fizjologiczne maczniaka jeczmienia (E. graminis f.sp. hordei) wystepujace w Polsce w latach 1980-1982. Hod. Roslin Aklim. Nasienn. 28, 219-230.

Frauenstein, K., Meyer, H. and Wolfram, H. 1979. Pathotypen von Erysiphe graminis f.sp. tritici und F. graminis f.sp. hordei in Europa. Arch. Phytopathol. u. Pflanzenschutz. 15, 391-399.

Gacek, E. 1983. Analiza patogenicznosci grzybow porazajacych jeczmien. Biul. IHAR, 150, 11-19.

Priestley, R.H. 1981. Choice and deployment of resistant cultivars for cereal disease control. In "Strategies for the Control of Cereal Disease" (Eds. J.F. Jenkyn and R.T. Plumb), pp.65-72. Blackwell Scientific Publications.

Ralski, E. and Mikolajewicz, T. 1969. Studia nad odpornoscia Jeczmienia na maczniaka prawdziwego Erysiphe graminis f.sp. hordei. Hod. Roslin Aklim. Nasienn. 13, 145-157.

Schwarzbach, E. 1981. Report on the consultancy mission to Poland 1981. FAO Rome. 11 pp.

Wolfe, M.S. 1978. Some practical implications of the use of the cereal variety mixtures. In "Plant Disease Epidemiology" (Eds. P.R. Scott and A. Bainbridge), pp.201-217. Blackwell Scientific Publications, Oxford.

Wolfe, M.S. and Schwarzbach, E. 1975. The use of virulence analysis in cereal mildews. Phytopath. Z. 82, 297-307.

INCIDENCE OF POWDERY MILDEW (ERYSIPHE GRAMINIS) ON BARLEY AND THE STRUCTURE OF THE BARLEY CROP IN SPAIN

J.L. Molina-Cano

La Cruz del Campo, S.A., Malting and Brewing Company,
Dept. of Barley Breeding, Av. Andalucia, 1, E-41007 Sevilla, Spain

ABSTRACT

The structure of the barley crop in Spain is presented, showing the acreage of both spring and winter types and the percentage of each variety sown.

The incidence of powdery mildew on barley yields in Spain is generally slight, however some tentative data on the influence of the disease both under irrigated and dryland conditions are presented.

INTRODUCTION

Although powdery mildew does not have a strong influence on the losses produced on the barley crop in Spain, because of the dry climate prevalent over most part of the country, it certainly has some importance in particular regions in the north and also on irrigated crops throughout the growing area.

This paper summarises the structure of barley production and also the current situation regarding mildew in Spain.

THE STRUCTURE OF THE BARLEY CROP IN SPAIN

Barley is the first crop in Spain taking into account the surface area of this cereal grown each year. In 1985 more than 4 million hectares were sown, of which, approximately, half were of winter and half of spring varieties. Although it is grown mainly on drylands, a small percentage is sown under irrigation.

Fig. 1 gives the distribution of the barley crop among the different regions of Spain. From these data it can be concluded that Castilla-Leon, Castilla La Mancha and Aragon cultivate more than 75% of the total barley area. These three regions have a typical continental climate with hard winters and hot summers, and a very rapid rise in temperature at the end of the spring. The average rainfall is about 400 mm per year, with an uneven distribution both among and within years, the winter months being more rainy. There is usually a lack of rainfall at the end of spring.

The varietal structure is presented in Table 1. It is clear from these data that more than 70% of the acreage is sown to old varieties, e.g. Beka, Pallas, Albacete and Hatif de Grignon. This is because these varieties are well-adapted to arid and cold environments, whereas modern varieties, which are bred normally in central and northern European countries, are adapted to high rainfall and fertility conditions. The breeding programmes under way in Spain are trying hard to improve this situation.

INCIDENCE OF POWDERY MILDEW ON SPANISH BARLEYS

It is not surprising that under the dry and hot conditions of the major part of the barley-producing areas of the country, powdery mildew is only a minor disease. In this connection, the survey of European Cereal Diseases carried out by Zadoks and Rijsdijk (1984), shows no data on losses caused by this disease on barley in Spain. Moreover, a review of Plant Breeding Abstracts from 1977 to 1985, showed that no paper was published during this period on powdery mildew in Spain.

With the lack of data on this problem, I will give here only preliminary information based on my personal experience as a barley breeder. This information should therefore be considered as "soft" information as defined by Zadoks and Rijsdijk (1984), i.e. based only on personal impressions.

The map depicted in Fig. 2 shows the Iberian Peninsula divided into agroclimatic regions according to Thran and Broekhuizen (1965). Within those relevant as barley-growing areas two scores are given, for mildew incidence (on a 1-4 scale), and for dryland or irrigated (followed by "R") conditions.

From these data it can be concluded that powdery mildew incidence on dryland crops is important only in the northern regions, whose rainfall is above 600 mm. It should be pointed out, however, that these are regions where barley is scarcely grown, probably on an area of less than 150,000 ha. On the other hand, the mildew incidence under irrigation is more important on the upper half of the country, but here, again, the irrigated surface devoted to barley is less than 10% of the total barley acreage of these regions.

Thus, as a general although provisional conclusion, powdery mildew has, quantitatively speaking, only a slight influence on barley losses in Spain. This last statement does not mean, obviously, that mildew studies are not important to this country, on the contrary, Spanish barley breeders are eager for information about this topic.

REFERENCES

European Brewery Convention. Barley & Malt Committee. Report on Field Trials 1985. Zoeterwoude (The Netherlands). p 23.

Plant Breeding Abstracts (1977 to 1985). Commonwealth Bureau of Plant Breeding and Genetics. Cambridge.

Thran, D. and Broekuizen, S. (1965). "Agro-ecological atlas of cereal growing in Europe. I. Agro-climatic atlas of Europe." Pudoc, Wageningen and Elsevier Publ. Comp., Amsterdam, London and New York. 125 maps and 38 pp.

Zadoks, J.C. and F.H. Rijsdijk (1984). "Agro-ecological atlas of cereal growing in Europe. III. Atlas of cereal diseases and pests in Europe." Pudoc, Wageningen. 168 pp.

TABLE 1 Main barley varieties grown in Spain in 1984 and 1985 (1)

Spring barley: 49% in 1985			Winter barley: 51% in 1985		
Area:	2-row	2,044,400 ha	Area:	6-row	2,110,700 ha
Harvest:	2-row	5,845,200 tons	Harvest:	6 row	4,834,700 tons

Variety	Year 1984 (%)	Year 1985 (%)	Variety	Year 1984 (%)	Year 1985 (%)
Pallas	18.3	18.6	Albacete	35.0	25.0
Beka	12.1	11.7	H. Grignon	22.4	22.2
Hassan	8.8	8.4	Dobla	11.3	10.9
Alpha*	8.5	8.1	Monlon	6.8	7.1
T. Union	6.4	6.7	AO-1	5.9	5.9
Menuet	6.7	6.4	Barbarrosa	3.5	3.5
Union	6.7	6.4	Alsekal	1.7	2.0
Georgie	4.0	3.8	Vegal	1.4	1.5
Athos	3.1	3.2	Plaisant	1.2	1.2
Kym	2.3	2.2	Ager	1.2	1.2
Koru	2.2	2.1	Tina	1.1	1.1
Sonja	-	2.0	Hop	1.0	1.0
Zephyr	2.2	2.0	Robur	0.9	0.9
Logra	1.8	1.8	Others	6.6	16.5
Igri*	-	1.8			
Others	18.3	15.6			

* winter barley

Note: The percentage of area sown to each variety has been calculated from the certified seed production.

(1) From: European Brewery Convention. Barley & Malt Committee. Report on Field Trials 1985. Zoeterwoude (The Netherlands). p 23.

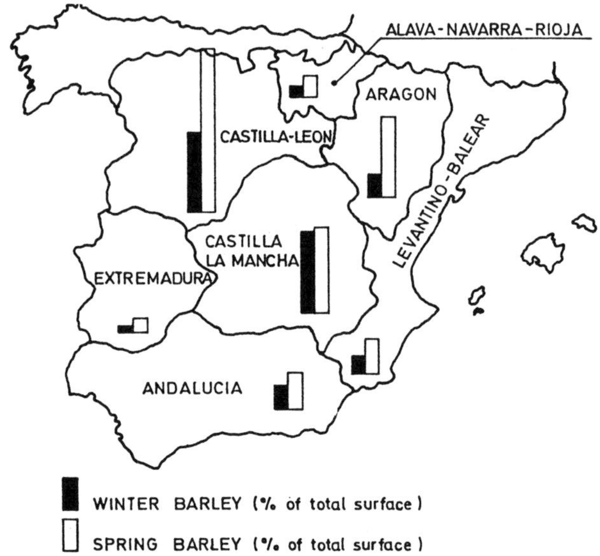

Fig. 1 Distribution of the barley acreage in 1985 among the different regions of Spain.

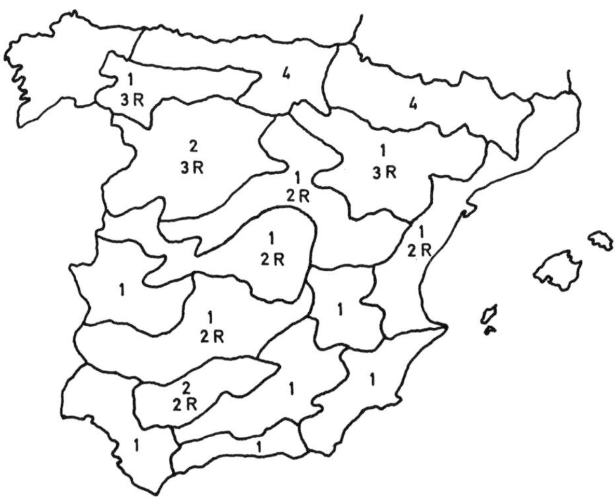

Fig. 2 Mildew (*Erysiphe graminis*) incidence on the main agroecological regions of Spain.

Note: Scale 4 (very important) to 1 (negligible)
R: under irrigated conditions.

CULTIVARS OF SPRING BARLEY AND POWDERY MILDEW (*ERYSIPHE GRAMINIS* F.SP. *HORDEI*) IN SWEDEN

L. Wiik

Swedish University of Agricultural Sciences
Department of Plant Protection, P.O. Box 44, S-230 53 Alnarp, Sweden

ABSTRACT

In yield loss trials with spring barley, mildew was the major disease. Using propiconazole sprays, the most effective treatment was a single spray at GS 34.

The most common spring barley varieties used during the last 10-20 years are listed, together with the average mildew attack on the varieties grouped by their specific resistance genes.

There was some indication of an increase in insensitivity to triazoles from 1985 to 1986, but no evidence of any loss in efficacy. Preliminary data from the use of a Wind Impaction Spore Trap in Sweden are given, using both differential varieties and near-isogenic lines.

INTRODUCTION

In Sweden the acreage of cereals is approximately 1500 000 ha. Of this roughly 650 000 ha are spring barley; winter barley is grown only to a limited extent, less than 5 000 ha, in the far south. Different varieties especially from Svalöf AB and W. Weibull AB are grown all over Sweden.

Sources of resistance to powdery mildew in barley were investigated by A. Wiberg (e.g. Wiberg, 1974), and by B. Leijerstam at the Swedish University of Agricultural Sciences, Experimental Division of Plant Resistance, as a link in the approval-scheme of Swedish varieties (barley and wheat). Sources of resistance to powdery mildew are also tested by plant breeders in Sweden and researchers in other countries (e.g. Torp *et al*., 1978; Wolfe *et al*., 1985). A minor research task concerned the role of winter barley for transmission of powdery mildew to spring barley (Grönt0ft, 1980).

Race non-specific resistance to barley mildew was investigated by B.E. Nilsson (1976). Nordic investigations on this topic are proceeding further (Jørgensen and Knudsen, 1983).

Currently, in Sweden, a mobile spore trap is being used to learn more about the powdery mildew population; this will continue to 1989.

YIELD LOSS STUDIES

Spraying at different growth stages with propiconazole (Tilt 250 EC,

0.5 l/ha) has been used in yield loss studies.

In 1980 we obtained relatively large yield increases from fungicide treatment of wheat mildew in spite of low levels of infection (Wiik, 1981). We therefore made a new powdery mildew assessment form with eleven classes, taking into account low infection levels. This form has also been used in recent years for spring barley. Assessments were made 2-7 times at each leaf level in the barley field trials which were sprayed at seven different occasions. A computer-program was written to calculate the AUDPC (Area Under Disease Progress Curve).

There were four trials in farmers' fields in the southern part of Sweden in each of the years 1983-1986. The varieties in the 16 trials were Alva (7), Golf (1), Flare (3), Ida (4) and Roland (1). Growth stages were determined according to the decimal code (Zadoks et al., 1974) and results are presented in Table 1.

TABLE 1 Mean yields and mildew infection in 16 field trials in farmers' fields, 1983-1986.

Treatment	G.S.*	Date	Yield increase kg/ha	Mildew index
Untreated			5440	136
Tilt**	23	29/5	+310	55
"	29	5/6	+450	59
"	34	12/6	+560	41
"	42	19/6	+480	51
"	51	26/6	+420	67
"	58	3/7	+260	93
"	66	10/7	+190	120
"	23+42	29/5+19/6	+700	12
"	29+51	5/6+26/6	+690	14
LSD, 5%			150	29

* Growth stages, decimal code
** Tilt 250 EC, 0.5 l/ha, propiconazole

Powdery mildew was the major disease; net blotch, scald and rust were absent or present only to a minor extent. On average, for single sprays, the highest yield increase was obtained from treatment at growth stage 34; earlier and later fungicide treatments were inferior. The data are still under evaluation.

MOST FREQUENTLY USED CULTIVARS

The race specific genes of the cultivars used most frequently during the last 10-20 years are given in Table 2. Different varieties with the same race specificity often differ considerably in mildew susceptibility.

TABLE 2 Most frequently used cultivars in Sweden during the last 10-20 years.

Cultivar	Race specific gene(s)[1]	Approved[2]
Agneta	Ml-a9	1978
Alva	Ml-(LA)	1977
Ansgar	Ml-k, Ml-a7	1973
Birgitta	-	1963
Birka	Ml-k, Ml-a9	1981
Cilla	-	1968
Edda II	-	1949
Etu	-	1971
Eva	-	1973
Flare	Ml-g, Ml-(LA)	1983
Frida	Ml-g, Ml-a12	1984
Golf	Ml-g, Ml-(LA)	1985
Gunilla	-	1970
Hellas	-	1967
Hulda	Ml-k, Ml-a9, Ml-(LA), ?	1985
Ida	Ml-k, Ml-a9, Ml-g	1979
Ingrid	-	1956
Jenny	Ml-a13, Ml-a(Ru2)	1980
Kajsa	Ml-a9	1977
Kara	Ml-a9, Ml-g, ?	1984
Lina	Ml-a7, Ml-a?	1982
Lofa	Ml-(LA)	1970
Mette	Ml-k, Ml-a9, Ml-(LA), Ml-g, ?	1984
Mona	(Ml-k), Ml-a9	1970
Nery	Ml-(LA)	1978
Pernilla	-	1979
Roland	Ml-a9	1982
Rupal	Ml-a13	1973
Simba	Ml-(LA)	1975
Särla	-	1965
Tellus	Ml-g	1970
Welam	Ml-a9	1976
Wing	Ml-k, Ml-a7	1970
Yrjar	-	1979

[1] According to B. Leijerstam, SLU; P. Lundin, W. Weibull AB; J. Meyer, Svalöf AB

[2] According to National Plant Variety Board, Descriptive list of cultivars

ACREAGE AND POWDERY MILDEW

The acreage (%) in bold, according to Official Statistics of Sweden: Agriculture and Fisheries, and powdery mildew (%) in normal figures, according to Sortval 1975–1986, Aktuellt från Lantbruksuniversitetet by A. Bengtsson, are shown in Table 3. Comparison of mildew infection among varieties in any one year are valid, but not between years because the calculation of the values differed in some years. From Table 3 the acreage of cultivars with no resistance gene has been smaller and the use of more complex host resistance more frequent during recent years.

TABLE 3 Acreage (%) and powdery mildew (%) (from 1974) of cultivars of spring barley in their race specificity groups.

	Year														
	70	71	72	73	74	75	76	77	78	79	80	81	82	83	84
NONE															
Birgitta	7	7	7	8	7	5	3	2	1						
					10	11	10								
Cilla	3	9	8	5	5	3	3	2	2						
					14	13	13	12	12	10					
Edda II	5	6	6	6	5	5	5	5	4	2	2				
Etu				1	1	2	3	4	5	5	6	4	4	4	4
					30	30	30	33	33	36	36	14	15	28	29
Eva						1	4	4	5	4	4	3			
					13	13	12	11	9	9	8	4	6	8	
Gunilla				2	3	5	6	6	8	8	9	9	10	8	7
					21	21	21	22	22	20	19	7	9	16	26
Hellas	8	7	5	5	4										
					16	16									
Ingrid	57	50	48	40	37	31	22	18	12	8	5	3	2	1	1
					17	17	16	16	15	12	12	5			
Pernilla														7	13
										19	19	8	9	12	19
Särla	7	11	10	11	9	9	8	7	6	5	4	3	2	2	1
					23	23	23	25	21	18	18	7	7	9	
Yrjar											1	1	2	1	1
Total acreage	87	90	84	78	71	61	54	48	43	32	31	23	20	23	27
Total mildew/n					18	18	18	20	19	18	19	8	18	15	25
LAEVIGATUM															
Alva										4	6	8	11	11	13
									6	11	11	8	7	8	9
Lofa															
					1	1	1	1	1	1	1	1			
Nery											1	1	2	4	4
									4	4	4	1	1	2	4
Simba									4	5	4	3	2		
					4	4	3	3	3	3	3	2	2	3	
Total acreage									4	9	11	12	15	15	17
Total mildew/n					3	3	2	2	4	5	5	3	3	4	7

TABLE 3 continued

	70	71	72	73	74	75	76	77	78	79	80	81	82	83	84
MONTE CRISTO															
Agneta											**1**	**5**	**8**	**8**	**6**
										32	32	13	15	31	36
Birka											**1**	**0**	**0**	**2**	**8**
												4	4	6	10
Kajsa										**3**	**4**	**2**			
									29	27	27				
Mona		**0**	**2**	**4**	**8**	**9**	**8**	**8**	**5**	**3**	**1**				
					1	6	12	13	12	16	16	11			
Roland															**1**
													3	5	8
Welam									**4**	**11**	**13**	**12**	**3**	**2**	**1**
								6	6	6	6	5	5	6	
Total acreage		**0**	**2**	**4**	**8**	**9**	**8**	**8**	**9**	**17**	**20**	**19**	**11**	**12**	**16**
Total mildew/n					1	6	12	10	16	20	20	8	7	12	18
LYALLPUR															
Wing		**1**	**6**	**12**	**9**	**5**	**2**								
					1	8	8								
WEIHENSTEPHAN															
Tellus				**0**	**3**	**13**	**23**	**31**	**33**	**34**	**32**	**26**	**15**	**11**	**6**
					12	12	12	11	11	9	9	5	5	7	10
COMBINATIONS															
Flare															
													3	4	6
Frida															
															4
Golf															
															4
Ida										**0**	**3**	**15**	**31**	**33**	**26**
									7	5	5	4	3	6	9
Lina															
													1	1	1
Mette															
															0
Total acreage										**0**	**3**	**15**	**31**	**33**	**26**
Total mildew/n									7	5	5	4	2	4	4
ALL															
Total acreage	87	**91**	**92**	**94**	**91**	**88**	**87**	**87**	**89**	**92**	**97**	**95**	**92**	**94**	**92**

In Table 4 the average mildew attack (%) in the different race specific gene groups has been calculated during different periods in Swedish cultivar trials.

TABLE 4 Average mildew attack (%) during different periods in Swedish cultivar trials.

Gene(s)	1974-1976	1974-1984	1978-1984
None	18	18	17
Lyallpur	6		
Monte Cristo	6	13	14
Weihenstephan	12	9	8
Laevigatum	3	4	4
Combinations			4

METHODS

1. Ciba Geigy method

Sensitivity in mildew populations to the EBI-fungicide propiconazole was estimated in the years 1984-1986 by I. Svedinger. Seedlings are grown in glass tubes under fixed climatic conditions for eight days. Fungicide is added at eight concentrations 0, 0.1, 0.3, 1.0, 3.0, 10.0, 30.0 and 100.0 ppm and the seedlings inoculated with a powdery mildew isolate. Assessment of the powdery mildew attack is made after 10 days (barley) or 14 days (wheat).

2. WIST (Wind Impaction Spore Trap)

During 1985 and 1986 a mobile spore trap (Wolfe, 1981; Limpert and Schwarzbach, 1981) has been used in the far south of Sweden. During 1986, spores were trapped in two regions. The tours were 50 km each and the ten days old seedlings were exposed for about one hour. In 1985, trapping was started on 2/8 and stopped on 11/10. One tour was made each week, giving a total of eight in 1985. In 1986 we started earlier on 2/6, and average results from 24 tours are presented here.

Some adaptations have been made to the WIST method used in Sweden. Instead of first collecting spores and then testing them, we test directly. This necessitates the inclusion of replicated varieties, near-isogenic lines and a variety treated with different fungicide concentrations for each tour. The results appear promising even with low spore catches. Assessments of the number of colonies are made ten days after exposing seedlings in the WIST.

RESULTS AND DISCUSSION

Fungicide insensitivity

The average ED_{50} values in ppm achieved from the Ciba-Geigy method were, respectively, 3.9 in 1984, 4.5 in 1985, and 3.8 in 1986. It should be pointed out that the variation, even among reference isolates, was large. In wheat the average ED_{50} values were lower, 1.5-4.0 ppm, but also in this case the variation was large.

The numbers of powdery mildew colonies assessed on ten leaves are presented in Table 5 (WIST data).

TABLE 5 Monitoring for fungicide insensitivity with the WIST (No. of colonies/10 leaves spring barley).

TILT, ppm	1985 East	1986 East	1986 South
0	3.7	2.7	4.6
0.1	5.0	2.5	3.8
0.3	4.4	2.2	4.0
1.0	3.4	2.7	3.8
3.0	3.2	2.0	2.8
10.0	2.5	2.3	4.2
30.0	0.7	1.6	2.8
LSD, 5%	2.5	0.9	2.1

In 1986 spores were trapped in two adjacent regions, here designated east (of Alnarp) and south (of Alnarp). Fungicide insensitivity increased apparently from 1985 to 1986 but in the spring barley field trials in 1986 there was no observed decrease in the efficacy of triazoles against powdery mildew.

Spore trapping on different varieties and near-isogenic lines

In Table 6 the average number of colonies per ten leaves on different varieties are shown.

The agreement between the two years is good. The average was calculated from tours when powdery mildew was caught. In 1985 mildew was caught in five tours out of eight and 1986 in 15 tours out of 24; correction for leaf size was not made.

In Table 7 results from different near-isogenic lines of the variety Pallas are presented.

TABLE 6 WIST 1985 and 1986: no. of colonies/10 leaves. One tour = 35 leaves/variety, East.

Variety	Gene(s) M1-	1985 2/8-11/10 5/8 tours	1986 2/6-13/10 15/24 tours
Golden Promise	none	3.1	2.6
Ingrid	none	3.9	3.7
Nery	-(La)	1.6*	1.5
Alva	-(La)	2.2	1.8
Rupal	-a13	0.2	0.1
Triumph	-a7,-Ab	3.6	3.3
Birka	-a4,-a9	0.7	0.8
Ida	-a4,-a9,-g	0.7	0.7
LSD, 5%		3.6	1.3

* Missing value 1 tour

TABLE 7 WIST, 1986: no. of colonies/10 leaves; isogenic lines, Pallas.

Isogenic line	Gene(s) M1-	South	East
P01	-a	0.1	0.3
P07	-a9,-k	1.4	0.6
P10	-a12	1.2	1.0
P11	-a13,-(Ru3)	0	0
P14	-(41/145)	4.4	2.0
P15	-(Ru2)	2.2	1.4
P16	-k	5.4	2.1
P17	-k	3.9	2.1
P21	-g,-CP	4.2	2.1
P22	-o5	0.2	0.1
P23	-(La)	5.3	2.2
LSD, 5%		2.2	1.1

Unfortunately, the recurrent parent Pallas (no resistance gene) was not used. The catches were largest on P14 (M1-(41/145)), P16 and P17 (M1-k), P21 (M1-g, M1-CP) and P23 (M1-La) and less on P07 (M1-a9, M1-k), P10 (M1-a12) and P15 (M1-(Ru2)) and even less on P01 (M1-a) and P22 (M1-o5). No colonies were caught on P11 (M1-a13, M1-a(Ru3)). Where comparison is possible with the Danish mobile nursery tests, agreement is satisfactory (Houmøller, 1985).

FUTURE DEVELOPMENT

The use of the mobile spore trap (WIST) will continue for at least three more years. The method will be extended during this time and complementary tests with mobile nurseries will be made. Evaluation of the methods used during the coming three years will perhaps indicate the benefit for plant breeders and the extension service. This may possibly lead to a further continuation after the three year evaluation.

ACKNOWLEDGEMENT

Bengt Leijerstam, Per Lundin and Jan Meyer are gratefully acknowledged for their help with designation of different genes in the Swedish varieties (Table 2), Lisa Munk and Jørgen Løhde for supplying the near-isogenic lines and Ingrid Svedinger and Ciba-Geigy, Nordic agrodivision for the fungicide insensitivity testing. I am most grateful to Prof. M.S. Wolfe for his helpful suggestions and critical reading of the manuscript.

REFERENCES

Anon., 1973-1985 (Published every year). Yearbook of Agricultural Statistics. Official Statistics of Sweden. National Central Bureau of Statistics.

Bengtsson, A. 1975-1986 (Published every year). Sortval 1975-1986. Aktuellt från lantbruksuniversitet. Mark, Växter.

Gröntoft, M. 1980. The role of winter barley for transmission of powdery mildew (Erysiphe graminis DC. f. sp. hordei) to spring barley. (Engl. summary). Examensarbeten 1980:2, Swedish University of Agricultural Sciences, Dept. of Plant and Forest Protection, Uppsala.

Houmøller, M.S. 1986. Undersøgelser af bygmeldug ved virulens-analyser i Danmark 1985. Statens Planteavlsforsøg, Meddelelse nr 1965, 88 årgang, 8 maj 1986.

Jørgensen, J.H. and Knudsen, J.C.N. 1983. The co-operative Nordic project on disease resistance: race non-specific powdery mildew resistance in barley. Acta Agric. Scand. Suppl. 23, 75-79.

Limpert, E. and Schwarzbach, E. 1981. Virulence analysis of powdery mildew of barley in different European regions in 1979 and 1980. Barley Genetics IV, Proc. 4th Int. Barley Genet. Symp., Edinburgh, 458-465.

Nilsson, B.E. 1976. An investigation of non-specific resistance to barley mildew. (Engl. summary). Ph.D. thesis. Institutionen för växpatologi, Lantbrukshögskolan, Uppsala 1976.

Torp, J., Jensen, H.P. and Jørgensen, J.H. 1978. Powdery mildew resistance genes in 106 northwest European spring barley varieties. Royal Veterinary and Agricultural University, Copenhagen, Denmark. Yearbook 1978, 75-102.

Wiberg, A. 1974. Sources of resistance to powdery mildew in barley. Hereditas 78, 1-40.

Wiik, L. 1981. Chemical treatment against powdery mildew and glume blotch: Field trials during 1980. (Engl. summary). Växtskyddsrapporter. Jordbruk 15, 74-82.

Wolfe, M.S., Slater, S.E. and Minchin, P.N. 1981. Mildew of barley. U.K. Cereal Pathogen Virulence Survey. 1980 Annual Report, 42-56.

Wolfe, M.S., Slater, S.E. and Minchin, P.N. 1985. Mildew of barley. U.K. Cereal Pathogen Virulence Survey. 1984 Annual Report, 38-48.

Zadoks, J.C., Chang, T.T. and Konzak, C.F. 1974. A decimal code for the growth stages of cereals. Weed Research 14, 415-421.

VIRULENCE ANALYSIS OF POWDERY MILDEW OF WHEAT COLLECTED FROM DIFFERENT CULTIVARS IN SWITZERLAND, 1980-1985

P.M. Fried and Ph. Streckeisen

Swiss Federal Research Station for Agronomy
8046 Zürich-Reckenholz, Switzerland

ABSTRACT

A limited survey of wheat mildew across Switzerland from 1980 to 1985 revealed that pathogenicity on Pm2 and Pm4b was considerably less common than pathogenicity on Pm1, Pm5, Pm6, Pm8 and Mli.

MATERIAL AND METHODS

From 1980 to 1985 we collected 565 leaf samples infected with powdery mildew of wheat (Erysiphe graminis f. sp. tritici) from the main spring and winter wheat cultivars and nurseries grown across the country. Conidia from single pustules were transferred to the susceptible cultivar Ardus for multiplication and storage (at 4°C on plantules raised on vermiculite in test tubes). Only 409 isolates survived the transfer (54, 59, 76, 27, 64, and 129 isolates/year in the years 1980-1985, respectively). They were tested on a differential set (Table 1), obtained from the Plant Breeding Institute, Cambridge (Bennett and Van Kints, 1982). The set was supplemented by some lines from Weihenstephan (Zimmermann et al., 1984). According to Heun and Fischbeck (in press), the breeding line 623-65 carries the resistance genes Pm8 + Blaukorn (= Pm4b). The differentials were raised in trays and inoculated 10 days after planting. Disease readings were taken

TABLE 1 Differential cultivars used for determination of the virulence frequencies of wheat powdery mildew in Switzerland, 1980-1985.

Differential cultivar	Resistance gene(s)	Differential cultivar	Resistance gene(s)
Hobbit *)	Pm-	Maris Dove	Pm2 + Mld
Anfield	Pm1	Sappo	Pm2 + Pm4b
Bounty	Pm2	Brigand	Pm2 + Pm6
Weihenstephan M1	Pm4b	Walter	Pm2 + Pm4b + Pm6
Hope	Pm5	Crossbow **)	Pm2 + Pm6 + Mli
Timgalen	Pm6	CWW 1645-5 **)	Pm2 + Pm6 + Pm8
Salzmuende 14-44	Pm8	Sicco	Pm5 + Mli + ?
Flanders	Mli	Weihenstephan 623-65	Pm8 + Blaukorn

*) susceptible control **) 1983-1985

approximately 8 to 10 days after inoculation, on a scale from 1 to 9 according to Newton and Cherewick (1947). More details about our testing procedures were presented earlier (Streckeisen and Fried, 1985).

RESULTS AND DISCUSSION

Although infected plant material was collected across the country and from all main cultivars, our sampling was not fully random. Nevertheless, we think that for our breeding programme Figure 1 provides useful information on the actual relative effectiveness of the genes or the gene combinations tested. High virulence frequencies were found on Pm1, Pm5, Pm6, Pm8 and Mli, whereas matching pathogenicity on Pm2 and Pm4b was relatively rare. This is in accordance with the results reported by Ramgraber (1984) and Zimmermann _et al_. (1984) for southern Germany.

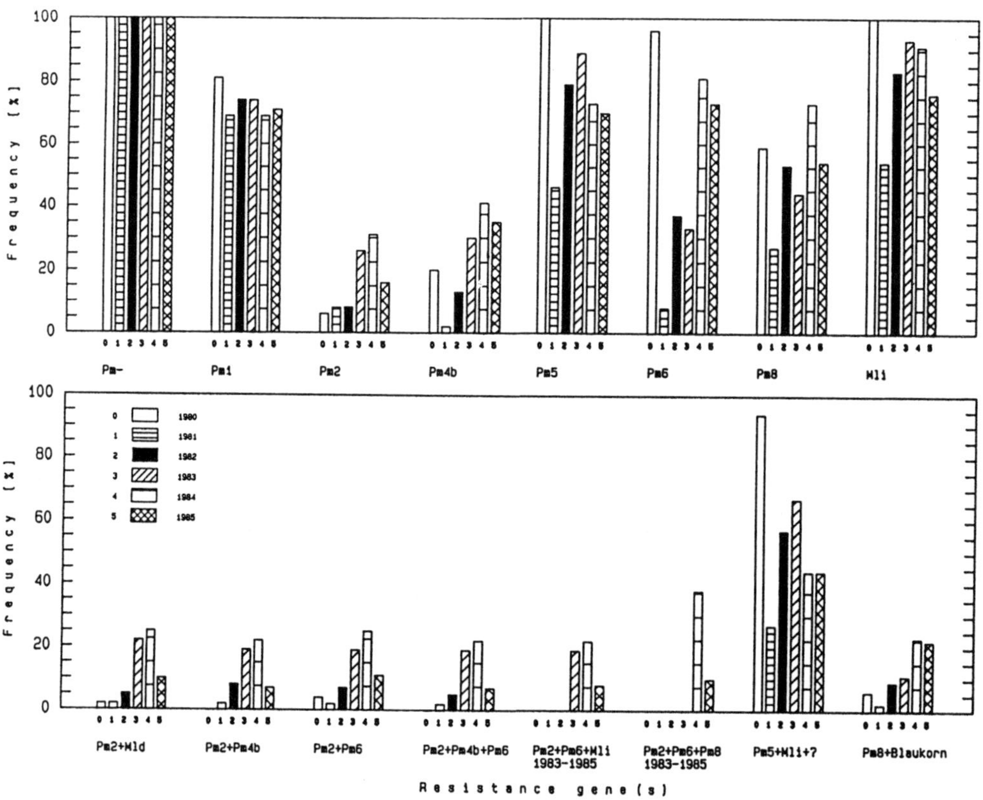

Fig. 1 Mean pathogenicity of wheat powdery mildew in Switzerland, 1980-1985

REFERENCES

Bennett, F.G.A. and Van Kints, T. 1982. Mildew of wheat. UK Cereal Pathogen Virulence Survey 1982 Annual Report, 7-26.

Heun, W. and Fischbeck, G. In Press. Identification of wheat powdery mildew resistance genes by analysing host-pathogen interactions.

Newton, N. and Cherewick, W.J. 1947. Erysiphe graminis in Canada. Canadian Journal of Research 25, 73-93.

Ramgraber, L. 1984. Beiträge zur Virulenzanalyse auf den Mehltau des Weizens, Diplomarbeit, Technische Universität München/Weihenstephan.

Streckeisen, Ph. and Fried, P.M. 1985. Virulenzanalyse im Weizenmehltau in der Schweiz in den Jahren 1981 bis 1983. Schweiz. Landw. Fo., 24, 261-269.

Zimmermann, G., Wenisch, K. and Strass, F. 1984. Further results of attempts to combine mildew resistance genes in winter wheat. Vorträge Pflanzenzüchtung 6, 85-102.

POPULATIONS OF THE BARLEY MILDEW PATHOGEN IN THE UK

M.S. Wolfe, Susan E. Slater and P.N. Minchin

Plant Breeding Institute, Maris Lane, Trumpington, Cambridge CB2 2LQ, England

ABSTRACT

Populations of the barley mildew pathogen have been surveyed in the UK since 1964. Current methods and recent trends in the populations are summarised.

Isolates pathogenic on varieties derived from Rupee were identified from the field for the first time in 1986, which facilitated characterisation of the resistances of several new varieties.

A low degree of pathogenicity was confirmed among isolates obtained from Atem at a site in northern England.

Other new varieties carry combinations of previously used resistance genes which select for pathogen phenotypes that have previously occurred for a long period at lower than expected frequencies in the pathogen population. This unfortunate trend is likely to be deleterious to the potential for variety diversification.

Isolates have been identified from Scotland with a low but stable level of insensitivity to fenpropimorph. As with variety resistance, it is evident that the pathogen is now responding to selection by mixtures of fungicides that have been previously used singly, which may prove deleterious to fungicide diversification.

INTRODUCTION

Barley mildew populations in the UK have been studied continuously since 1964, to provide breeders and advisers with up-to-date information on the response of the pathogen to commercial varieties of winter and spring barley. From 1967, the work was incorporated in the UK Physiologic Race Survey which was changed in name in 1977 to the UK Cereal Pathogen Virulence Survey (CPVS). In recent years, the studies have been expanded in two ways, first to provide similar information for response to fungicides, and second to provide a better understanding of the dynamics of the pathogen as a background to the development of strategies for disease control.

Each year, a summary of the data is distributed in the annual report of the CPVS which also contains similar information on the other important airborne pathogens of cereal crops. In addition, an annual list is now being circulated that contains the mildew resistance phenotypes of all barley varieties grown in the UK.

METHODS

The methods used have been under constant development since the surveys were started. Currently, they are as follows:

1. Sampling a variety or fungicide treatment

a. Leaf samples are taken directly from plants in a field or plot of known variety or fungicide treatment, either by ourselves or by advisers. The samples may be treated as bulks or single colony isolates.

b. The stick sampler, containing seedlings or leaf segments of test varieties, is dragged through a field or plot of known variety or fungicide treatment to dislodge spores that then act as inoculum. This method has replaced 'mobile nurseries' because only a single visit is required to the collection site.

2. Sampling the air spora of a region

a. A static nursery containing seedlings of susceptible control (Golden Promise) and differential test varieties is exposed weekly on a high roof in Cambridge. The numbers of colonies incubated weekly on Golden Promise indicate the course of the local epidemic. Bulk collections of these infections can be used to determine the frequencies of the principal pathogenicity characters in the population. These data can be checked against colony numbers on the differentials. The frequencies of simple combinations of pathogenicity characters can be estimated from the populations collected on the differentials. This is achieved by testing the populations from each differential on all of the other differentials.

Alternatively, single colonies can be isolated from the original collections. These can be tested later and separately on differential varieties and fungicide-treated seedlings.

The static nursery is not suitable for fungicide tests because the fungicide content of the test leaves changes considerably during the period of exposure.

b. The Wind Impaction Spore Trap (WIST), fitted to a car roof, is driven around a set route near Cambridge (200 km) about once every three weeks to expose seedlings of Golden Promise, untreated or treated with different fungicides, to the air spora. In this way, fungicide treated seedlings are exposed to a large air mass over a short period (1 hr).

The WIST is also used to sample the air spora over England and Scotland, usually at the beginning of July each year. This involves exposures over 80 km of seedlings of Golden Promise with some differential varieties and fungicide treated seedlings, over the 1900 km round trip.

Identification of resistance phenotypes

One of the major functions of the surveys is to identify, as early as possible, isolates of the pathogen that have previously unknown pathogenicity or fungicide insensitivity characters. These occurrences are clearly of great importance to breeders and to fungicide manufacturers. It is also important to identify isolates with complex combinations of characters; such isolates are essential for differentiating the resistance components of new varieties.

Isolates pathogenic on varieties with resistance derived from Rupee (Mla13, Ru2, Ru3) were obtained from the field for the first time in the UK in 1986, even though the matching varieties occupy only a small area. From tests with these and other isolates, Sherpa and Pipkin probably have Ru2 combined with another gene, but the greater quantitative resistance of Sherpa may be due to an additional factor, possibly from the parent, Maris Mink. Clansman, Digger and Sergeant probably all have Mla13 combined with Ru3; it is not known whether or not they also possess Ru2. At least one of the new isolates obtained is pathogenic on all five varieties, but to a lesser extent on Sherpa.

Isolates were obtained from infected fields of Atem (MlLa, mlo) in Northumberland. In a blind test assessed by E. Schwarzbach (see Schwarzbach, this volume), the isolates were shown to have a low but significant level of quantitative pathogenicity specific for Atem. The level was similar to that found in the same season in Austria and Poland.

Pathogen population structure

Table 1 lists the principal differential varieties used in the survey.

TABLE 1 Principal differential varieties for pathogen isolates.

BMR gp	Variety	BMR gp	Variety	BMR gp	Variety
0	Gold. Promise	6a (Mlk)	H. 1063	10a(Mla13)	Digger
1a	W. 37/136	6b (Mla7)	Porter	4,6b	Doublet
1b	W. 41/145	6ab	Ark Royal	4,6ab	Klaxon
2 (Mlg)	Julia	6bc(Mla7,Ab)	Triumph	4,7	Vista
3 (Mla6)	Midas	7 (Mla1)	Delta	4,8	Kym
4 (MlLa)	Lofa	8 (Mla9)	Simon	5,6c	Natasha
5 (Mla12)	Hassan	9 (mlo)	Apex	5,10b(MlRu2)	Sherpa

Changes in frequency of the major pathogenicity characters in recent years are shown in Table 2. This table is based on data from winter barley, to indicate the frequencies of pathogenicity genes coming from the largely non-selective winter host population which normally provides the main source of inoculum for the spring crop.

TABLE 2 Changes in frequency of the major pathogenicity characters since 1978 determined from samples from winter barley varieties.

Year	BMV character										
	1b	2	3	4	5	6a	6b	6ab	6bc	7	8
1978	-	68	34	7	28	51	-	35	0	0	0
1979	-	39	20	13	16	44	-	34	0	0	0
1980	-	61	27	16	21	-	-	38	2	5	0
1981	-	44	31	45	23	-	-	7	4	1	0
1982	-	48	45	31	41	-	-	18	4	1	0
1983*	-	68	43	36	42	-	-	29	37	3	0
1984	56	47	51	15	13	-	-	52	52	1	1
1985	45	45	43	18	21	49	42	-	47	-	2
1986	26	28	23	8	24	26	23	-	24	1	2

* NIAB trials samples in 1983

The majority of pathogenicity characters decreased in frequency from 1985 to 1986 (Table 2), partly due to the decrease in area of Triumph, and partly due to the increased cultivation of currently resistant varieties. The decrease of BMV 2 appeared to relate only to the unidentified resistance gene in Julia; pathogenicity for Mlg (Goldfoil) has remained high since the 1960's. The slight increase in BMV 5 was probably due to infection of Natasha (BMR 5, 6c), recently introduced on a large scale.

WIST data from a wider area revealed more general increases in the frequencies of pathogenicity characters (Table 3), particularly in Scotland. The reason for the apparent increase in BMV 3 in Scotland is not known; BMR 3 varieties tended to decrease during this period. The increases in BMV 4, 5, 6bc and 4,9 presumably reflect the growing diversity of resistant varieties in Scotland as the popularity of Golden Promise (BMR 0) continued to decline.

Linkage disequilibrium in the pathogen population

Pathogenicity genes have often been in linkage equilibrium in the population, that is, the actual frequency of a combination of two genes was equal to the product of the individual frequencies of the two genes (Wolfe and Knott, 1982). However, from the mid-1970s until recently, particular genes were always in negative disequilibrium; they occurred less frequently than expected (Table 4, 1985 data).

TABLE 3 Distribution of major pathogenicity characters from 1984-86 determined from direct tests in the WIST (roof data for Cambridge).

Year	Region					
	Cambr.	E. Mids	N. Eng.	Lothns	E. Sctl	N. Sctl
BMV 3						
1984	-	62	73	48	36	23
1985	27	27	36	53	55	46
1986	19	31	55	56	125	86
BMV 4						
1984	-	37	56	27	35	51
1985	22	19	28	47	55	15
1986	29	22	50	47	77	44
BMV 5						
1984	-	18	51	27	39	40
1985	26	27	23	54	74	21
1986	31	46	58	68	148	32
BMV 6bc						
1984	-	91	21	21	15	16
1985	20	39	17	33	62	23
1986	28	44	66	81	54	61
BMV 4,9						
1984	-	4	4	9	0	5
1985	4	12	10	20	2	2
1986	2	5	22	8	10	13

TABLE 4 Colony counts relative to those on Golden Promise for isolates trapped on differential varieties exposed on a roof in Cambridge (1985).

Source	BMV character						
	1a	1b	3	4	5	6a	6bc
BMR 1a	70*	60	50	28	21	52	53
1b	29	53*	48	53	31	30	38
3	25	48	67*	85	17	44	19
4	1	30	70	92*	7	0	1
5	5	26	50	48	25*	0	0
6a	44	8	66	7	5	74*	52
6bc	58	39	52	0	0	86	72*

From Table 4, among the isolates from BMR 4 and 5, BMV 6a and 6bc occurred at much lower frequencies than they did in samples from BMR 1a, 1b and 3. Conversely, among the isolates from BMR 6a and 6bc, BMV 4 and 5 occurred at much lower frequencies than in samples from other varieties. The separation of BMV 4 and 5 from BMV 6a, b and c virtually divided the pathogen population into two major sub-populations (Fig. 1). This division has been most useful in the diversification and mixing of varieties; the usefulness was maintained as long as the resistance gene were kept separate in different varieties.

In the last two years, however, there has been a considerable increase in varieties with combined resistances. The effects of this trend are shown in Table 5 and diagrammatically in Fig. 1. The combinations of resistance genes are now causing selection in the pathogen population against the previous negative linkage disequilibria. For example, Cameo (BMR 4,6a,8) and Kym (BMR 4,8) select for BMV 4 which, because of a positive disequilibrium, also carries along a relatively high frequency of BMV 5.

Simultaneously, the linkage disequilibrium associated with selection for BMV 6a and 8 results in a high frequency of BMV 6b in the populations on these varieties. Interestingly, the high frequency of BMV 6a and 6b does not lead to a high frequency of the commonly occurring pathogenicity for Triumph (BMV 6b,c). This suggests that pathogenicity for Cameo and Kym evolved from an older, less common part of the population related to earlier selection of the pathogen on Akka during the late 1970's.

Auto (BMR 3,4,6a), Doublet (BMR 4,6b) and Klaxon (BMR 4,6a,6b) apparently select for the same fraction of the population since BMV 8 occurs at a high frequency, BMV 6bc is uncommon, and the effects of linkage disequilibrium are similar. Everest is also similar, but it probably carries BMR 6c in addition to BMR 4,6b, so that pathogenicity for Triumph is more common and for BMR 8 less so, than on the other three varieties.

The increase in area and increased infection of these varieties may encourage survival of Atem-virulent isolates because of the high frequency of BMV 4. Previously, isolates virulent on Atem would have failed on the large area of Triumph, because of the apparent selection against BMV 4 on the latter variety.

Blenheim, Corniche, Kingpin and Natasha all have BMR 5,6c and so they also cross the boundary between the previous two sub-populations. In this case however, the linkage disequilibrium between BMV 4 and 5 appears to have been reversed, and BMV 4 occurs only at low frequencies. The disequilibrium between BMV 6a and 6b also appears to have been affected in that the frequency of the unselected BMV 6a was considerably lower than in

TABLE 5 Frequencies of pathogenicity characters in leaf samples from new varieties with combined resistances.

	BMV character													
	3	4	5	6a	6b	6bc	7	8	4,6b	4,6ab	4,6bc	4,7	4,8	5,6c
Auto	39	25	25	89	43	8	0	48	8	48	19	0	43	4
Doublet	3	26	62	70	67	0	0	29	42	78	31	0	24	11
Klaxon	12	40	20	69	57	6	0	24	24	51	27	0	24	5
Everest	19	29	34	60	78	26	0	10	35	57	50	0	13	6
Cameo	27	38	30	90	51	1	0	69	15	54	28	0	79	2
Kym	32	34	24	59	30	3	3	52	8	40	16	1	66	2
Vista	10	33	49	19	11	0	130	5	3	2	14	137	29	2
Regatta	16	32	12	63	27	0	106	11	3	18	31	106	18	1
Blenh'm	7	8	57	15	56	39	0	0	0	0	2	0	0	55
Cornich	4	3	70	15	42	31	3	0	0	0	0	0	1	57
Kingpin	11	3	52	33	50	41	0	0	0	3	1	0	3	64
Natasha	12	3	50	12	40	35	0	0	2	1	1	0	0	62

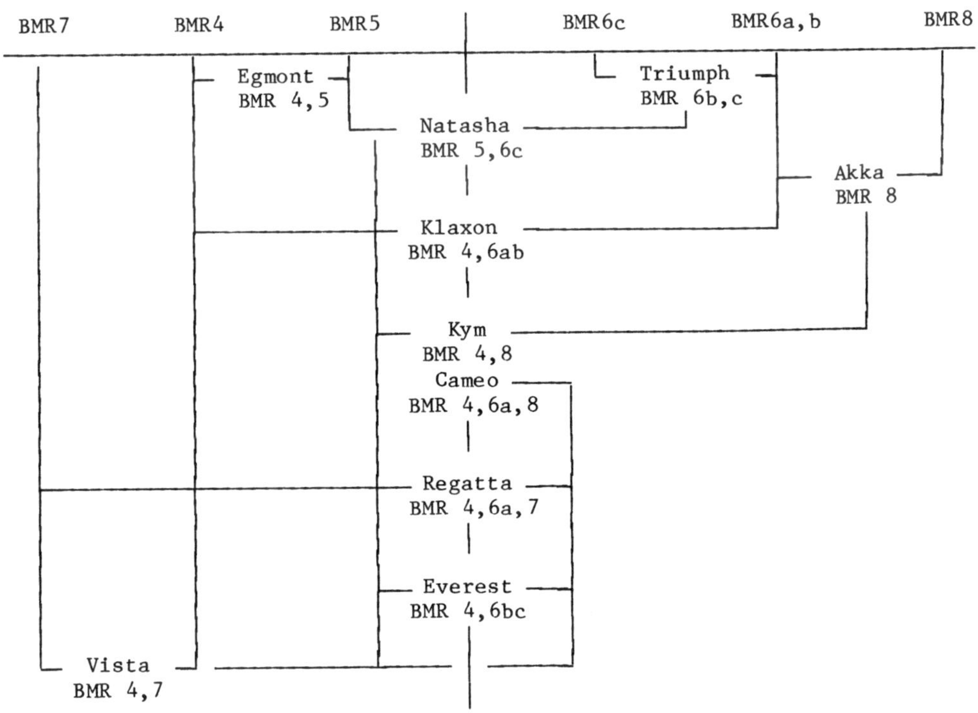

Fig. 1 Diagrammatic representation of recent trends in barley breeding affecting pathogen evolution.

the other populations. In the pathogen populations on this group of varieties, therefore, there appears to have been relatively strong selection against unnecessary pathogenicity genes.

Introduction of varieties with combined resistances is unfortunate, first, because their effectiveness is likely to be ephemeral. Second, the potential for diversification either between or within fields is reduced. Third, the long-standing separation of BMV 4 and 5 from BMV 6a, 6b and 8, could be lost. Fourth, the probable increase in the BMV 4 population may endanger the resistance of Atem (BMR 4,9). Fifth, if there is a large increase in the absolute frequency of recombinant phenotypes of the pathogen, then varieties with combined resistances and those with the separate components may become highly susceptible (Wolfe and Barrett, 1976).

Pathogen response to fungicide

Since the large-scale introduction of triazole fungicides for mildew control in 1980, there has been a steady increase in the insensitivity of the pathogen population to them. In 1984 and 1985, we started to find, commonly, isolates insensitive to seedlings treated at the field rate with triadimenol.

In 1986, there was an increase in higher levels of insensitivity to triazoles in East Anglia although performance of the pathogen against low doses of the fungicide decreased (Table 6). These changes probably reflect a swing in the pathogen population towards a higher frequency of the highly insensitive genotype, which has a relatively flat dose response curve, and a lower frequency of the less insensitive genotypes which have steeper response curves.

TABLE 6 Change in triazole insensitivity over years in barley mildew in East Anglia: direct WIST counts of colonies relative to untreated control.

Dose	Year					
	81	82	83	84	85	86
0.025 (1/15)	23	51	83	85	64	52
0.075 (1/5)	-	27	54	54	70	62
0.125 (1/3)	-	-	-	24	43	45
0.250 (2/3)	-	-	-	-	-	74
0.375 (1/1)	-	-	-	-	-	44

The shift between classes of insensitivity towards higher levels was also evident within the 1986 season from samples taken at different times.

The increase in triazole insensitivity was widespread in 1986 (Table 7). This general change was probably brought about by the rapid increase in the use of Ferrax. This was partly confirmed by the general increase in ethirimol insensitivity (Table 7). Data from indirect tests of isolates from the roof nursery also indicated a considerable increase in the response to Ferrax, particularly late in the season. In comparing the response of the pathogen to 1/5 field rate of Ferrax and Milstem, there tended to be many more colonies on the Ferrax-treated leaves which presumably is due to the half rate of ethirimol in the mixture relative to that in commercial Milstem. It suggests that although the pathogen is responding to Ferrax, it is still being controlled by high rates of ethirimol, but this may not be maintained under selection.

The potential problem with Ferrax was highlighted by previous evidence from local analyses of populations, that insensitivity to both triazoles and ethirimol was positively and significantly associated.

TABLE 7 Regional changes with time of pathogen insensitivity to triadimenol and to ethirimol (WIST data from triadimenol and ethirimol treated seedlings).

	Triadimenol 1:3			Ethirimol 1:5		
Source	1984	1985	1986	1982	1985	1986
E. Anglia	24	43	45	-	16	15
E. Midls	80	60	62	2	17	25
N. Engl	35	27	48	2	3	11
Lothians	22	17	55	4	25	27
E. Scotl	16	28	44	5	12	31
N. Scotl	12	11	85	6	15	16
Mean	32	31	57	4	15	21

Table 8 summarises the results of indirect tests for fungicide response among isolates obtained from seedlings treated in different ways and exposed in different regions. Colony numbers on the triadimenol-treated leaf segments were highest for the populations collected on triadimenol-treated plants, but were low, relatively, only for the populations collected on fenpropimorph-treated plants.

In a similar way, colony counts on the ethirimol-treated segments were highest for the populations from ethirimol-treated plants. From other

TABLE 8 Indirect tests of isolates obtained from different regions on seedlings untreated or treated with different fungicides. All tests completed in autumn 1986.

	Fungicide and rate relative to commercial application							
	Triad.	Ethir.	Frrx.	Trdph.	Fnprop.	Ptrl.	TltT	Drn.
Source	1:1	1:4	1:15	1:20	1:100	1:20	1:15	1:20
Controls								
CC 1	0	25	0	7	4	0	0	0
83/209	0	1	0	17	2	1	21	27
83/3	69	28	20	3	0	0	19	25
85/379	63	0	0	13	0	0	21	37
Untrtd.								
Essex	87	34	12	44	0	0	24	77
E. Mids	35	21	12	23	1	0	34	33
N. Engl	60	30	12	22	0	0	24	34
Loths.	52	2	1	0	0	0	2	0
E. Sctl								
N. Sctl	48	3	4	0	0	0	1	1
Triad. 1:1								
Essex	122	8	20	38	0	0	39	49
E. Mids	54	43	16	19	2	1	28	42
N. Engl	87	89	26	19	0	0	27	49
Loths.	74	1	1	0	0	0	7	26
E. Sctl	45	24	13	0	0	0	4	2
N. Sctl	84	3	14	2	0	0	3	8
Ethir. 1:5,1:15								
Essex	46	11	11	3	1	0	26	25
E. Mids	164	75	14	11	0	0	73	96
N. Engl	44	58	45	23	0	0	22	34
Loths.	31	11	64	0	0	0	0	0
E. Sctl	81	25	12	0	0	0	0	0
N. Sctl	42	12	15	4	12	5	26	6
Tridem. 1:20								
Essex	57	24	8	10	0	0	24	29
E. Mids	58	1	2	0	0	0	0	14
N. Engl	53	0	0	0	0	0	0	0
Loths.	79	3	3	0	0	0	6	1
E. Sctl	68	3	1	0	0	0	2	1
N. Sctl	41	7	12	2	0	0	0	0
Fenprp. 1:100								
Essex	26	13	0	0	0	0	35	42
E. Mids	-	-	-	-	-	-	-	-
N. Engl	-	-	-	-	-	-	-	-
Loths.	28	13	22	20	19	34	37	19
E. Sctl	27	0	0	89	47	38	76	52
N. Sctl	2	44	26	18	41	80	73	12

sources they tended to be noticeably lower except, perhaps from the triadimenol-treated plants. As expected, the colony counts on Ferrax-treated leaf segments were also highest for the populations from the ethirimol-treated plants since ethirimol provides the only effective bottleneck against the increase of isolates with combined insensitivity to triazoles and ethirimol.

A response to tridemorph was evident in samples from England, but from Scotland it was associated with a response to fenpropimorph. The fungicide in the tridemorph trap seedlings may have been too concentrated to reveal this effect. Isolates trapped on the fenpropimorph-treated seedlings grew to some extent on all morpholine-treated material. One of these isolates was re-tested on three occasions and gave a consistent positive response to fenpropimorph. These results suggest that there may be one phenotype, relatively common in England, that has a low level of insensitivity to tridemorph but which is sensitive to fenpropimorph. A different phenotype occurs less commonly in Scotland that gives a low level of cross-insensitivity to tridemorph, fenpropimorph and fenpropidin.

As expected, leaf segments treated with Tilt Turbo or Dorin (triazole-tridemorph mixtures), gave high colony counts with populations that had high counts for the individual fungicide components, although Tilt Turbo appeared less effective than Dorin against the fenpropimorph isolates from Scotland. Indeed, the data from Scotland indicate that genotypes exist with some insensitivity to triazoles, ethirimol, and tridemorph and other morpholines. In this sense, the use of mixtures of fungicides containing components that have been previously exposed separately, is analogous to recombination of previously exposed resistance genes in host varieties. In either case, the initially large extra effect in controlling mildew is likely to be short-lived (Wolfe and Barrett, 1976).

Molecular biology

Survey investigations at PBI are now being supported by work on restriction fragment length variation in the DNA of the barley and wheat mildew pathogen. Variation in DNA fragments extracted from infected leaf segments has been assessed largely from HindIII restriction digests by hybridisation with an intermediate repetitive DNA sequence, isolated and purified by cloning. Among an initial small group of randomly chosen isolates, variation proved to be extensive. Further tests were made, therefore, with a group of isolates again from different years and locations, but all of which carry the relatively uncommon character of pathogenicity for Mla9. Again, most isolates differed considerably, although two were indistinguishable except on a single band.

To focus more closely on variation within populations, 55 isolates were tested from the pathogen population established on Atem (see above). DNA from these isolates was also digested with HindIII and hybridised with the same probe. Considerable variation was revealed suggesting a diverse origin for this population. On the other hand, the isolates pathogenic on Mla13, Ru2 and Ru3 were identical to each other. However, they differed considerably from pathogenically similar continental isolates, suggesting a different origin from them. Clearly, the sensitivity of the technique for comparisons among isolates underlines its potential for improving our understanding of the migration of populations. The banding patterns of all of the isolates tested so far are now being characterised precisely to facilitate the use of the data in a computer program to analyse relationships among the isolates by similarities and differences.

REFERENCES

Wolfe, M.S. and Barrett, J.A. (1976). The influence and management of host resistance on control of powdery mildew on barley. Barley Genetics III. Proceedings of the Third International Barley Genetics Symposium, Garching, pp. 433-439.

Wolfe, M.S. and Knott, D.R. (1982). Populations of plant pathogens: some constraints on analysis of variation in pathogenicity. Plant Pathology 31, 79-90.

AN INVESTIGATION OF THE SENSITIVITY OF *ERYSIPHE GRAMINIS* F. SP. *TRITICI* TO VARIOUS ERGOSTEROL INHIBITING FUNGICIDES

J.T. Fletcher, S.T. Cooper and A.L.H. Prestidge

Agricultural Development and Advisory Service, Olantigh Road, Wye, Ashford, Kent TN25 5EL, England

ABSTRACT

The fungicide sensitivity of populations of *Erysiphe graminis* f. sp. *tritici* was examined in 1985 in southern England in a small survey and also at a high point (177 m above sea level) near to Wye, Kent. More than half of the populations examined showed decreased sensitivity to triadimenol but there was no recognisable change with fenpropimorph. In two randomised and replicated experiments, fungicide rates and fungicide mixtures were compared for their effects on the sensitivity of mildew populations. In 1985, a half rate spray of triadimefon resulted in the greatest incidence of forms less sensitive to triadimenol. There was no obvious difference between normal and 1.5 normal rates or between any of the rates of fenpropimorph on triadimenol or fenpropimorph sensitivity. In 1986, triazole and morpholine fungicides were compared with mixtures of the two. Two triadimenol sprays resulted in a large decrease in the sensitivity of the population when it was assessed shortly after the second spray. The triadimenol and tridemorph mixture had a similar but less marked effect. None of the other treatments markedly influenced triadimenol or fenpropimorph sensitivity.

INTRODUCTION

Work in England has demonstrated the decreasing sensitivity of *Erysiphe graminis* f. sp. *hordei* to the triazole group of the ergosterol biosynthesis inhibiting fungicides (Fletcher and Wolfe, 1983; Wolfe *et al.*, 1983; Wolfe *et al.*, this volume). This has resulted in an increase in the use of morpholine fungicides and although these also inhibit ergosterol biosynthesis, they do so at a different site from the triazoles. So far, no significant change in sensitivity to the morpholines has been reported.

In 1985, studies were started in Kent on the sensitivity of populations of *E. graminis* f. sp. *tritici* to the same group of fungicides (see also Wolfe and Van Kints, this volume). No marked reduction in performance had been reported from the field at that time although fungicide comparisons had indicated that some triazole fungicides controlled mildew better than others.

MATERIALS AND METHODS

Measurement of fungicide sensitivity

The sensitivity of pathogen populations to triadimenol was measured using the mobile nursery technique. Seeds of the mildew susceptible cultivar Kanzler were treated with triadimenol (as Baytan) to give the following rates:- 0, 4, 8, 12.5, 25, 37.5 and 62.5 g ai/100 kg seeds. Lines of treated seeds were sown in Fisons Universal Compost in trays measuring 250 x 220 x 60 mm. The seeds were arranged in the tray to give seven lines across, with a gradient of increasing seed treatment concentration away from one end which had the untreated line. There were approximately 10 seedlings in each of the seven lines. When the first leaf was fully expanded the trays were taken to the exposure site and left for two to four days. They were then returned to the greenhouse and seven to nine days later mildew was assessed on leaf 1 using the standard ADAS disease assessment key for wheat powdery mildew. A mean mildew figure was obtained for each line and the FD^{50} value calculated. The FD^{50} value is defined as the fungicide rate which had half the amount of mildew in comparison with that of the untreated line.

Sensitivity to morpholine fungicides was measured on cv. Kanzler seedlings, grown in Fisons Universal Compost in Plant Pack pots which measured 70 x 70 x 80 mm. Following germination and the development of the first leaf, the pots were drenched with fenpropimorph (as Corbel) at the following rates:- 0, 6.2, 12.5, 25 and 100 ppm. The volume of each pot was 250 ml, and 50 ml of the fungicide drench or water was applied to slightly dry compost to ensure that all the fungicide treatment was retained within the pot. Once the drenches had been applied all further watering was from the base. One day after drenching, the pots were arranged in a tray which contained 24 pots in total, four from each treatment. The pots were arranged so that the fungicide drench treatment increased in concentration away from the untreated pots which were at one end. Exposure times, assessment and calculation of the FD^{50} values were the same as described for the triadimenol assay.

Surveys

In 1985, wheat was surveyed in southern England by exposing mobile nurseries in 15 farm crops. Also, the sensitivity of the population was monitored at a high point on Wye Downs (177 m above sea level) from May 22 to August 14.

Experiments

In 1985, the effects on fungicide sensitivity of spray treatments with triadimefon and fenpropimorph used 0.5, 1.0 and 1.5 times the commercial rate and compared with untreated plots was measured in a replicated (4) experiment using the mildew susceptible cv. Mission. Each replicate consisted of a plot 17.5 x 2.25 m bordered by guard plots which were identically treated. All records were taken from the control plot. The fungicide sprays were applied on May 29 at GS39. In 1986 in an experiment with cv. Kanzler, identical plot sizes and layouts were used but with five replicates of each treatment. The following fungicides and fungicide mixtures were compared:- tridemorph (Calixin), fenpropimorph (Corbel), fenpropidin (Patrol), triadimenol (Bayfidan), propiconazole + tridemorph (Tilt Turbo), triadimenol + tridemorph (Dorin) and there was an unsprayed control. The fungicides were applied on May 16 and June 23.

RESULTS

1985 survey

The results of the sensitivity measurements from 15 farm crops (Table 1) indicate a variation between the sensitivities of the populations to triadimenol but more than half showed greatly reduced sensitivity.

TABLE 1 FD^{50} values of populations of *Erysiphe graminis* f. sp. *tritici* to triadimenol in 1985.

		Number of populations in each FD^{50} category						
Location	Survey date	0	4	8	12.5	25	37.5	62.5
Canterbury 1	June 24-27	0	0	0	2	2	4	0
Canterbury 2	July 16-28	0	0	2	4	0	2	0
Chichester	July 2-4	0	0	1	1	2	3	0
Reading	May 29-June 3	0	0	0	0	1	2	0
Wolverhampton	June 21-24	0	1	0	1	0	0	0
TOTAL		0	1	3	8	5	11	0

There was no detectable variation in the sensitivity to fenpropimorph; FD^{50} values were between 25 and 35 ppm.

1985 high point survey

Measurements of triadimenol and fenpropimorph sensitivities were made on eleven occasions between May 22 and August 14 (Table 2).

Triadimenol sensitivity decreased considerably between June 12 and July 17, the period when most crops in the area were sprayed. Fenpropimorph sensitivity was also least during this period, but by August 14 populations were sensitive to both fungicides.

TABLE 2 FD^{50} values of population of Erysiphe gramins f. sp. tritici measured at a high point on Wye Downs.

Date	triadimenol	fenpropimorph
May 22	37.5	25.0
May 29	25.0	12.5
June 5	25.0	12.5
June 12	62.0	25.0
July 3	62.5	25.0
July 10	62.5	25.0
July 17	45.0	12.5
July 24	25.0	12.5
July 31	25.0	12.5
August 7	37.5	1.9
August 14	25.0	6.25

Experiment - 1985

Measurements of sensitivity were made on May 21, 28, June 4, 11, 18, 25, July 2, 9, 16 and 23 (Table 3). In the triadimefon-treated plots the highest triadimenol FD^{50} values were recorded in the half rate spray treatment for a period of about four weeks after the fungicide was applied. In contrast the values in the full rate and one and a half rate treatments remained comparatively low. Triadimenol FD^{50} values for the fenpropimorph spray treatments were low throughout and the mildew population in the high rate treatment was particularly sensitive to triadimenol. Sensitivity to

TABLE 3 Sensitivity of Erysiphe graminis f. sp. tritici populations to triadimenol (T) and fenpropimorph (F) from unsprayed plots and those sprayed with 0.5, 1.0 or 1.5 times commercial rates of triadimefon and fenpropimorph.

	FD^{50} values													
	Untreated		Triadimefon						Fenpropimorph					
Assessment	0		0.5		1.0		1.5		0.5		1.0		1.5	
date	T	F	T	F	T	F	T	F	T	F	T	F	T	F
May 5	*12.5	-	12.5	-	12.5	-	12.5	-	12.5	-	12.5	-	11	11
May 28	12.5	11	12.5	12	12.5	11	12.5	11	12.5	11	12.5	11	11	11
June 4	11.0	25	19.0	20	10.0	18	12.5	15	12.5	18	12.5	23	10	24
June 11	20.0	20	50.0	31	21.0	17	18.0	25	16.0	21	22.0	8	10	24
June 18	12.5	39	-	6	24.0	39	10.0	49	24.0	39	11.0	36	19	26
June 25	37.5	16	55.0	37	23.0	43	18.0	22	21.0	50	34.0	24	21	23
July 2	34.0	28	28.0	35	31.0	24	25.0	17	23.0	29	26.0	20	15	24
July 9	28.0	28	52.0	26	31.0	24	40.0	14	30.0	45	29.0	32	20	46
July 16	15.0	28	12.5	25	26.0	26	15.0	15	18.0	26	20.0	17	17	14
July 23	16.0	21	21.0	27	19.0	20	30.0	9	10.0	16	22.0	14	19	20

* each figure is a mean of two assessments

fenpropimorph was variable but within a narrow range. The least sensitive population was detected in the half rate fenpropimorph spray treatment. This treatment was the only one which appeared to be causing a response to selection in the pathogen population.

Regular disease assessments were done throughout the experiment and mildew levels remained extremely low, never exceeding 3% of the leaf area covered. There were no other diseases at significant levels. Yields ranged from 7.4 to 8.2 tonnes per hectare with only those from the half rate spray treatments with both fungicides not being significantly different from the unsprayed control.

Experiment - 1986

Measurements of sensitivity were made on May 5, June 4, 18, July 2 and 16 (Tables 4 and 5). The most marked shift was recorded on July 2 when the triadimenol sensitivity in the triadimenol-treated plots decreased markedly.

TABLE 4 Triadimenol sensitivity of *Erysiphe graminis* f. sp. *tritici* populations from plots sprayed with various fungicides and fungicide mixtures.

Treatment	FD50 values at each assessment date May 5	June 4	June 18	July 2	July 16
untreated	22.8	12.2	9.7	31.3	9.8
tridemorph	15.1	21.1	10.9	29.4	16.7
fenpropimorph	36.9	16.1	18.8	21.4	12.1
fenpropidin	17.8	37.9	13.8	46.0	14.1
triadimenol	22.5	32.7	17.4	408.3	30.8
propiconazole + tridemorph	19.3	20.8	18.9	26.6	12.6
triadimenol + tridemorph	18.4	28.8	21.1	97.0	18.3

TABLE 5 Fenpropimorph sensitivity of *Erysiphe graminis* f. sp. *tritici* population from plots sprayed with various fungicides and fungicide mixtures.

Treatment	FD50 values at each assessment date May 5	June 4	June 18	July 2	July 16
untreated	6.2	8.7	21.9	12.5	12.3
tridemorph	6.2	7.5	27.5	17.5	14.7
fenpropimorph	6.2	6.2	22.5	15.6	12.6
fenpropidin	6.2	6.2	27.5	12.5	13.1
triadimenol	6.2	8.7	27.5	13.7	14.0
propiconazole + tridemorph	6.2	7.5	22.5	15.0	12.8
triadimenol + tridemorph	6.2	8.5	27.5	10.9	13.3

To a lesser extent the same effect was recorded in the triadimenol + tridemorph treatment and to a much smaller extent in the fenpropidin treatment.

Fenpropimorph sensitivity showed no difference between treatments but FD^{50} values from the June 18 assessment were higher than at any other time. At the final assessment (July 16) the mildew population was sensitive to both fungicides.

Mildew developed early in this experiment but after an initial increase never exceeded 14% of the leaf area covered (leaf 3, June 18, GS69). A late attack of brown rust (*Puccinia recondita*) is thought to have been responsible for a highly significant increase in yield in all spray treatments except for tridemorph alone.

DISCUSSION

The survey and high point monitoring results indicate that populations of *E. graminis* f. sp. *tritici* in south-east England show decreased sensitivity to triadimenol but not to fenpropimorph. The use of triadimenol seed treatment (Baytan) is widely practised in the area and it is not unusual for three further fungicide applications to be made (GS31, 39 and 59). The selection exerted by the later applications probably accounts for the high FD^{50} values recorded for triadimenol at the high point in June and early July. A small increase in the values for fenpropimorph may reflect a similar trend and could be the first indication of changes in sensitivity of the population to this fungicide.

The apparent reversion to more sensitive forms is at present resulting in a slow step-wise shift towards greater insensitivity to triadimenol. Further selection of better adapted forms could result in a significant increase in this shift in the coming season.

The results of the field experiments indicated that the use of reduced rates of triadimenol are likely to increase the incidence of populations less sensitive to triadimenol. A corresponding trend was not seen with fenpropimorph treatment. Fungicide mixtures decreased the incidence of less sensitive triadimenol forms although the two used in the 1986 experiment gave different results. Tilt Turbo appeared to be more effective than Dorin in delaying the increase in less sensitive forms. The two products differ in two respects, first, the triazole component, propiconazole and triadimenol respectively, and second, the quantity of tridemorph at 225 ml and 375 ml per litre respectively. Possible differences in the sensitivity of the mildew population to propiconazole and triadimenol cannot be ruled out (de Waard et al., 1986) and the proportion of the tridemorph component may also be important. Further work needs to be done on mixtures to investigate the influence of different amounts of the two components on the incidence of decreased sensitivity to either one. The response to each fungicide should be monitored separately especially among the triazoles, rather than assuming that the response will be the same for all members of each group.

From these results it appears that in 1986 the wheat mildew situation was very similar to that of barley mildew in 1982. In 1983 there was widespread failure of triazole fungicides to control barley mildew and the same may be true for wheat mildew in the next year or two.

REFERENCES

Fletcher, J.T. and Wolfe, M.S. 1981. Insensitivity of Erysiphe graminis f. sp. hordei to triadimefon, triadimenol and other fungicides. Proceedings of the 11th British Crop Protection Conference Vol. 11, 633-640.

de Waard, M.A., Kipp, E.M.G., Horn, N.M. and van Nistelrooy, J.G.M. 1986. Variation in sensitivity to fungicides which inhibit ergosterol biosynthesis in wheat powdery mildew. Netherlands Journal of Plant Pathology 92, 21-32.

Wolfe, M.S., Minchin, P.N. and Slater, S.E. 1984. Dynamics of triazole sensitivity in barley mildew, nationally and locally. Proceedings of the 12th British Crop Protection Conference Vol. 11, 464-470.

POPULATIONS OF THE WHEAT MILDEW PATHOGEN IN THE UK

M.S. Wolfe and Thea M.C. van Kints
Plant Breeding Institute, Trumpington, Cambridge CB2 2LQ, England

ABSTRACT

Populations of the wheat mildew pathogen have been surveyed in the UK since 1961. Current methods and recent changes in the population are summarised.

Pathogenicity for varieties with WMR 2 and 2+6 remains high. Pathogenicity for WMR 7 was low in the population generally, but there is concern that it may be increasing, or about to increase, rapidly. The intermediate resistance of Talent is still effective in the field; the resistance of spring wheat varieties is also effective.

Despite high levels of pathogenicity on seedlings of WMR 2 varieties, the adult plants still tend to be moderately to highly resistant in the field. This was particularly noticeable for Fenman. In more detailed tests, there was no clear evidence of pathogenicity specific for varieties within the WMR 2 group.

Insensitivity to triadimenol fungicide appears to be still increasing in the pathogen population.

INTRODUCTION

The methods used for the wheat mildew surveys are similar to those for the barley mildew surveys (see Wolfe et al., this volume) except that the surveys on the wheat pathogen have been less extensive. A comparison of the uniformity of the data obtained from WIST sampling compared with sampling with a static nursery on a high roof, showed that the latter method was significantly more reliable, at least for virulence characters. For fungicide surveys, the static nursery suffers the disadvantage of changing fungicide concentration in the test seedlings during their long exposure.

PATHOGEN RESPONSE TO VARIETIES

The differential varieties currently used for testing are listed in Table 1.

The frequencies of infection on the main differentials exposed on the roof are given in Table 2 for 1985 and 1986. As in previous years, WMV 2 was very common and may now be fixed in the population. The combination WMV 2+6 was also common, but could now be declining as a result of the increasing diversity of the host population. Pathogenicity for Mission (WMR 4+8) increased and was higher than for Mercia in both years. This difference indicates the effect of the background resistance in the latter variety. Although the values are small, the doubling of pathogenicity on Slejpner is almost certainly a real effect. The intermediate resistance of Talent inherited by Brock remained effective.

TABLE 1 Differntial varieties for wheat mildew with their resistance genes and WMR grouping.

WMR group	Gene	Variety
0	-	Cerco, Moulin, Jerico, Minaret, Alexandria
1	Pm1	Anfield
2	Pm2	Galahad, Longbow, Fenman, Norman, Avalon
3	Pm3a,3b,3c	Asosan, Chul, Sonora
4	Pm4a,4b	Khapli, Armada
5	Pm5	Hope
6	Pm6	Timgalen
-	Pm7	Transec
7	Pm8	Ambassador, Corinthian
8	Mli	Aquila, Mercia
9	Pm2,Mld	Maris Dove
2+4		Sappo
2+6		Brimstone, Gawain
2+7		Hornet
2+Talent		Brock
4+8		Mission
7+?		Dauntless, Slejpner
2+4+6		Rendezvous
2+6+7		CWW 1645/5
2+6+8		Parade
5+8+?		Broom, Tonic
Sona		Solitaire, Wembley
Axona		Axona

TABLE 2 Pathogenicity, relative to Cerco, on the main differential varieties exposed on a high roof in Cambridge.

Year	Differential group and variety						
	0	2	2+6	8	4+8	7+?	2+Talent
	Moulin	5 vars.*	Brimstone	Mercia	Mission	Slejpner	Brock
1985	88	88	62	28	37	4	19
1986	94	90	54	26	51	9	24

* mean of 5 WMR 2 varieties.

Pathogenicity values for the five WMR 2 varieties in Table 2 are given separately in Table 3, but averaged for 1985 and 1986 since there was little difference between the years. Within this group, Galahad is consistently the most resistant and Avalon the most susceptible. The closely related varieties Longbow, Norman and Fenman are intermediate.

TABLE 3 Seedling pathogenicity values, averaged for 1985 and 1986, for five WMR 2 varieties (Galahad, Longbow, Norman, Fenman, Avalon) and three others (Brock, Brimstone, Mission), together with adult plant infection percentages for 1986.

	WMR 2 varieties					Non-WMR 2 varieties		
Source	Galahad	Longbow	Norman	Fenman	Avalon	Brock	Brimstone	Mission
Seedlings	66	82	86	99	112	22	44	58
Adults	14	20	14	6	27	9	31	21
Seed/Adult*	5	4	6	17	4	2	1	3

* seedling scores divided by adult plant scores and rounded to nearest whole number.

Also in Table 3, seedling data for Brock, Brimstone and Mission have been included, together with adult plant infection data for both the WMR 2 and non-WMR 2 varieties. Seedling values for the WMR 2 varieties are notably higher than for the others and are less well correlated with the adult plant infection scores. In particular, Fenman was considerably more resistant at the adult plant stage than expected from the seedling values.

Such high values at the seedling stage on the WMR 2 varieties may reflect the long period of adaptation of the pathogen to this resistance. It is particularly interesting to note, therefore, that the adult plants of these varieties were no more infected than those of the more recently introduced non-WMR 2 varieties. This effect is demonstrated by the ratios obtained by dividing the seedling colony count for each variety by its percentage leaf infection at the adult stage (Table 3).

In a separate experiment, isolates collected from adult plants of the five WMR 2 varieties were tested on seedlings of the same group to test for adaptation within WMR 2. There was a slight, but not significant, trend towards adaptation to "own" host; the largest deviation was obtained with isolates from Fenman which were more pathogenic to Fenman than expected from the null hypothesis. Further tests are needed to determine the significance of this observation.

Table 4 summarises data from tests of leaf samples taken from key varieties and tested on seedling leaf segments of those varieties. The pathogen showed specificity for each of the varieties in the Table with the single exception of Brock. It also appears that Anfield and Wembley may be more closely related in terms of resistance than had been thought previously.

TABLE 4 Mean pathogenicity values relative to Cerco of single colony isolates collected on differential seedlings on the roof.

Source	Winter differentials Hornet	Mercia	Mission	Brock	Spring differentials Anfield	Broom	Wembley
Hornet	62	46	6	69	6	1	12
Mercia	7	52	38	37	27	9	18
Mission	5	31	97	40	33	16	17
Brock	8	33	42	31	19	7	14
Anfield	3	48	30	31	76	9	40
Broom	0	49	60	53	29	49	15
Wembley	6	46	71	57	83	8	54

Pathogenicity for Hornet (WMR 2+7) was uncommon in the samples from all other varieties, although, from past experience, it is highly likely that it will increase rapidly if the variety is grown more widely. In the samples from Hornet itself, it was notable that pathogenicity for Mission was unusually low. Less surprisingly, pathogenicity for spring varieties was generally less in samples from Hornet than in those from other varieties, presumably because of the more recent introduction of Hornet.

Also from Table 4 there was evidence of a tendency for unnecessary genes for pathogenicity to accumulate on spring wheat varieties. This would be explained by the later epidemic development on spring compared with winter wheats. However, this phenomenon may be of little importance at present since there is very little spring wheat grown in the UK and pathogenicity for spring wheat varieties, with the exception of Jerico and Minaret (both WMR 0), is uncommon.

FUNGICIDE INSENSITIVITY

Cerco seedlings grown from seed treated with different concentrations of triadimenol were exposed in the WIST in Cambridgeshire (80 km). After incubation the colonies were counted and expressed as percentages relative to those on untreated Cerco seedlings (Table 5).

The proportion of the pathogen population insensitive to 0.04 g a.i. of triadimenol has changed little during the last three years. However, insensitivity to the other concentrations increased, markedly so for the field rate (0.375 g a.i.) in 1986. Tests have not yet been made with morpholine fungicides.

TABLE 5 Mean colony numbers (weekly exposures, May to August each year) on Cerco seedlings grown from triadimenol-treated seed relative to those on untreated seedlings.

Year	Seed treatment (g a.i. kg^{-1}) 0.040	0.125	0.250	0.375	Number of journeys
1983	51	18	-	-	11
1984	69	33	-	-	17
1985	66	32	14	5	15
1986	70	49	25	14	16

POWDERY MILDEW ON BARLEY IN CANADA

Duane E. Falk and Jim Louter

Assistant Professor and Research Associate
Dept. of Crop Science, University of Guelph, Guelph, Ontario, Canada

ABSTRACT

Following the breakdown of Mla1 in barley cultivars in Ontario, a WIST survey in 1986 revealed 24 virulence combinations among 118 isolates, based on 8 virulence genes. The sources of inoculum are not yet known.

INTRODUCTION

Canada is not a member of the E.E.C., nor even close to the European Continent, but we do have similar difficulties with mildew on wheat and barley in the province of Ontario and eastern Canada. Powdery mildew is generally not a problem on wheat or barley in western Canada.

In Ontario, barley mildew was effectively controlled for a number of years by background resistance and the Mla1 gene in several cultivars. However, in 1984 the frequency of virulence on Mla1 increased dramatically and, by 1986, the resistance was completely ineffective. Currently there are no registered 6-row barley cultivars in Ontario with good mildew resistance.

Sampling with a WIST in 1986 turned up 24 different virulence combinations in a sample of 118 single colony isolates (see Table 1). Virulence was present for all tested resistance genes except Mla6 and Mla13. Virulences on Mla10, Mla7 + LG and Mlg were relatively rare while virulences on Mlh, Mla1, Mlp and Mlk were more common. Recombination appeared to be random. More than half of the isolated races had three or more virulence genes. Moseman's isogenic series in Manchuria was used to determine the virulence genes in the isolates.

We have not yet pin-pointed our source(s) of primary inoculum. We know that the pathogen is capable of over-wintering on the winter barley as well as being blown in from the Mid-West of the USA during the summer. We do not know what genes have been deployed in the USA nor do we know what genes our winter barley cultivars possess. Our most recently released winter barley has become quite susceptible to mildew in the three years since its release.

Since we now have some idea of the genetic variability of the pathogen, our next step is to determine what sources of useful resistance are available to use in adapted and foreign materials. We are currently screening our breeding material and a number of introductions for resistance genes which are not matched by any virulence genes in our pathogen population.

We anticipate using the Pallas isogenic series from Denmark in future WIST surveys and so should be able to relate our data directly to that of European researchers.

Any of our barley breeding materials or pathogen isolates that would be useful to other scientists are available by writing to the Dept. of Crop Science at the University of Guelph.

TABLE 1 Virulence gene combinations corresponding to known resistance genes from WIST in 1986 (118 colonies).

Corresponding resistance	Frequency	Overall frequencies	
Ml-a	0.02	Ml-h	0.94
Ml-h	0.13	Ml-p	0.63
Ml-a, Ml-h	0.10	Ml-k	0.43
Ml-a, Ml-a (Mu)	0.01	Ml-a	0.38
Ml-h, Ml-p	0.13	Ml-a(Mu)	0.21
Ml-h, Ml-k	0.05	Ml-a(LG)	0.04
Ml-h, Ml-a (Mu)	0.02	M-a^{10}	0.01
Ml-a, Ml-h, Ml-p	0.05	Ml-g	0.01
Ml-a, Ml-k, Ml-p	0.01	Ml-a6	-
Ml-a, Ml-h, Ml-k	0.01	Ml-a13	-
Ml-a, Ml-h, Ml-a (Mu)	0.01		
Ml-a, Ml-a (Mu), Ml-p	0.01		
Ml-h, Ml-a (Mu), Ml-p	0.03		
Ml-h, Ml-k, Ml-p	0.13		
Ml-h, Ml-k, Ml-a (Mu)	0.01		
Ml-h, Ml-a (LG, Ml-p	0.02		
Ml-g, Ml-h, Ml-p	0.01		
Ml-a, Ml-h, Ml-k, Ml-p	0.11		
Ml-h, Ml-k, Ml-a (Mu), Ml-p	0.05		
Ml-h, Ml-k, Ml-a (LG), Ml-p	0.01		
Ml-a, Ml-h, Ml-a (Mu), Ml-p	0.02		
Ml-a, Ml-h, Ml-k, Ml-a (Mu), Ml-p	0.03		
Ml-a^{10}, Ml-h, Ml-k, Ml-a (Mu), Ml-p	0.01		
Ml-h, Ml-k, Ml-a (LG), Ml-a (Mu), Ml-p	0.01		

List of participants

EEC countries

CEC	Dr. Vittoria Alliata, 200 rue de la Loi, B-1049 Brussels. Tel.: (32-2) 2358386; Telex: COMEU B 21877
Denmark	Dr. J.H. Jørgensen, Risø National Laboratory, Postbox 48, DK 4000 Roskilde. Tel.: (45-2) 371212; Telex: 43116
	Mr. M. Houmøller, Institute of Plant Pathology, Lottenborgvej 2, DK-2800 Lyngby. Tel.: (45-2) 872510; Telex: 37677 PVC DK
England	Prof. M.S. Wolfe, Plant Breeding Institute, Maris Lane, Trumpington, Cambridge CB2 2LQ. Tel.: (44) 223 840411; Telex: 818280 PBICAM G
	Dr. J.T. Fletcher, ADAS, South East Region, Olantigh Road, Wye, Ashford, Kent TN25 5EL. Tel.: (44) 233 812761
France	Dr. Claude de Vallavieille-Pope, Plant Pathology, National Institute of Agronomy, F-78850 Thiverval-Grignon. Tel.: (33-1) 30544510; Telex: 697388 F
	Mr. D. Andrivon, as above.
Germany, FRG	Dr. Pag, BML, Referat 313, 5300 Bonn-Duisdorf. Tel.: (49) 228 5293672; Telex: 886844
	Prof. Dr.Dr.h.c. G. Fischbeck, TU München, Chair of Agronomy and Plant Breeding, D-8050 Freising-Weihenstephan. Tel.: (49) 8161 713422; Telex: 522854 TUMUE D
	Dr. M. Heun, as above. Tel.: (49) 8161 713183
	Mr. F.G. Felsenstein, as above. Tel.: (49) 8161 713562
	Dr. E. Limpert, as above. Tel.: (49) 8161 713562
Ireland	Dr. Brendan Dunne, Plant Pathology Dept., An Foras Taluntais, Oak Park, Carlow. Tel.: (353) 503 31425; Telex: 33038 AFTO EI
Italy	Dr. M.T. Ialongo, Patologia Vegetale, Via Bertero 22, 00156 Rome. Tel.: (39) 8274218
Netherlands	Dr. A.G. Balkema-Boomstra, Stichting voor Planten veredling, SVP, Postbus 117, 6700 Wageningen. Tel.: (31) 8370 19112
Spain	Dr. J.L. Molina-Cano, La Cruz del Campo, S.A., Dept. Barley Breeding, Avda. Andalucia 1, SP-41007 Sevilla. Tel.: (34) 54 574000; Telex: 72239

Non-EEC countries

Austria	Dr. B. Zwatz, Bundesanstalt für Pflanzenschutz, Trunnerstraße 5, A-1020 Wien. Tel.: (43) 222 241511; Telex: 135259 BANST A
Canada	Dr. D.E. Falk, Dept. of Crop Science, University of Guelph, Guelph, Ontario N1G 2W1. Tel.: (519) 824 4120
Finland	Dr. R. Karjalainen, Dept. Plant Pathology, Univ. Helsinki, Viikki 21, SF-00710 Helsinki. Tel.: (93) 378011
Germany, GDR	Dr. C.O. Lehmann, Akademie der Wissenschaften der DDR, Zentralinst. für Genetik und Kulturpflanzenforschung, DDR-4325 Gatersleben.
Poland	Dr. E. Gacek, Exp. Breeding Station, I.H.A.R. 47-233 Bakou, Kluczbork. Tel.: (48) 8192; Telex: 0732327
Sweden	Mr. L. Wiik, Dept. Plant and Forest Protection, Swedish Univ. Agric. Sciences, Box 44, S-23053 Alnarp. Tel.: (46) 40 415000
Switzerland	Dr. P.M. Fried, Forschungsanstalt für Landwirtschaft, Reckenholtzstr. 191-211, CH-8046 Zürich. Tel.: (41) 1 578800

Further experts

Dr. I.M. Smith, EPPO, 1 rue le Notre, F-75016 Paris. Tel.: (33-1) 45207794; Telex: 614148 OEPP F

Doz. Dr. E. Schwarzbach, Probst. Saatz. A-2301, Groß-Enzersdorf. Tel.: (43) 2215 21933

NB: All telephone numbers are in the form for international dialling.